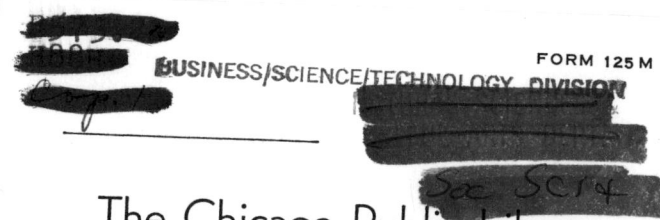

UNIVERSITY OF EDINBURGH
Pfizer Medical Monographs

5

This volume
comprises the papers read
at the fifth Symposium of
the Pfizer Foundation of the
Post-Graduate Medical School,
University of Edinburgh, 1969.
Further volumes will
be published
annually

Pfizer Medical Monographs 5

HUMAN POPULATION CYTOGENETICS

EDITORS
PATRICIA A. JACOBS, W. H. PRICE
AND PAMELA LAW

The Williams and Wilkins Company
Baltimore
1970

© University of Edinburgh 1970
All rights reserved
The Williams and Wilkins Company
428 E. Preston Street
Baltimore, MD 21202
First published in Great Britain
by Edinburgh University Press
Printed in Great Britain

Library of Congress Catalog
Card Number 71-113375

TO W. M. COURT BROWN

Opening address *p.*1
SIR HAROLD HIMSWORTH

Observations on meiosis
in normal males and females *p.*9
R. G. EDWARDS

The behaviour of structural
aberrations at male meiosis *p.*23
M. HULTÉN and J. LINDSTEN

Robertsonian translocations *p.*63
JOHN L. HAMERTON

Reciprocal translocations in
human populations *p.*81
J. LEJEUNE, B. DUTRILLAUX
and J. DE GROUCHY

The inheritance of randomly
ascertained chromosome
abnormalities *p.*89
PATRICIA A. JACOBS

Chromosome abnormalities and
spontaneous abortions *p.*103
D. H. CARR

Applications of quantitative
karyotypy to chromosome
variation *p.*119
H. A. LUBS and F. H. RUDDLE

Chromosome studies of normal
newborn infants *p.*143
P. S. GERALD and S. WALZER

Chromosome patterns in a
general neonatal population *p.*153
J. H. TURNER and N. WALD

Incidence studies of constitutional
chromosome abnormalities in
the post-natal population *p.*159
P. G. SMITH and
PATRICIA A. JACOBS

Population cytogenetics and
environmental factors *p.*191
H. J. EVANS

Human chromosomes and
natural selection *p.*217
L. S. PENROSE

The population cytogenetics of
other mammalian species *p.*221
C. E. FORD

The operation of selection *p.*241
J. H. EDWARDS

Progress of an automatic system
for cytogenetic analysis *p.*263
NIEL WALD et al.

Instrumentation and organisation
for chromosome measurement and
karyotype analysis *p.*281
D. RUTOVITZ

The principles of a software
system for karyotype analysis *p.*297
C. JUDITH HILDITCH

Opening Address
SIR HAROLD HIMSWORTH

Medical Research Council

OPENING ADDRESS

¶ SOME twenty years ago, I became Secretary of the British Medical Research Council; a body the remit of which stretches from the heights of psychiatry on the one hand to the depths of molecular biology on the other. This involved my moving from the sylvan groves of academic life into the windswept arena of national administration. But, with that courtesy and good manners that characterises the higher ranks of the British Civil Service, whether they are implementing an increase in the income tax or attempting to reconcile irreconcilables, I was made welcome. Among those with whom I lunched on those occasions was Sir Frank Lee with whom I had been associated during the War.

'Tell me', he said, 'how do you do a job like yours? The span is colossal, and all of it expert. How do you do it?'

I need hardly say that by that time I was beginning to wonder myself. But I had an inspiration. I said 'Do you paint?'

He replied 'Yes, but very badly'.

I said 'So do I. And probably worse than you. But do you know when a picture is well painted?'

'I've got you', he said.

Well, there, ladies and gentlemen, you have epitomised the qualifications and disqualifications of a person like myself on an occasion like this. I am no expert but you cannot do a job like that of Secretary of a Research Council without developing a feeling for jobs that are being done well and for those that are not.

And you would be quite correct in deducing from that two things:

The first is that I should not be here this morning if I did not believe that those engaged in the field of 'Human Population Genetics' were doing a very good job indeed and that the future was full of exciting promise.

The second is that I am fully aware of my incompetence to meet you on your special ground. But there is one ground that I want to meet you on.

One cannot spend 20 years as the chief executive officer of a central research organisation without forming some idea of the pattern that developing biomedical research tends to follow, and acquiring some views on the factors that determine its progress. This morning, therefore, I propose to look briefly at cytogenetics from outside, with the object of trying to identify the factors that have contributed to your past success and which are likely to be equally important in determining the future progress of your subject. Before embarking on this, however, it might be as well if we looked together at a line of research that is sufficiently far from your interests that we can both approach it as objective outsiders.

The field I have in mind is nutrition and the example I have chosen is the train of research that led through to the discovery of the B vitamin and its role in intracellular metabolism.

The story starts with the identification of the disease beri-beri and the characterisation of its pathology both in its cardiac and neuritic types. There knowledge halted until an event that the obtuse would call 'chance' happened. You remember Pasteur's observation, 'In the field of research chance favours only the prepared mind.' The event was, of course, a disastrous outbreak of beri-beri in the Japanese Navy. Fortunately, in the person of Takaki, there was a prepared mind watching and this event was rapidly followed by the epidemiological investigation showing that the outbreak had followed the substitution of highly milled for unmilled rice in the sailors' diet: then we were off. It was soon shown that an analogous condition could be produced in experimental animals by a diet of milled rice and that this could be prevented by adding the polishings removed by milling to the diet. Experimental pathology had now become feasible. And biochemistry came hard on its heels. The active factor was isolated from the rice polishings, analysed and then synthesised.

Thiamine, or Vitamin B, had been discovered. But the story did not stop there. The question could now be asked, 'How does Vitamin B work in the metabolism of the cell?' Its role in co-enzymes was elucidated. And the story is continuing today – at the molecular level. It is a classical story.

But what were the factors in its success? Well, may I point out certain features?

First: It is evident that many disciplines from the clinical to molecular biology were necessary to achieve this result. None alone would have sufficed.

Second: These disciplines were not mixed up in a random fashion. Progress was marked by their coming into play in an orderly sequence. Clinical medicine, pathology, epidemiology, experimental pathology, biochemistry, organic chemistry, intracellular chemistry – and they could only be deployed effectively in that sequence. Of course work in these fields was not in abeyance until the developing sequence was reached. It was developing its own knowledge. My point is, however, that research in these fields became intellectually relevant to the subject of nutrition in proportion to the development of the sequence.

Third: This sequence constituted a frame of reference within which each discipline could check the intellectual relevance of its activities. Thus the organic chemist, at every step, could check if he was on the right track by testing the substances he was producing on the experimental condition in animals. And if this was satisfactory he could check further by getting the clinician to use it on the real patient. I ask you to consider how far the biochemistry of the vitamins could have got if it had not been linked to pathology.

Now consider another field, nearer your own.

OPENING ADDRESS

Twenty years ago, I first came in contact with radiobiology. Until then my acquaintance with radiations had been limited to signing request forms for x-rays. It was not surprising, therefore, when I entered on my new post that one of the first things I did was to tour the radiobiological centres of this country. And I became more and more disconcerted. I couldn't see any shape or direction in the work. It seemed to be stretching pseudopodia out tentatively and at random in all directions. And the situation was made no easier by everybody asserting that he was doing fundamental research. It then dawned upon me that nobody had ever studied patients exposed to fatal doses of ionising radiations or done a post-mortem on one. They had no frame of reference within which to check the relevance of their ideas or to see the significance of their results. Well, I tried to found a monkey hospital but before this got far reports began to come in on the appearance of leukaemia in the survivors of the atomic bomb explosions at Nagasaki and Hiroshima. Prompted by these, people recalled the rumours about an increased incidence of leukaemia in radiologists. And then Court Brown and Abbatt came out with their work on the increased incidence of this condition in patients with ankylosing spondylitis treated with radiation. Of course, this was only one facet of the whole subject. But the whole atmosphere altered. There was now a sense of direction and in consequence a quickening of confidence that spread over into contiguous studies.

Now let us come to your own field of cytogenetics and – as we are in Edinburgh let me take some examples from the work of the institution that Michael Court Brown created.

Consider the work that established that the Y chromosome plays the key role in male sex determination. Until 1959 it was believed, on the basis of work on *Drosophila*, that the Y chromosome was inert. Male sex, it was believed, was determined by autosomal genes which, in the female, were checked by her having two rather than one X sex chromosome. There the matter stuck until Jacobs and Strong ascertained the karyotype of cases of Klinefelter's syndrome. According to the theory that the Y chromosome was inert these, with their XXY composition, should have been females. But they were not. They were males – and the positive effect of the Y chromosome in determining the male sex was thus established. Again consider the question of a man found to have a mental defect when there is more than one X chromosome associated with the Y. Or consider the association of unusual height and psychopathic personality associated with a YY configuration.

In each of these cases neither the karyotype nor the clinical features – sex type, mental defect, tallness – taken alone are much more than mere descriptions. Taken together, however, they at once acquire a striking significance. Not only do they give a strong indication of a link between structure and function but they indicate the structure from which the function derives. All very obvious, you may say. But consider the implications for future research.

Clearly the means by which changes in the nuclear material produce their effects must be through the agency of intracellular biochemistry and biophysics. When we can establish the precise linkages between chromosome structure and intracellular chemistry the advance in our understanding will be incalculable. At one time that seemed no more than a dream. But would I be wrong in seeing in the links that have been established between the karyotype form and the clinical picture the beginning of sequences of research comparable to that which led from beri-beri to the concept of deficiency diseases and the characterisation of vitamins and their metabolic significance?

Again take the salutary shock that was delivered to our tacit assumptions by the discovery that mongolism was associated with an *extra* chromosome. We had got so used to regarding the mongol as deficient that I am sure that, before this discovery, we should all have expected something to be *missing* from the karyotype. But doesn't the actual finding suggest that the essential process is inhibitory and point to a possible link with something comparable to the repressor genes. In any case, we are beginning to formulate questions. We can't help it now. Findings of this kind are intellectual irritants when previously our situation was bordering on passive acquiescence.

That is the approach through the clinical syndrome. But there is another contiguous approach to more subtle conditions. That through epidemiology. Court Brown was already beginning to speculate on the association of karyotype and morbidity. Of course, the natural association of cytogenetics with classical genetics ensured that the epidemiological approach came readily to mind. And the necessary foundations are now being laid in the surveys of the incidence of the various karyotypes in the general population. Clearly this was necessary to establish which were compatible with normality. But already it has thrown up a series of disconcerting observations. It has revealed that noxious influences like infections can produce nuclear disorganisation. That others like radiations can produce new clones of cells. Further, that with age the sex chromosomes are affected so that, in man, an increasing number of cells are found that have lost either an X or a Y chromosome. Then there are the discrepancies – and in research a discrepancy is always an opportunity. If the incidence of the XYY karyotype in newborn male babies is 1 in 1,000, why is it apparently so uncommon in adult man? Does it predispose to early death or – dare I say it – does the second Y 'disappear' early as the Y of the ordinary XY disappears from a proportion of cells with advancing age?

[5

Then, take the cytogenetics of cancer. The evidence that a cancer may be made up of a series of lines of cells with different karyotypes – and that these may have different susceptibilities to drugs, hormones or physical agents – may well be one of the most fruitful ideas not only for medicine but for normal cell biology. It was Claude Bernard who said that drugs were instruments for the dissection of function. May this not be a way in to dissect out the functional chemistry of the cell that is related to changes in the structure of its nuclear controlling mechanism?

By this time I may have lost you. But this is what I am driving at:

Just as the biochemistry of the vitamins was an indispensable component of the sequence of knowledge that ran from beri-beri down to molecular biology so, in my opinion, is cytogenetics an indispensable component of a whole series of other sequences. But no subject can realise its full significance in isolation. It is only by developing its links with the context in which it naturally mixes that it can realise its full significance. Divorce it from its context and cultivate it for its own sake and it will not be long before it begins to show the symptoms of intellectual anaemia. It is on the interplay of knowledge, between disciplines, along such natural sequences that conceptual progress depends. As Herbert Spencer put it over a century ago : 'A more general science as much owes its progress to the presentation of new problems by a more special science, as the more special science owes its progress to the solutions that the more general science is thus led to attempt.'

Your sequences start with the abnormal – and never forget that the abnormal is still an experiment even though it is done by God rather than yourself. At one extreme of your sequences are clinical medicine and epidemiology. Pathology and cytogenetics follow on. And that is largely where the bulk of our achievement has got to. But looking to the future, and seeing the probing advances already in progress. The next in line is biochemistry, biophysics and so to the molecular level. That will be the broad form of your intellectual sequences when with further development they become established throughout their full range.

And what is the moral of all this? It is in my opinion that the progress of your subject depends to a large degree on men throughout the whole range of relevant disciplines keeping alive their awareness of, and their receptivity to, the significance of findings throughout the whole length of relevant studies. It was Alvin Weinberg, a nuclear physicist, who concluded that in the biomedical field the terms fundamental and applied had little meaning. In the biomedical field research at all levels is equally fundamental to the appreciation of intellectual significance.

And that brings me to the man who must have been in all our minds this morning: Michael Court Brown. As you know, he organised this conference before he died. If you like, I am here today as a token of my respect for him. My predecessor, Edward Mellanby, used to divide research workers into discoverers and investigators. Court Brown had a large element of the authentic discoverer in him. His interests were so comprehensive and his receptivity to potential significance so wide-ranging that he saw ahead with very real confidence. I know what his plans were at his untimely death and I know their promise. He had a whole range of relevance under his view. It is perhaps because our basic philosophies of scientific development were so much in accord that I count having known him as one of the events of my professional life.

Observations on Meiosis in Normal Males and Females

R. G. EDWARDS

Physiological Laboratory
Cambridge

¶THIS WORK has been carried out in collaboration with Dr S. A. Henderson. The majority of the data involves the study of the oocyte since this is our major area of research. The earliest stages of prophase of the first meiotic division occur in the foetal or neonatal ovary in mammals. After diplotene, oocytes enter dictyotene shortly before or after birth of the mother, and this stage persists until the mature oocytes commence their pre-ovulatory development in the adult ovary. The surge of LH in mid-cycle stimulates meiosis to be resumed in the oocytes. After dissolution of the nucleus (germinal vesicle), the stages between diakinesis and metaphase of the second meiotic division (metaphase-II) follow rapidly. In most species, oocytes are ovulated while in metaphase-II, and anaphase-II and telophase-II occur after fertilisation. In the male, the sequence is quite different, for all stages of meiosis occur in the testis of the adult.

Chromosomal analysis of oocytes and spermatocytes. Analysis of the conformation of bivalents is best made at diakinesis and metaphase-I. Oocytes must therefore be collected following endogenous LH secretion, or after the injection of gonadotrophins into the mother. Techniques of superovulation in animals provide some oocytes, especially in laboratory animals, but a different approach is much more useful. This is to liberate oocytes from the follicles into a suitable culture medium; meiosis is then resumed. First noted on rabbit oocytes in 1935 [1, 2], later work has extended to other species including man [3, 4]. Pyruvate in the culture medium is evidently

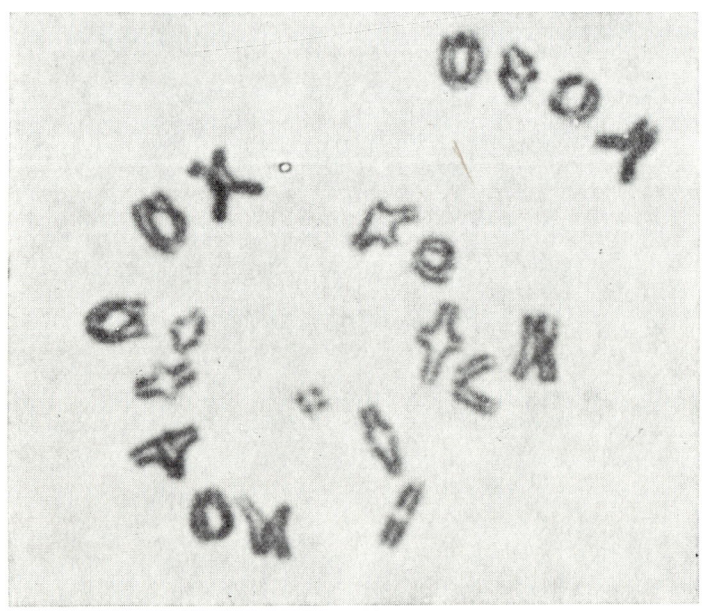

FIGURE 1
Diakinesis in a mouse oocyte with 28 chiasmata. Courtesy *Nature*

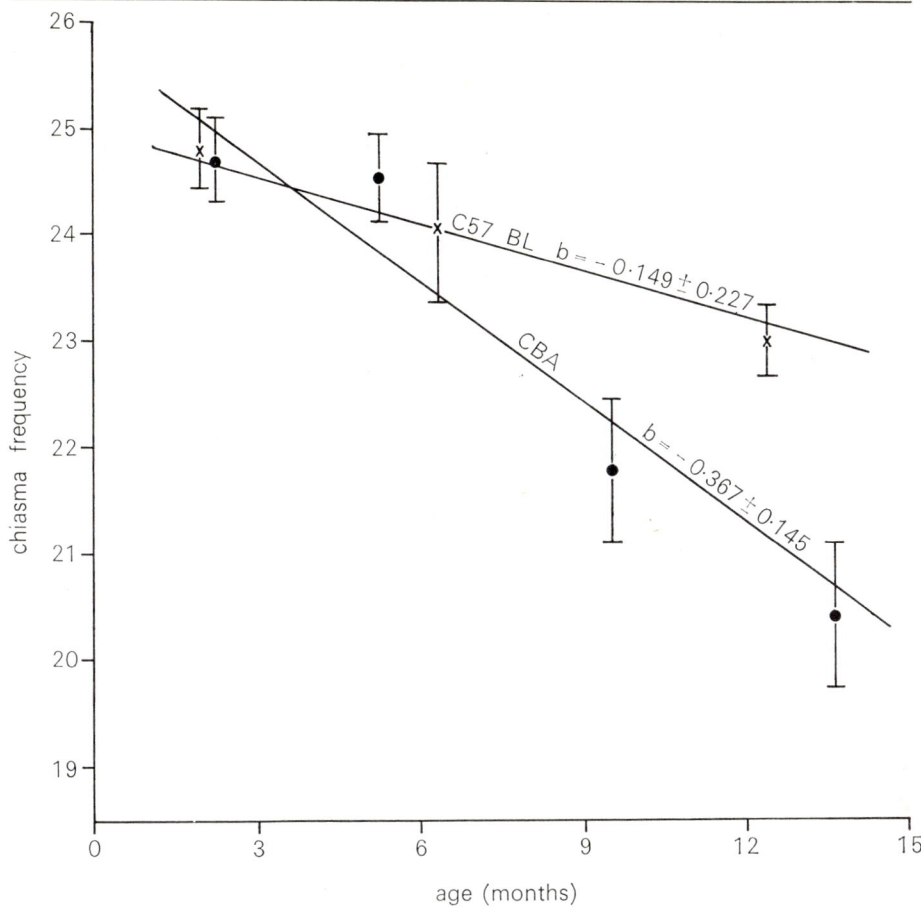

FIGURE 2
Decline in chiasma frequency with increasing maternal age in CBA and C57BL mice. In male CBA mice age 2 or 9 months, mean chiasma frequency was 24 and 26 respectively. Courtesy *Nature*

essential for maturation [5]. Oocytes from any particular species mature synchronously *in vitro*; their exact stage of maturation at a specific time can thus be predicted with accuracy.

Until more is known about maturation *in vitro*, it is wise to run parallel studies in oocytes maturing *in vivo* [6] to ensure that results are similar. Recent observations on diakinesis and metaphase-I indicate that few, if any, differences exist between the chromosomal configurations of mouse oocytes matured *in vitro* and *in vivo* [7]. Human oocytes can be obtained in relatively large numbers from excised ovaries, and up to 80 per cent of these will mature *in vitro*. With the development of an excellent technique for flattening the oocytes for chromosomal examination [8], detailed studies can now be carried out on them without difficulty.

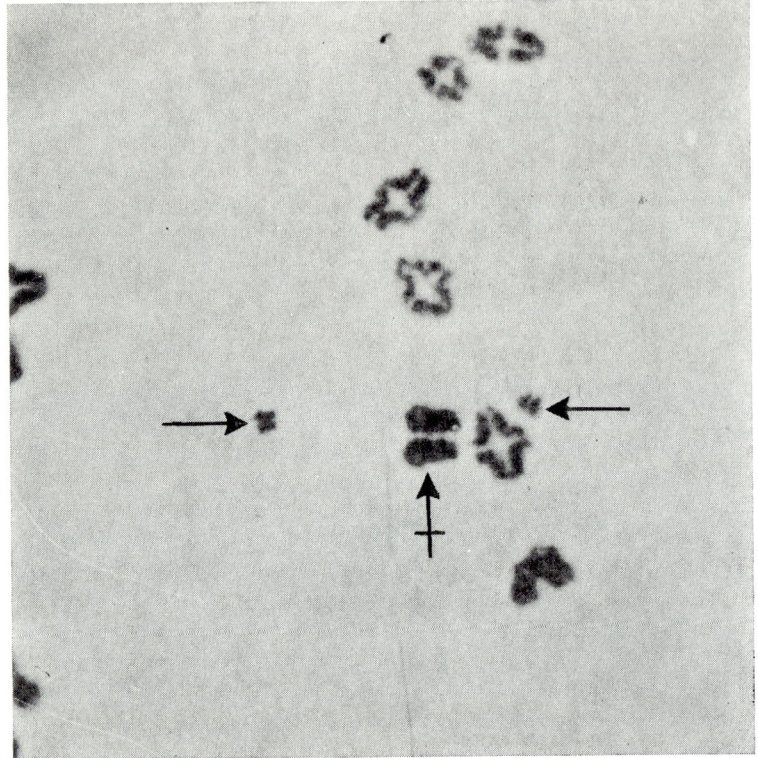

FIGURE 3
Univalents showing secondary lateral association (⊻) or completely separate (↓).
Courtesy *Nature*

We carried out a detailed study of mouse oocytes in diakinesis and metaphase-I, and examined similar stages in spermatocytes [7]. We have added little to this data in the past year, and I will summarise our data briefly. Clear preparations of oocytes in diakinesis and metaphase-I were obtained (Fig. 1). The number of chiasmata per oocyte declined significantly with increasing maternal age in oocytes of *CBA* mice (Fig. 2), and less rapidly in *C57BL* mice. There was no decline in chiasma frequency in spermatocytes with increasing paternal age; if anything, there was a slight increase. Chiasma frequency was thus much lower in oocytes of older females than in spermatocytes from older males. The decline in chiasma frequency in oocytes with increasing maternal age was accompanied by the localisation of chiasmata to the terminal parts of the bivalents, and by the presence of univalents. Some of the univalents were found to be widely separated from their partners (Fig. 3); others displayed some secondary association, occasionally with the wrong partner. The incidence of univalents was also correlated with maternal age. Almost three quarters of the

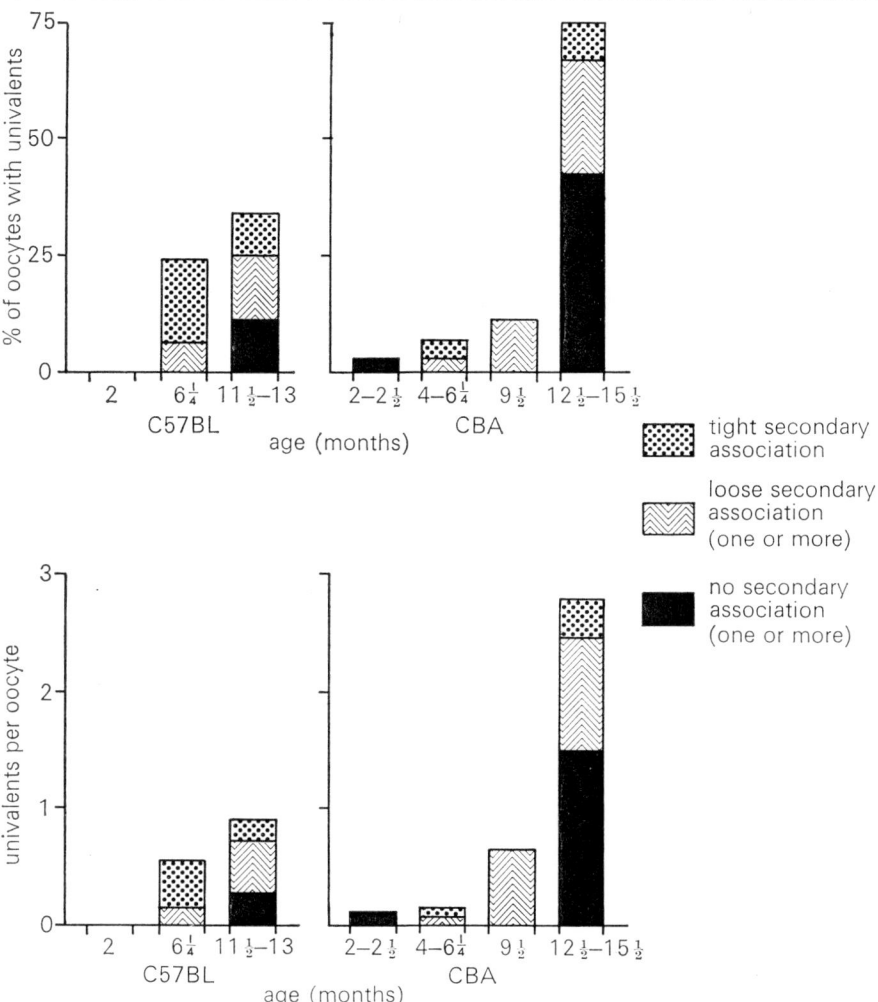

FIGURE 4
Incidence of univalents in oocytes of CBA and C57BL mice with increasing maternal age.
Courtesy *Nature*

oocytes taken from *CBA* females more than one year old possessed two or more univalents (Fig. 4). The citrate pre-treatment of oocytes might have caused some univalents to form or separate, although if so, the frequency of this effect was more common in older oocytes.

Human oocytes matured *in vitro* have also been studied (Figs 5 and 6). The intervals between various stages of development are shown in Figure 7 when medium 199 plus 15 per cent (v/v) foetal calf serum was employed. The use of a medium richer in pyruvate [9] in combination with follicular

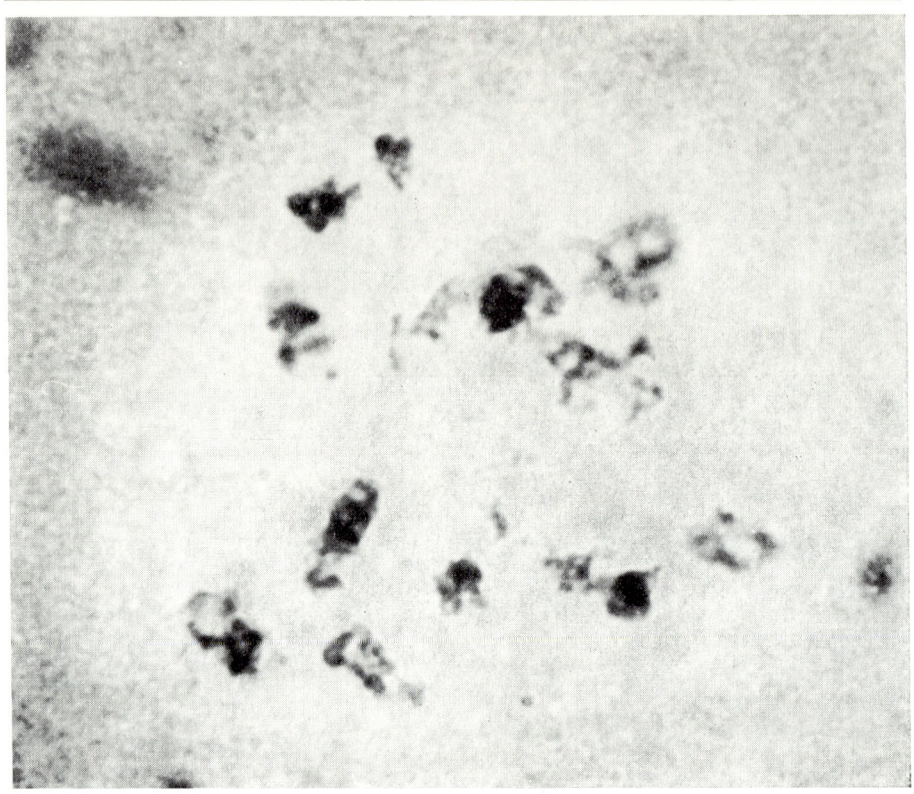

FIGURE 5
Early diakinesis in a whole mount of a human oocyte. The nucleolus can still be seen, and some of the chromosomes are in association with it. Courtesy *Lancet*

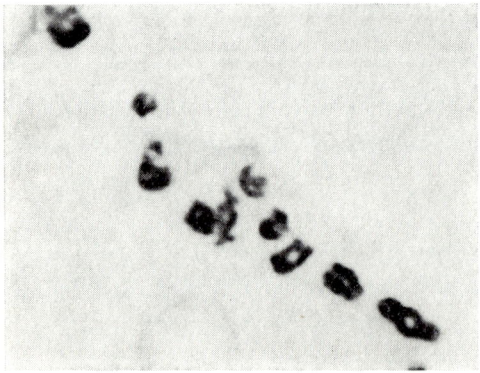

FIGURE 6
Detail of metaphase-I in a human oocyte

Maturation of human eggs *in vitro*

	Hours in culture				
	0–24	25–27	28–31	32–36	> 36
Germinal vesicle	97	44			
Diakinesis/Metaphase I		52	76	18	6
Anaphase I			7	4	
Metaphase II		3	4	16	77	93

FIGURE 7
Timing of meiosis in human oocytes cultured in medium 199 and 15% foetal calf serum. After 28 hours in culture, eggs still in the germinal vesicle stage were excluded. Changes in the culture medium might alter these timings

fluid has given us timings more rapid than those shown in Figure 7. The effect of different media on human oocyte maturation requires analysis.

Human oocytes displayed similar chromosomal anomalies to those found in mouse oocytes. Univalents showing weak association [10] or completely separate, were found at diakinesis and metaphase-1 (Fig. 8). At least one oocyte, taken from a 35-year-old woman, contained a widely separated

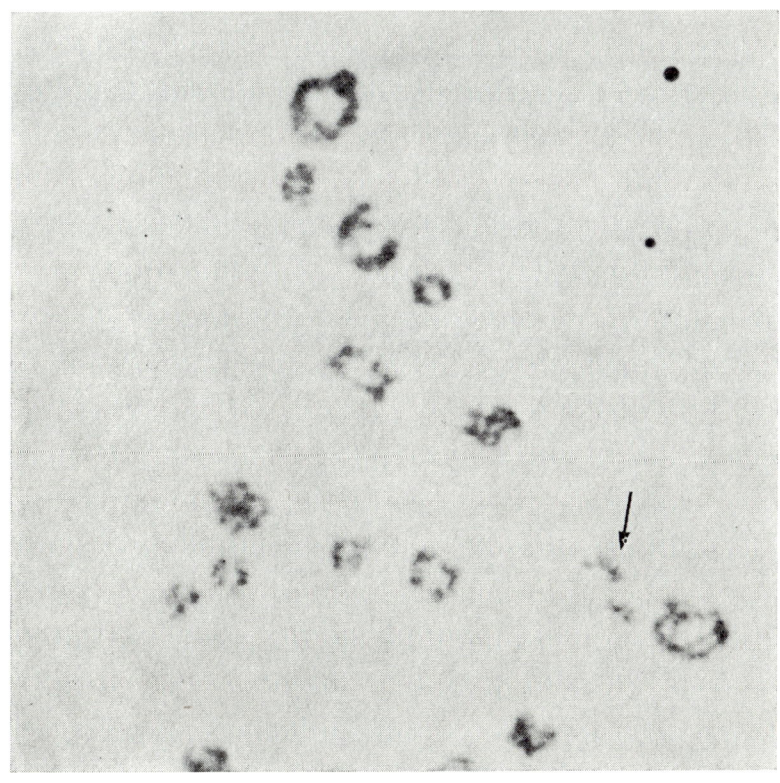

FIGURE 8
Detail of a human oocyte, showing two chromosomes with loose secondary association (↓)

univalent pair in the 21–22 group (Fig. 9). Chiasma frequency in 15 oocytes was estimated as between 37 and 48, although insufficient oocytes have been scored to correlate chiasmata frequency with maternal age. Our estimates are lower than those reported by Kjessler [11] in human spermatocytes, although we must stress that our results are still preliminary. Yuncken [12] reported a chiasma count of 52 in a human oocyte matured *in vitro*, and our estimates of the chiasma frequency in published photographs of oocytes recovered from women given gonadotrophins [13] were between 46 and 51.

Some human oocytes were studied during anaphase-I or early metaphase-II (Figs. 10 and 11). In one oocyte, out of 33 examined at similar times, one bivalent had failed to separate at anaphase (Fig. 10). This condition was rare in comparison with errors of bivalent association at diakinesis. We have studied very few oocytes at metaphase-II (Fig. 11). Now that we can fertilise human eggs *in vitro* [14], it should be possible to examine anaphase-II and telophase-II. We have seen telophase-II, the first ever to be recorded [15]; there were no lagging chromosomes in this first egg to be examined.

The similarities between observations on human oocytes and on mouse oocytes, in material so far examined, are thus the presence of univalents and the lower frequency of chiasmata as compared with spermatocytes.

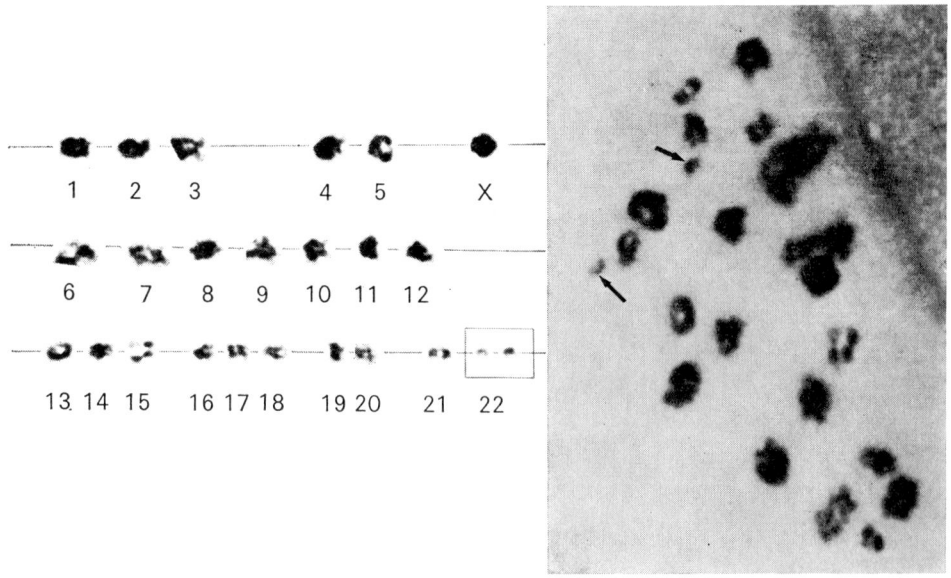

FIGURE 9

Metaphase-I in a human oocyte, showing two univalents in the 21–22♯ group. This oocyte could have been the precursor of a mongol foetus

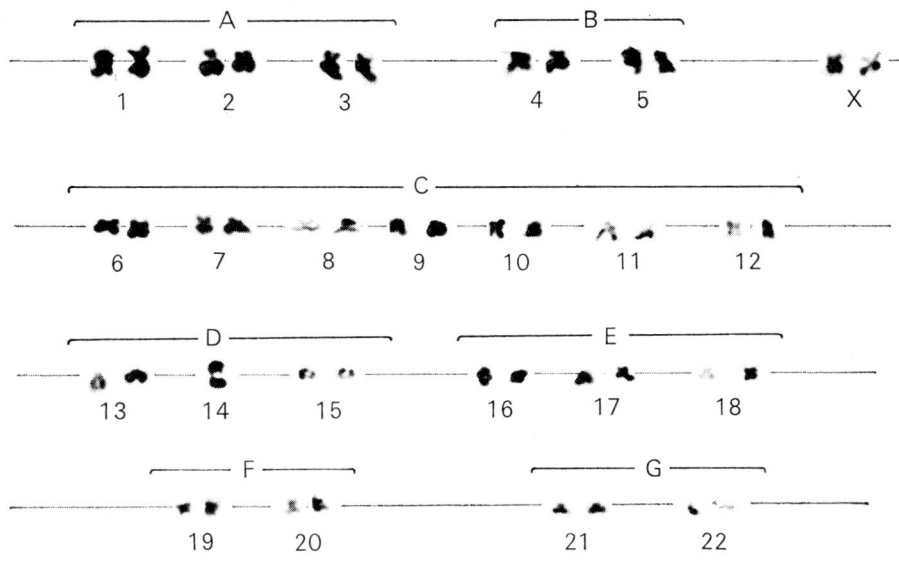

FIGURE 10
Anaphase-I in a human oocyte. One bivalent in the D group has evidently failed to separate

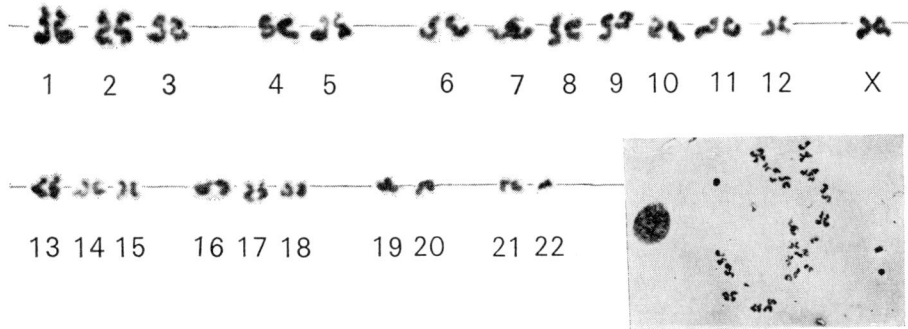

FIGURE 11
Metaphase-II in a human oocyte

The origin of monosomy and trisomy in man. How do these observations relate to the origin of chromosomal anomalies in man? Chromosomal studies of spontaneous abortuses have revealed that at least 20 per cent or more are chromosomally unbalanced [16], but this might be an underestimate since the more grossly abnormal embryos may die very early in pregnancy.

Triploid foetuses are found, and could obviously arise through polygyny or polyandry. Most imbalanced abortuses ($>$60 per cent) are trisomics or monosomics, and since the incidence of many autosomal trisomics is correlated with maternal age, the anomalies presumably arise in the egg or embryo. Sex-chromosome imbalance is due to errors in either spermatogenesis or oogenesis, as shown by the use of Xg to trace the origin of the sex chromosomes and by analysis of the frequencies of different Xg phenotypes [17].

Our observations on univalents in mouse oocytes, which were more often found among the smaller chromosomes, could explain the cause of abnormal segregation at anaphase-I. It is likely that enough abnormal embryos would arise to account for the incidence of imbalanced human conceptuses reported in the literature. So far formal proof of our hypothesis is lacking because there have been no studies on the chromosome complement of embryos of older mice or women. Extraneous factors could interfere with the pairing of chromosomes in oocytes. Univalents showing secondary association might become orientated normally on the anaphase-I spindle, unless external factors disrupted their regular orientation, and so temporarily increased the incidence of chromosomally imbalanced embryos.

The decline in chiasma frequency is evidently a primary event occurring during prophase of meiosis-I in the foetal ovary, and is not due to the loosening or movement of chiasmata ('terminalisation') during the prolonged dictyate stage. This conclusion is based on evidence that the recombination frequency between some linked genes in the mouse declines considerably with maternal age and not with paternal age [18, 19]. Crossing-over must have occurred during synapsis in the foetal gonad, and the decline in recombination frequency with maternal age must therefore have been due to events occurring in the foetal ovary. Terminalisation is excluded because this process would not lead to the decline in recombination frequency.

Why are oocytes with low chiasma frequency ovulated towards the end of the reproductive life of the female? The most likely explanation is that a 'production line' of eggs exists in the foetal ovary, which is reflected in the regular progression of eggs through ovulation during adult life [7]. Oocytes formed early in the foetal ovary have a high chiasma frequency and are ovulated early in adult life. Those formed later in the foetus have lower chiasma frequencies (perhaps due to environmental changes in the foetal ovary) and are ovulated in later life. In plants, developmental or nutritional factors can affect chiasma frequency in gametes formed successively [20], and sufficient time elapses during formation of oocytes in the human foetal ovary for a production line to develop there. Evidence of environmental changes in the foetal ovary may be provided by the detectable waves of oocyte degeneration at different embryological stages [21]. Embryological evidence also indicates a sequence of oocyte formation in man [22]. Thus, histological, embryological, genetical and indirect evidence supports the idea of a production line in the foetal ovary and, perhaps, the ordered ovulation of oocytes along this production line during adult life. The primary sequence could be laid down during the formation or migration of primordial germ cells, those entering the foetal ovary first being the first to enter meiosis, etc.

This theory has been criticised. Movement of chiasmata might occur through 'terminalisation', but the genetic and cytological evidence presented above argues against this explanation for a decline in chiasma frequency with age. Another criticism is that, if our theory is correct, embryos of an old female should fail to undergo development because they are intrinsically abnormal. This possibility can be tested by transferring embryos from old females into the uteri of young females. In hamsters, only 5 per cent of embryos from older mothers developed into foetuses when they were transferred into young recipients, as compared with almost 50 per cent of embryos from young females so transferred [23]. This evidence supports the production line theory. Many ova recovered from older mice are grossly abnormal [24], which could be further support for the theory. But experiments on ovum transfer in mice were judged to show that no differences existed in development viability between embryos from young and old mice [25], and have been considered as evidence against the 'production line' theory. However, this experiment is totally inconclusive, for there was indiscriminate use of mice of different inbred and hybrid strains as embryo donors or recipients. There are considerable differences. between strains in the decline of chiasma frequency with age [7], hence we now know that it is invalid to mix data from different strains in this way.

Our theory can also explain why trisomics and monosomics for the X chromosomes arise through non-disjunction in the egg or spermatozoa [17]. The XY bivalent in the testis shows an end-to-end association, and the X and Y chromosomes are often found as univalents in squash preparations [26]. The cytogenetic behaviour of the sex bivalent in the testis thus has similarities with that of autosomal and sex bivalents in oocytes. It is therefore not surprising that sex chromosome trisomics can arise through non-disjunction in either parent, whereas autosomal trisomics arise mostly through the oocyte.

Other suggestions have been advanced to explain the origin of trisomics, although with little supporting evidence; we have discussed some of these elsewhere [7, 27]. Non-disjunction has been attributed to persistence of the nucleolus preventing separation of the bivalents associated with it [28], for the X chromosomes, and those in groups D, E and G, are involved in known trisomic conditions. But there is no experimental evidence to confirm this suggestion. Delayed fertilisation resulting from sporadic or reduced frequency of coitus in older couples has been evoked as a cause of non-disjunction at metaphase-II [29]. The supporting data is mainly statistical, rests on many assumptions and does not sufficiently explain the age dependence. The decreased frequency of coitus is not exclusively confined to older women; age-dependency for frequency of coitus is insufficiently related to the age factor in mongolism [30]; and other statistical data fail to support the hypothesis. Cannings and Cannings [30] suggest that the long period of degeneration of the oocyte occurring in older women or a shortening of the life of spermatozoa in the female tract may be more important. Their suggestion rests on ground equally as tenuous as German's [29]. Moreover, experimental data from animals indicate that neither delayed ovulation nor delayed fertilisation are important factors in causing trisomy. Delayed ovulation of eggs in rats resulted in 18 out of 390 embryos being chromosomally abnormal (cp. 6/410 in controls), most of them being mosaics [31]. Similarly, the data on chromosome anomalies in embryos after delayed fertilisation of animal eggs [32–35] reveal that triploidy or mosaicism may result but not trisomy.

Thyroid disease and thyroid antibodies in the mother are considered of primary importance in the aetiology of Down's syndrome [36]; a specific mutant protein is postulated to combine with homologous chromosomes or chromatids, and so prevent anaphase separation during the first or second meiotic divisions. It is difficult to place any value in such a hypothesis. In Addison's and Cushing's disease, autoantibodies reacting with ova are found in a few patients [37].

External factors could influence chromosomal separation. Perfusion of intact canine testes with progesterone causes meiotic chromosomes from pachytene onwards to stick together [38]. Generalised stickiness could disrupt the extrusion of the first and second polar bodies in eggs, or cause separation of the bivalents, especially those showing secondary association. The meiotic chromosomes of male mice treated with lysergic acid diethylamide (LSD) showed an increase in the incidence of chromosome breaks, gaps and fragments [39]. Infections could be another source of non-disjunction, according to some statistical evidence [40, 41]. An infective cause of mongolism would explain the clustering reported in various studies, but an association with disease is denied as often as it is reported [42, 43]. The lack of supporting data on meiosis or on embryos leaves considerable doubt about the validity or applicability of these theories.

ACKNOWLEDGEMENT

I am indebted to the Ford Foundation for financial assistance.

REFERENCES

[1] Pincus, G. and Enzmann, E., *J. exp. Med.,* **62,** 665, 1935.
[2] Chang, M. C., *J. exp. Zool.,* **128,** 379, 1955.
[3] Edwards, R. G., *Nature, Lond.,* **208,** 349, 1965.
[4] Edwards, R. G., *Lancet,* **2,** 926, 1965.
[5] Donahue, R. P. and Stern, S., *J. Reprod. Fert.,* **17,** 395, 1968.
[6] Edwards, R. G. and Gates, A. H., *J. Endocrinol.,* **18,** 292, 1959.
[7] Henderson, S. A. and Edwards, R. G., *Nature, Lond.,* **218,** 22, 1968.
[8] Tarkowski, A. K., *Cytogenetics,* **5,** 394, 1966.
[9] Bavister, B. D., *J. Reprod. Fert.,* **18,** 544, 1969.
[10] Edwards, R. G. and Henderson, S. A., *Proceedings 6th Fertility and Sterility Conference, Tel Aviv,* 1968.
[11] Kjessler, B., *Karyotype, meiosis and spermatogenesis in a sample of men attending an infertility clinic.* Basle: Karger, 1966.
[12] Yuncken, C., *Cytogenetics,* **7,** 234, 1968.
[13] Jagiello, G., Karnicki, J. and Ryan, R. J., *Lancet,* **1,** 178, 1968.
[14] Edwards, R. G., Bavister, B. D. and Steptoe, P. C., *Nature, Lond.,* **221,** 632, 1969.
[15] Bavister, B. D., Edwards, R. G. and Steptoe, P. C., *J. Reprod. Fert.* **20,** 159, 1969.
[16] W. H. O. Conference. Standardisation of procedures for chromosome studies in abortion. *Cytogenetics,* **5,** 361, 1966.
[17] Race, R. R. and Sanger, R., *Br. med. Bull.,* **25,** 99, 1969.
[18] Bodmer, W. F., *Nature, Lond.,* **190,** 1134, 1961.
[19] Reid, D. H. and Parsons, P. D., *Heredity,* **18,** 107, 1963.
[20] Rees, H. and Naylor, B., *Heredity,* **15,** 17, 1960.
[21] Baker, T. G. and Franchi, L. L., *J. Cell Science,* **2,** 213, 1967.
[22] Van Wagenen, G. and Simpson, M. E., *Embryology of the ovary and testis. Homosapiens and Macaca mulatta.* New Haven: Yale University Press, 1965.
[23] Blaha, G. C., *Anat. Rec.,* **150,** 413, 1964.
[24] Boot, L. M. and Mühlbook, O., *Acta physiol. pharmac. néerl.,* **3,** 463, 1954.
[25] Talbert, G. B. and Krohn, P. L., *J. Reprod. Fert.,* **11,** 399, 1966.
[26] Ford, C. E. and Hamerton, J., *Nature, Lond.,* **178,** 1020, 1956.
[27] Edwards, R. G. and Edwards, Ruth E. In *Modern Trends in Human Genetics.* Ed. A. E. Emery. London: Butterworth, 1969.
[28] Evans, J. H., *Nature, Lond.,* **214,** 361, 1967.
[29] German, J., *Nature, Lond.,* **217,** 516, 1968.
[30] Cannings, C. and Cannings, M. R., *Nature, Lond.,* **218,** 481, 1968.
[31] Butcher, R. L. and Fugo, N. W., *Fert. Steril.,* **18,** 297, 1967.
[32] Austin, C. R., *Nature, Lond.,* **213,** 1018, 1967.
[33] Gates, A. H. and Beatty, R. A., *Nature, Lond.,* **174,** 356, 1954.
[34] Shaver, E. L. and Carr, D. H., *J. Reprod. Fert.,* **14,** 415, 1967.
[35] Vickers, A. D., *J. Reprod. Fert.* (in press).
[36] Burch, P. R. J., *Nature, Lond.,* **221,** 173, 1969.
[37] Irvine, W. J., Chan, M. M. W., Scarth, L., Kolb, F. O., Hartog, M., Bayliss, R. I. S. and Drury, M. I., *Lancet,* **2,** 883, 1968.
[38] Williams, D. L., Runyan, J. W. and Hagen, A. A., *Nature, Lond.,* **220,** 1145, 1968.
[39] Skakebaek, N. E., Philip, J. and Rafaelsen, O. J., *Science,* **160,** 1246, 1968.
[40] Stoller, A. and Collman, R. D., *Lancet,* **2,** 1221, 1965.
[41] Stoller, A. and Collman, R. D., *Nature, Lond.,* **208,** 903, 1965.
[42] Stark, C. R. and Fraumeni, J. F., *Lancet,* **1,** 1036, 1966.
[43] Leck, I., *Lancet,* **2,** 457, 1966.

The Behaviour of Structural Aberrations at Male Meiosis
Information from Man

M. HULTÉN *and* J. LINDSTEN

Department of Endocrinology and Metabolism
Karolinska sjukhuset, Stockholm

¶ STUDIES on mitotic cells are often sufficient to elucidate, if not the origin, at least the nature of chromosome abnormalities. In many instances, however, the diagnosis of such abnormalities remains at best presumptive on the basis of mitotic studies alone. In these cases meiosis may be of diagnostic value. Meiotic studies are also of theoretical and practical significance for an understanding of the modality of pairing and for the assessment of gamete output in subjects with normal and abnormal chromosome constitutions. The literature on human male meiosis is still very limited (reviews in Fraccaro et al. [1], and Luciani [2]).

The present paper is mainly devoted to a discussion of the appearance of the male meiotic chromosomes in one trisomic mongol, two D/G translocation carriers, two D/G translocation mongols, two D/D translocation carriers, two carriers of a small extra unidentified chromosome, and one carrier of an A/G reciprocal translocation which could not be identified in the mitotic karyotype.

MATERIAL AND METHODS

The cases studied were known from previous analysis of cells from lymphocyte cultures to have abnormal mitotic karyotypes. The testicular biopsies were obtained through a small incision made under local anaesthesia using Citanest[R] (prilocaine chloride) or under short-term general anaesthesia using Fluothane[R] (halothane) and nitrogen monoxide (the D/G translocation mongols). No complications to the biopsy were noted. Permission to perform the biopsy in the patients with mongolism had been obtained from both parents and from the physician at the institution for the mentally retarded where the patients were staying.

Four subjects with a normal male karyotype were used as controls. One was a 23-year-old healthy man with normal testicular histology previously published by Hultén et al. [3]. His spermiogram was normal five months after the testicular biopsy (100 million sperm per ml, 33 per cent morphologically abnormal). Two controls were orchidectomised because of prostatic cancer. They were 52 and 72 years old, respectively, and had normal testicular histology [4]. The fourth was the 46-year-old healthy non-carrier father of the D/G translocation carrier and mongol described below. He too had a normal spermiogram and testicular histology (51 million sperm per ml, 27 per cent morphologically abnormal).

Meiotic preparations were made using a modification of the air-drying technique of Evans et al. [5].

There are several practical and technical limitations involved in this type of study. The access to subjects with known chromosome aberrations willing to give a testicular biopsy is limited, and so is control biopsy material from healthy males of various ages with normal spermiograms and testicular histology. Furthermore, only small biopsies can be obtained. In our hands, the air-drying technique has yielded cells suitable for analysis mainly at the first spermatocyte stage. The spermatogonial metaphases are few

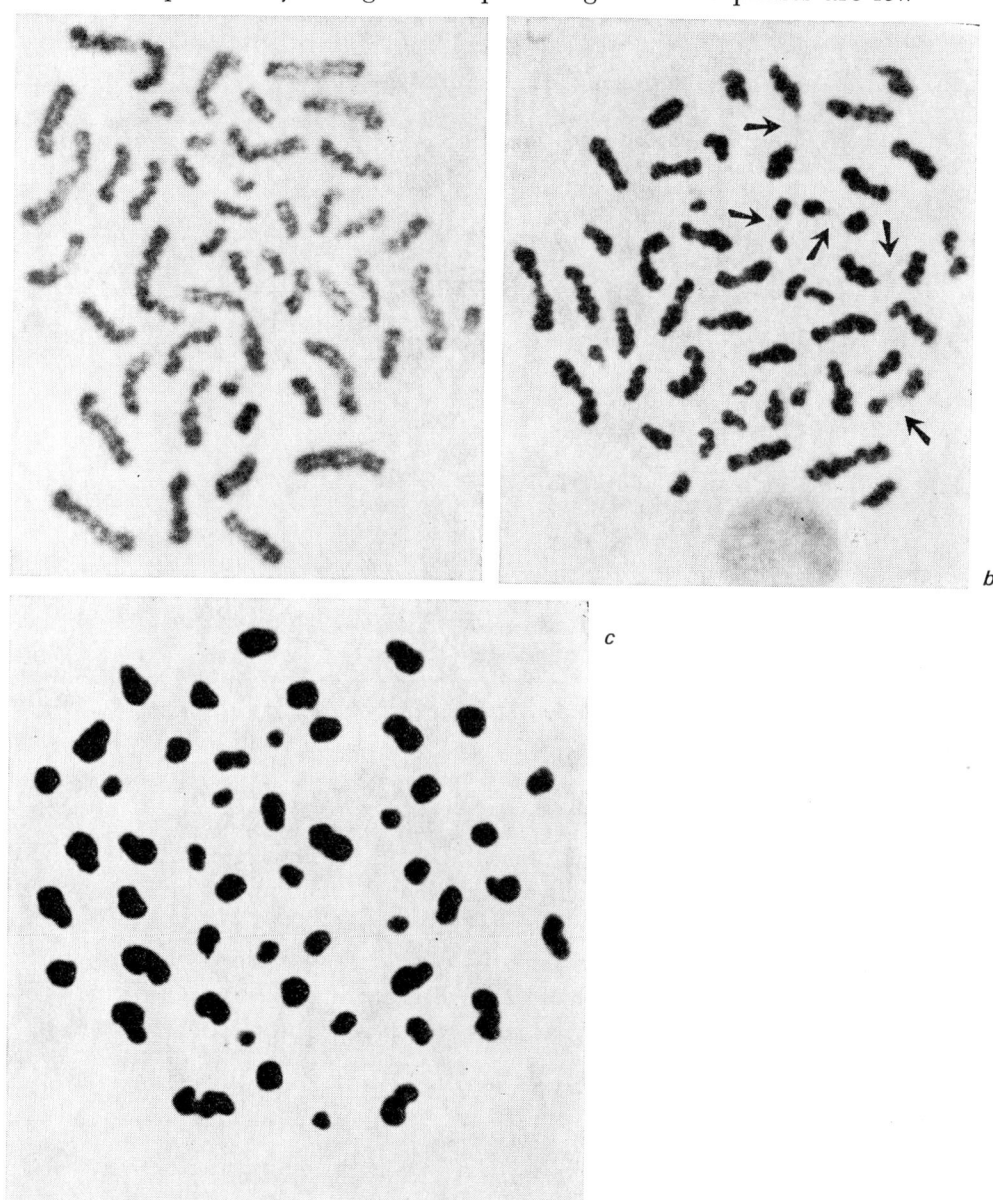

FIGURE 1

Two spermatogonial metaphases with 46 chromosomes (*a, b*) and one with 50 'chromosomes' (*c*). Note the extended secondary constrictions in *b* (arrows); from Hultén et al. [22]

TABLE 1. Number of analysed spermatogonial metaphases

Subject	Chromosome constitution	Number of cells with the chromosome number shown below								Total
		≤44	45	46	47	48	49	50	51	≥52	
C.M.	47,XY,21+	–	–	–	3	2	1	4	–	–	10
B.K.	45,XY,D–,G–,t(DqGq)+	–	4	–	1	2	–	–	–	–	7
J.K.	45,XY,D–,G–,t(DqGq)+	1	1	1	–	–	1	–	–	–	4
A.Å.	46,XY,D–,t(DqGq)+	–	–	1	1	–	–	–	–	–	2
L.K.	46,XY,D–,t(DqGq)+	–	1	8	3	3	2	1	2	–	20
H.A.	45,XY,D–,D–,t(DqDq)+	–	17	–	1	2	–	–	–	–	20
Å.S.	45,XY,D–,D–,t(DqDq)+	4	19	1	–	2	–	1	–	–	27
S.O.	47,XY,mar+	1	2	3	25	2	1	–	1	1	36
T.E.	47,XY,mar+	–	–	–	–	–	–	–	–	–	–
K.K.	46,XY,t(Gq;A)	–	–	6	–	4		1	–	–	11
Control 1 23 years	46,XY	2	2	19	3	–	–	1	–	–	27
Control 2 46 years	46,XY	–	–	15	3	1	1	1	–	–	21

but usually well spread. However, the secondary constrictions of chromosomes nos. 1, 9 and 16 demonstrate varying degrees of extension (Fig. 1). Therefore, the 'chromosome' count in the four control subjects varied between 44 and 50 (Table 1). This is a serious limitation when dealing with presumptive gonosomic mosaics. The chromosomes in second metaphase are generally not well spread. Even if they are spread, their fuzzy and curly appearance does not permit a detailed analysis. Chromosome no. 9 also demonstrates variously extended secondary constrictions in second metaphase (Fig. 2). Because of these limitations we have not been able to obtain information relevant to the problem of chromosome segregation during the first meiotic division. Thus, the discussion below will be limited to cells in diakinesis – first metaphase.

MEIOSIS IN G TRISOMY AND IN BALANCED AND UNBALANCED D/G TRANSLOCATIONS

To our knowledge meiosis has been studied in only seven patients with mongolism and trisomy 21. Mittwoch [6] studied one case without detecting any abnormalities (the chromosome constitution was not known). Miller et al. [7] in a short abstract reported the occurrence of univalents,

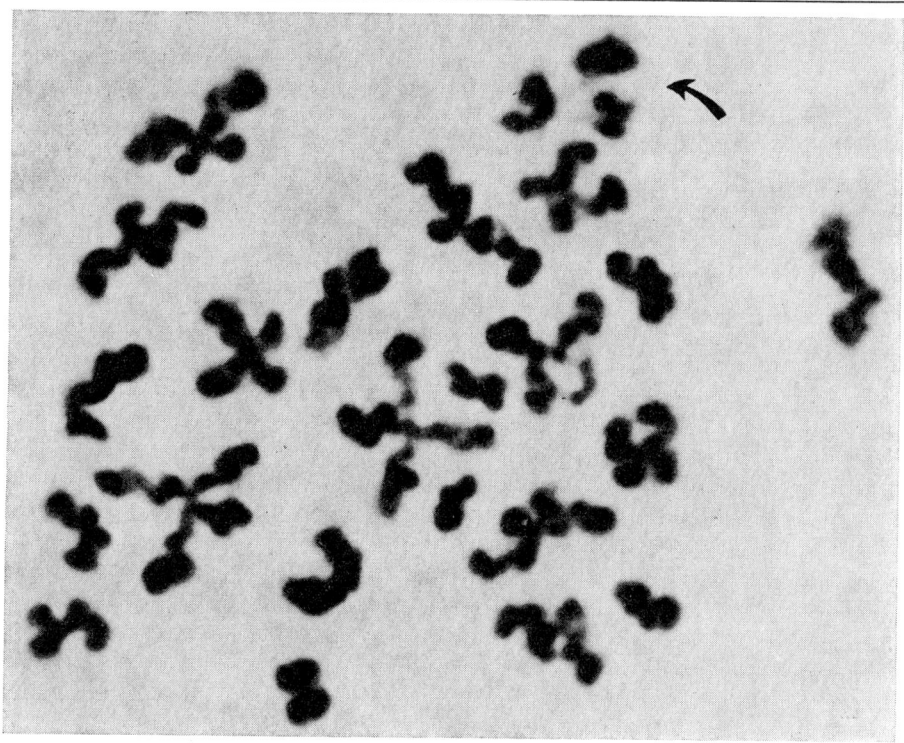

FIGURE 2
Second metaphase with 23 chromosomes from case B.K. Note the extended secondary constriction on probably chromosome no. 9 (arrow)

trivalents and multivalents in four mongols. A trivalent was found in 15 cells, and a univalent in one cell from one patient studied by Sasaki [8], while Finch et al. [9] found a univalent in the seven cells analysed from another patient. Hamerton [10] and Mikkelsen [11] each studied one D/G translocation carrier. Both showed two types of trivalents with different numbers of chiasmata, while Hamerton [10] also found two cells with three univalents and one cell with an unequal bivalent. Penrose [12] reported the finding of a quadrivalent in first spermatocytes from a mongol patient with a 21/22 translocation.

Trisomy 21. We have studied one 16-year-old typical mongol (C.M.) with trisomy 21 and with testes of ordinary size and consistency. It was not possible to obtain a semen specimen, but the testicular histology showed all stages of spermatogenesis even though considerably reduced.

Only ten spermatogonial metaphases could be counted, three had 47 (Fig. 3a), the others 48–50 'chromosomes', none had 46 chromosomes (Table 1). Thus, there was no evidence of a 46/47 mosaicism among these spermatogonial metaphases.

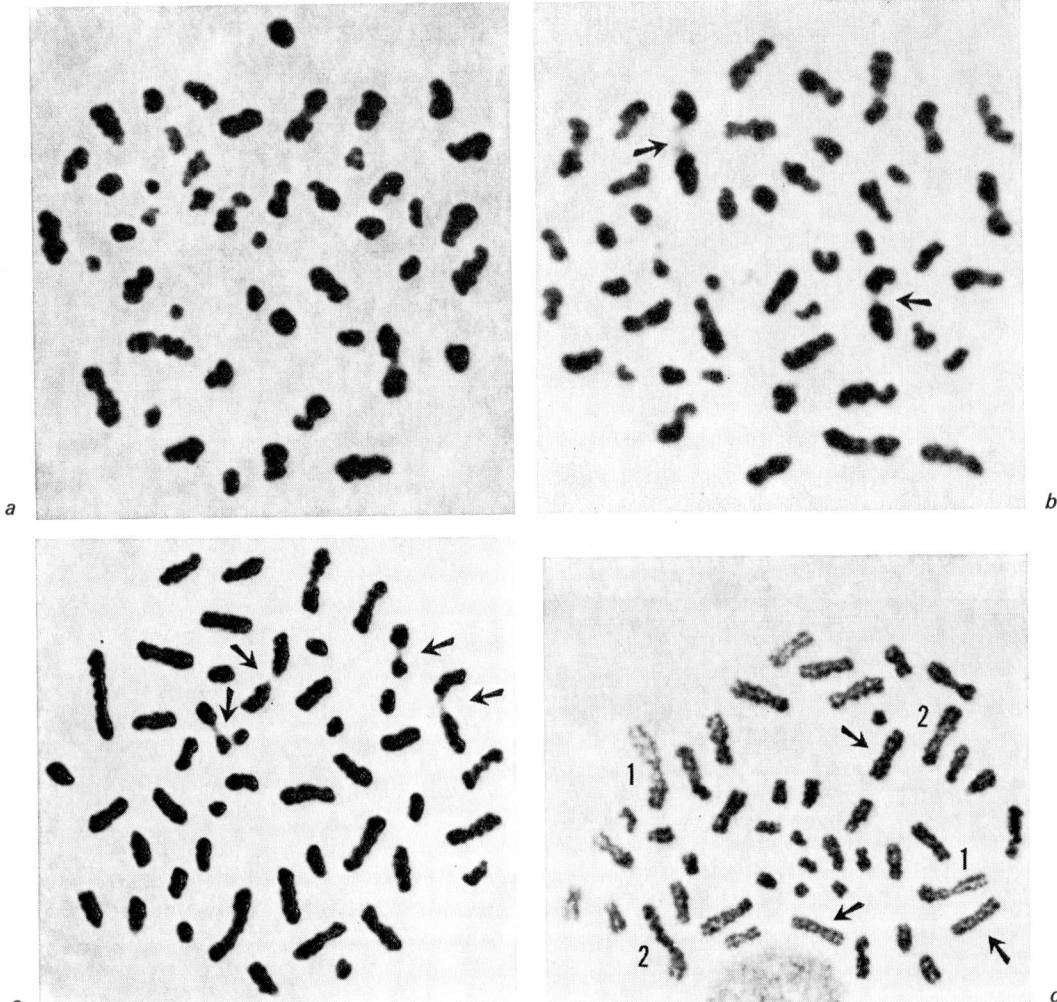

FIGURE 3
Four spermatogonial metaphases. (*a*) Trisomic mongol, 47 chromosomes. (*b*) D/G translocation carrier (B.K.), 45 chromosomes. (*c*) D/G translocation mongol (L.K.), 46 chromosomes. (*d*) D/D translocation carrier (H.A.), 45 chromosomes. Chromosomes nos. 1 and 2 are indicated by numbers, and the three chromosomes with the size of a no. 3 by arrows

The results of the analysis of the first spermatocytes are given in Table 2. Two apparently normal G bivalents plus a univalent were found in 44 out of the 88 cells studied (50 per cent). The reality of the G univalent was most clear in those cells with 25 structures and the X and Y separated (Fig. 4). An obvious trivalent was seen in 19 of the 88 cells (22 per cent).

TABLE 2. Patient C.M. with trisomy 21

Sex chromosomes	II + II + I	III : 2			III : 3	III ? or II + II ?	Total
		a	b	c			
XY	39	6	3	5	2	13	68
X + Y	5	3	–	–	–	8	16
X + Y ? or I ?	–	–	–	–	–	4	4
Total	44	9	3	5	2	25	88

Number of first spermatocytes divided according to the modality of pairing of the three homologous chromosomes and to the appearance of the sex chromosomes. II + II + I = cells with an obvious univalent with the size of a G chromosome; III : 2 and III : 3 = trivalents with 2 and 3 chiasmata respectively. For further explanation see text.

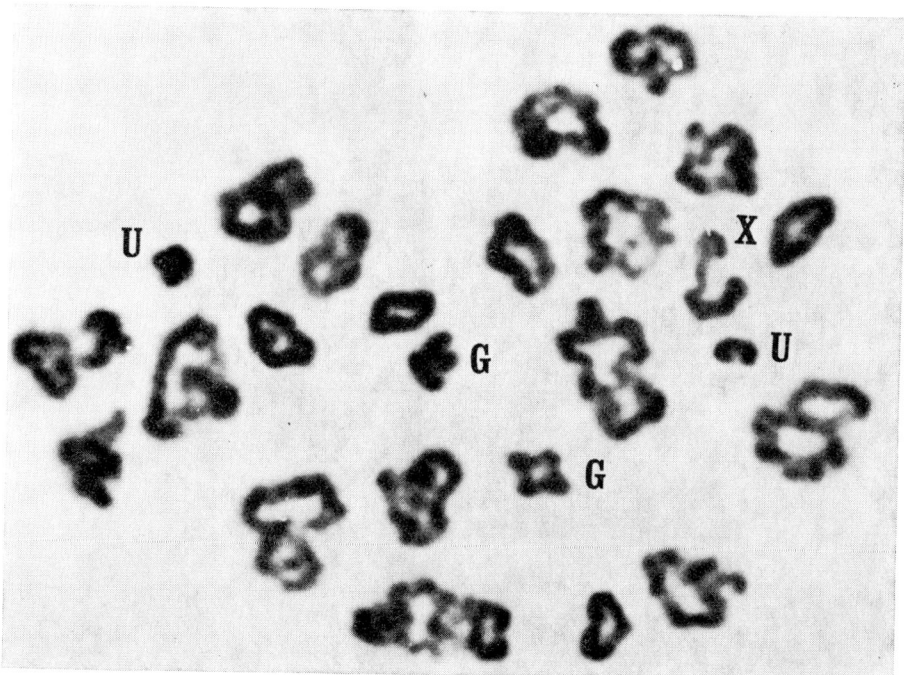

FIGURE 4

First spermatocyte from the trisomic mongol C.M. Twenty-five structures, two single-chiasma bivalents with the size of normal G bivalents (G) plus a univalent (U), separated X and Y (U) chromosomes

The trivalent had a different shape depending on the number and location of the chiasmata. The most common trivalent had a cross-like appearance with one chiasma on the long arm and another one apparently on the short arm (Fig. 5). Altogether three different types of trivalents were observed with either two chiasmata on the long arm, or one on the long and one on the short arm (Fig. 6).

The true frequencies of the various cell types might be somewhat different as the modality of pairing could not be identified with certainty in 25 of the 88 cells (29 per cent). It could not be decided with accuracy whether these cells in fact had a non-identifiable trivalent of a more complicated type, or had only two regular G bivalents. The latter interpretation seemed obvious in some cells, and therefore the extra chromosome 21 must have been eliminated, provided the patient is not a mosaic. An obvious univalent was found in four of the 25 cells but could not be distinguished from a Y chromosome separated from an X.

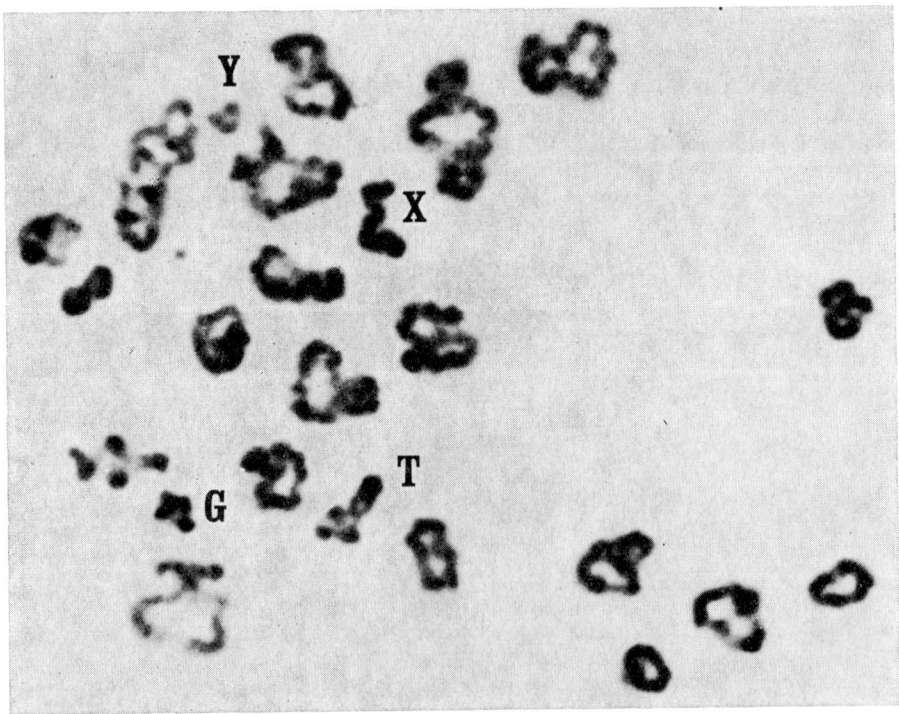

FIGURE 5
First spermatocyte from the trisomic mongol C.M. Twenty-four structures including one single-chiasma bivalent with the size of a normal G bivalent (G), one cross-shaped trivalent (T), and separated X and Y chromosomes

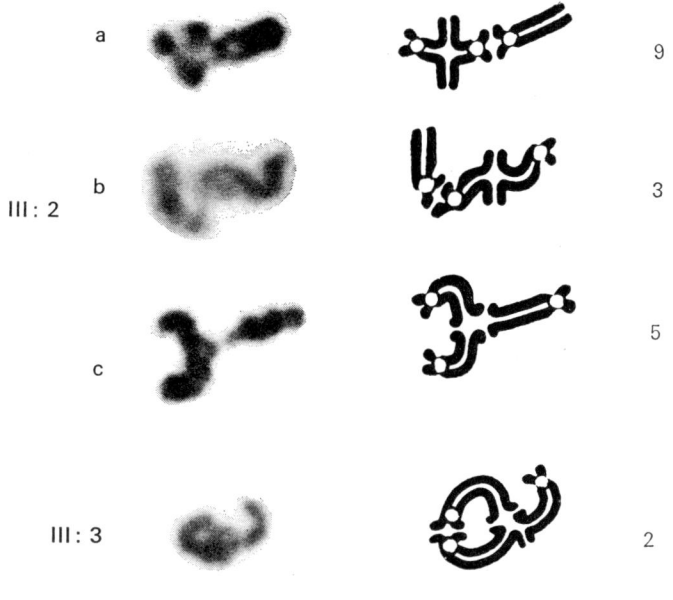

FIGURE 6

Cut-out trivalents (III) from first spermatocytes from the trisomic mongol C.M. The trivalents have been divided into two main groups according to the number of chiasmata (III : 2 and III : 3) illustrated in the drawings to the right of the photographs. The number of observed trivalents of a specific type is given to the right of the drawings

D/G translocation carriers. Two apparently healthy D/G translocation carriers were studied. Autoradiographic analysis of the DNA replication pattern of the chromosomes revealed that chromosome nos. 14 and 21 were the most likely to be involved in the translocation in both cases. The first one, B.K., whose brother had D/G translocation mongolism (see below and Fig. 7), was 18 years old, and had normal testes with an apparently normal histology but oligo-asteno-teratospermia (21 million sperm per ml., 66 per cent morphologically abnormal). The second, J.K., was 34 years old, childless, had only one testis (measuring 7·5 × 4·5 cm) and oligo-asteno-teratospermia (0·3–0·8 million sperm per ml, 88–93 per cent morphologically abnormal). Testicular histology showed reduced spermiogenesis and occasional sperm with pronounced degenerative changes. It has not yet been possible to study the chromosome constitution of his relatives.

Very few spermatogonial metaphases were found (Table 1). An example is shown in Figure 3.

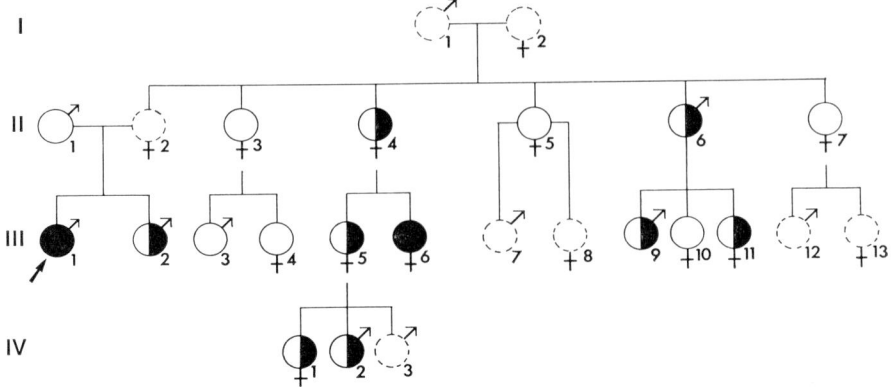

FIGURE 7
Pedigree of family K. with a segregating D/G translocation. The carrier B.K. is no. III : 2 and his mongol brother III : 1 (propositus indicated by an arrow). Filled symbols indicate D/G translocation mongols, semi-filled ones the carriers. Dashed symbols indicate subjects whose karyotypes are unknown

TABLE 3. Two D/G translocation carriers

Subject	Sex chromosomes	III : 2			III : 3		III : 4	U II + I	Non-identifiable	Total
		a	b	c	a	b				
B.K.	XY	25	9	3*	17*4	1†	—		12	71
	X + Y	—	—	—	—	—	—		—	—
J.K.	XY	18† 6	—		10	—	—	2	4	40
	X + Y	1	—	—	2	1	—	—	1	5
Total		44	15	3	29	5	1	2	17	116

Number of first spermatocytes divided according to the shape of the translocation figure and to the appearance of the sex chromosomes. III : 2, III : 3 and III : 4 = trivalents with 2, 3 and 4 chiasmata respectively. U II + I = unequal bivalent + univalent. For further explanation see text

* Non-terminalised chiasma between Gq in two cells
† Non- terminalised chiasma between Gq in one cell

Altogether 116 first spermatocytes were analysed (Table 3). Cells with three main different types of trivalent and cells with an unequal bivalent plus a univalent were observed in the two subjects in similar proportions. The two most common trivalents had, respectively, one chiasma at the middle of the long arm of the D chromosome material (III:2:a) and one at each end of these chromosomes, respectively (III:3:a) (Fig. 8). They were found respectively, in 38 and 25 per cent of the 116 cells. Other, less fre-

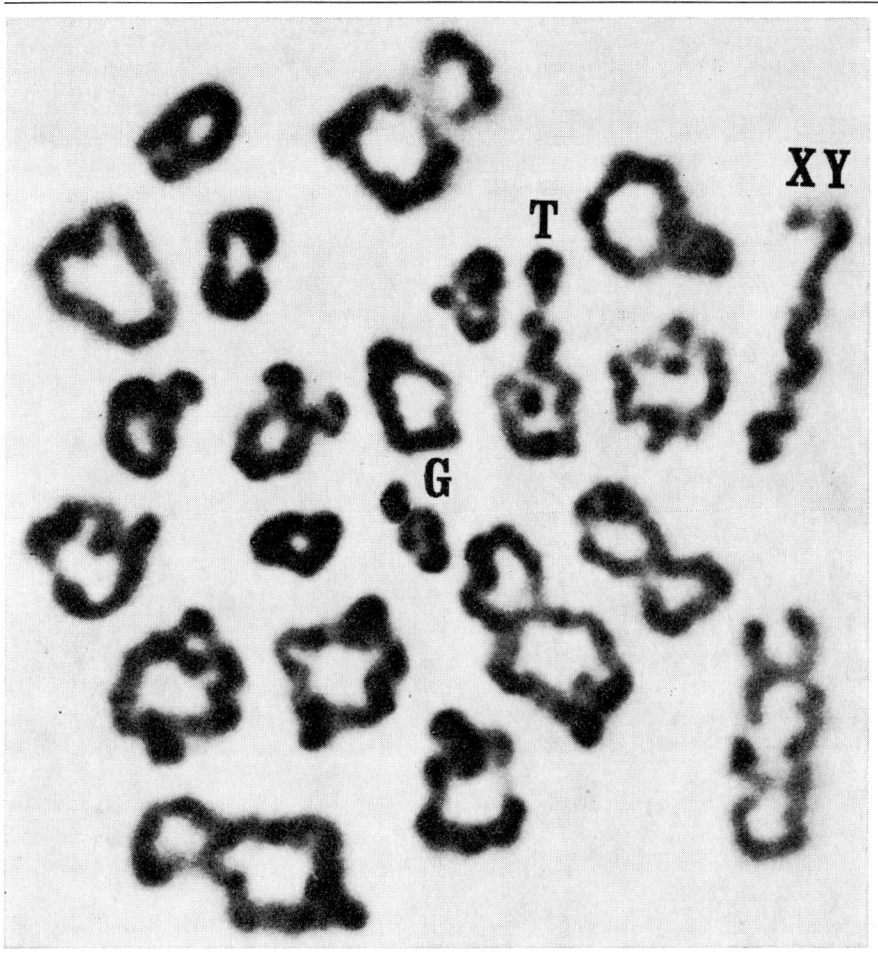

FIGURE 8
First spermatocytes from the D/G translocation carrier B.K. Twenty-two structures including one trivalent. (*a*) Trivalent with one chiasma at the distal end and another close to the centromere of the D chromosomes. (*b*) Trivalent with one chiasma in the middle of the long arm of the D chromosomes (see p. 34)

quent, types of trivalents were also seen (Fig. 9). The chiasma on the long arm of the G chromosome material was usually terminalised. Furthermore, no cell was found to have two chiasmata on the long arm of the G chromosome material. An unequal bivalent plus a univalent were found in two cells (Fig. 10). The modality of pairing could not be distinguished in a relatively large proportion of the cells (15 per cent). The technical standard of these cells was about the same as that of the others. Thus, there might be other and more complicated types of trivalents or more cells with an unidentifiable bivalent and lacking a univalent.

[33

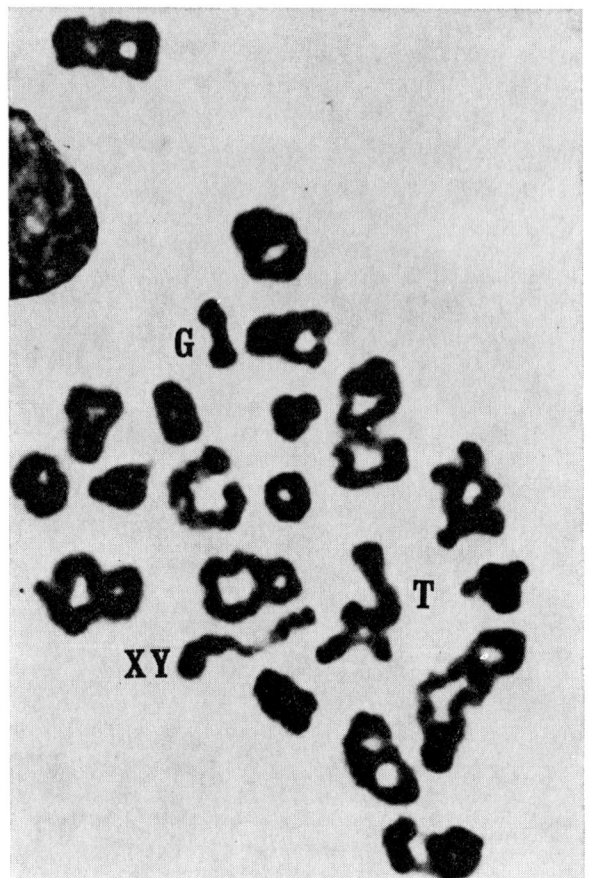

FIGURE 8b (caption p. 33)

D/G translocation mongols. Two D/G translocation mongols were studied, 18 (A.Å.) and 20 (L.K.) years old, respectively. Autoradiographic studies revealed that also in these cases chromosomes no. 14 and 21 were the most likely ones to be involved in the translocation. None of the patients could deliver a semen specimen. The testicular histology of A.Å. revealed spermatogenic arrest at the spermatid level. Occasional prespermatids were, however, seen. L.K. showed the same picture but in addition slight tubular sclerosis as well as peritubular and interstitial fibrosis. Some tubular regions were completely sclerotised. The brother of L.K. has been discussed above and the results of the chromosome analysis of his family is shown in Figure 7. The family of A.Å. is presented in Figure 11.

The number of counted spermatogonial metaphases is given in Table 1 and one cell with 46 chromosomes is shown in Figure 3. The chromosomes were often very contracted and the secondary constrictions invisible. This explains the great variation in the 'chromosome' counts.

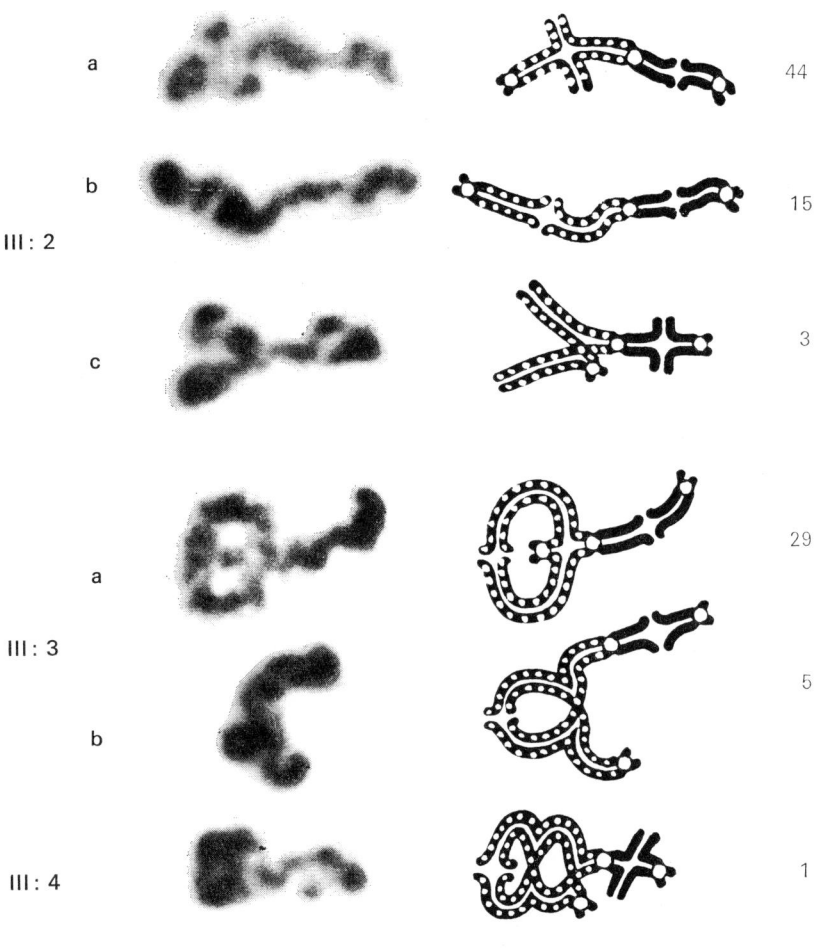

45, XY, D−, G−, t(DqGq) +

FIGURE 9
Cut-out trivalents (III) from first spermatocytes from the D/G translocation carriers. The trivalents have been divided into three main groups according to the number of chiasmata (III : 2, III : 3 and III : 4) illustrated in the drawings to the right of the photographs. The number of observed trivalents of a specific type is given to the right of the drawings

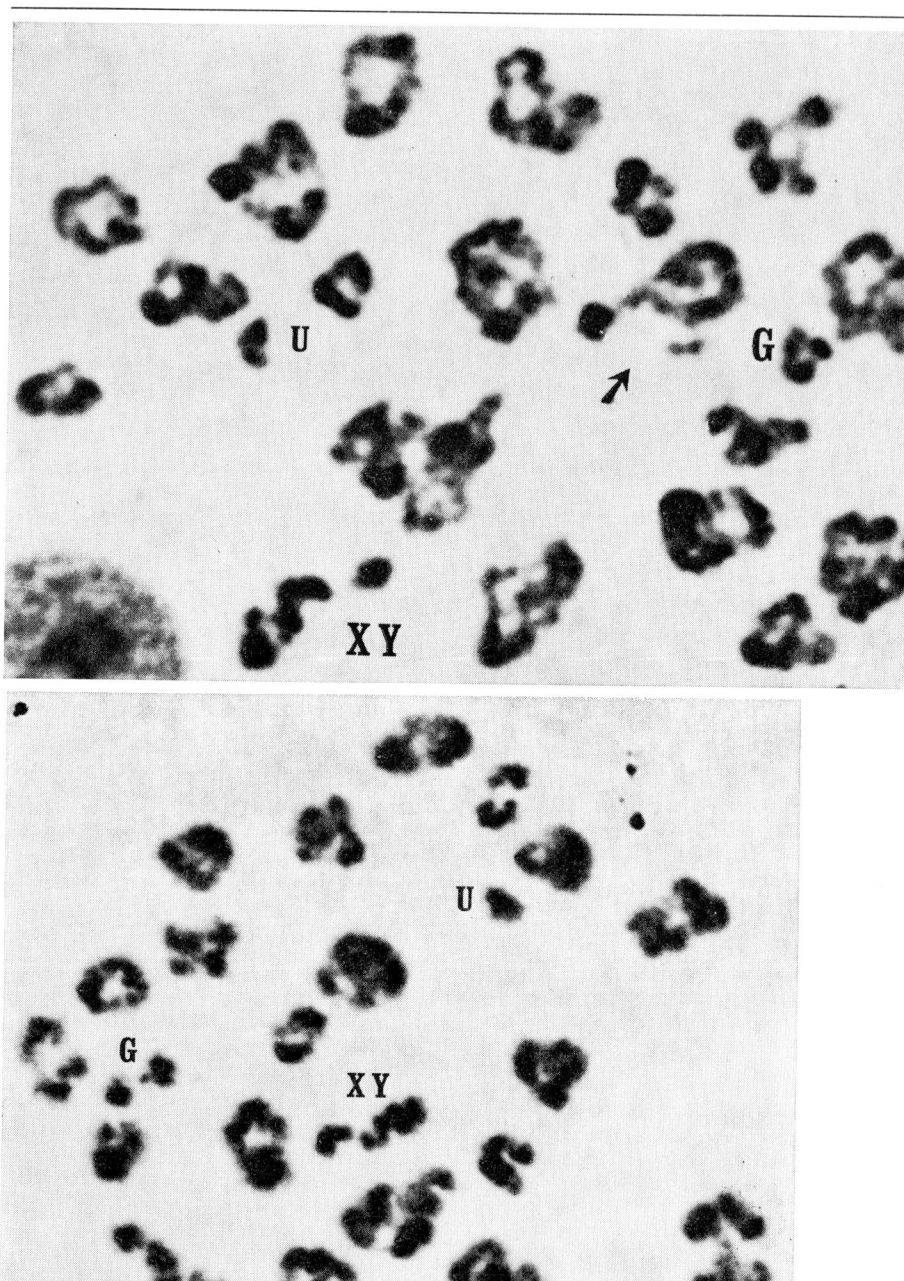

FIGURE 10

Two first spermatocytes from the D/G translocation carrier J.K. One presumptive unequal bivalent (arrow), one univalent (U) and one apparently normal G bivalent. (*a*) Two chiasmata on the long arm material of the D chromosomes. (*b*) One chiasma on the long arm material of the D chromosomes

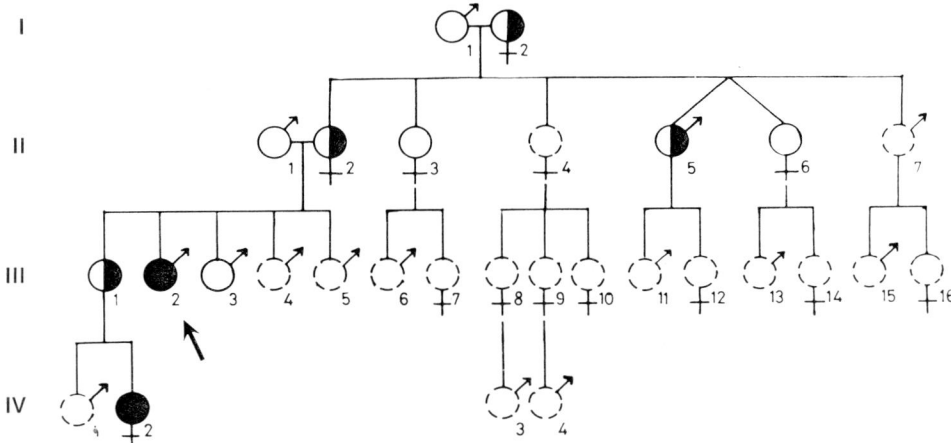

FIGURE 11
Pedigree of family Å. with a segregating D/G translocation. The mongol propositus A.Å. (III : 2) indicated by an arrow. Filled symbols indicate D/G translocation mongols, semi-filled ones the carriers. Dashed symbols indicate subjects whose karyotypes are unknown

The first spermatocytes were less numerous than in the D/G translocation carriers, and the interpretation of the modality of pairing presented great difficulties. A large number of different types of pairing was observed, and a tentative classification of these types is given in Table 4. Five out of a total of 83 cells (6 per cent) had two apparently normal G bivalents plus an unequal D/DG bivalent. Presumptive trivalents of the same types as those in the D/G translocation carriers were observed in 22 of the cells (25 per cent). However, an obvious univalent was found in only six out of these 22 cells. Whether this is due to previous elimination of the univalent or to misdiagnosis of a quadrivalent as trivalent, we do not know. Two additional types of apparent trivalents were seen which could be interpreted as having two chiasmata on the long arm of the G chromosome material. Presumptive quadrivalents were observed in 22 out of the 83 cells (25 per cent). Several different types were found with 3, 4 and 5 chiasmata, respectively (Fig. 12). The modality of pairing could not be evaluated with certainty in as many as 34 of the 83 cells (41 per cent).

Comments. The estimation of the number of chiasmata on the G bivalents in normal subjects is often difficult. The small size makes it sometimes impossible to differentiate between a single-chiasma cross-shaped and a two-chiasma ring-shaped G bivalent. Furthermore, in many cells one or both of the G bivalents appear as rods. These have been interpreted as having one terminalised chiasma on the long arm. However, depending on precocious separation some of these might represent two-chiasma bivalents.

TABLE 4. Two D/G translocation mongols

Subject	Sex Chromosomes	II+II+UII	III:2 a	III:2 b	III:3 a	III:3 b	III:3 c	III:4	IV:3 a	IV:3 b	IV:3 c	IV:4 a	IV:4 b	IV:4 c	IV:5 a	IV:5 b	Non-identifiable	Total
A.Å.	XY	4	4*	1	7†	—	2	2†	4	1	2	4	4	3	1	1	23‡	63
	X+Y	—	—	—	—	1†	—	—	—	—	—	—	—	—	—	—	2	3
	X+Y?	—	—	—	1	—	—	—	—	—	—	—	—	—	—	—	2†	3
L.K.	XY	1	1§	—	2+§	1§	—	—	2	—	—	—	—	—	—	—	7	14
	X+Y	—	—	—	—	—	—	—	—	—	—	—	—	—	—	—	—	—
	X+Y?	—	—	—	—	—	—	—	—	—	—	—	—	—	—	—	—	—
Total		5	5	1	10	2	2	2	6	1	2	4	4	3	1	1	34	83

Number of first spermatocytes divided according to the shape of the translocation figure and to the appearance of the sex chromosomes. II+II+UII = 2 apparently normal G bivalents plus one unequal D/DG bivalent; III: 2, III: 3 and III: 4 = trivalents with 2, 3 and 4 chiasmata respectively; IV: 3, IV: 4 and IV: 5 = quadrivalents with 3, 4 and 5 chiasmata respectively
* Two cells had a univalent † One cell had a univalent ‡ Four cells had a univalent § Non-terminalised chiasma between Gq

TABLE 5. One carrier of an homologous (Å.S.) and one of a non-homologous (H.A.) D/D translocation.

Subject	Sex Chromosomes	II+I	III:2 a	III:2 b	III:2 c	III:3 a	III:3 b	III:4	Non-identifiable	Total
Å.S.	XY	323	—	—	—	—	—	—	—	323
	X+Y	40	—	—	—	—	—	—	—	40
	X	2	—	—	—	—	—	—	—	2
H.A.	XY	—	4	7	—	10	5	4	2	32
	X+Y	—	—	5	1	—	2	—	—	8
	?	—	—	—	2	1	—	—	—	3

Number of first spermatocytes divided according to the shape of the translocation figure and to the appearance of the sex chromosomes. I = univalent; III: 2, III: 3 and III: 4 = trivalents with 2, 3, and 4 chiasmata respectively

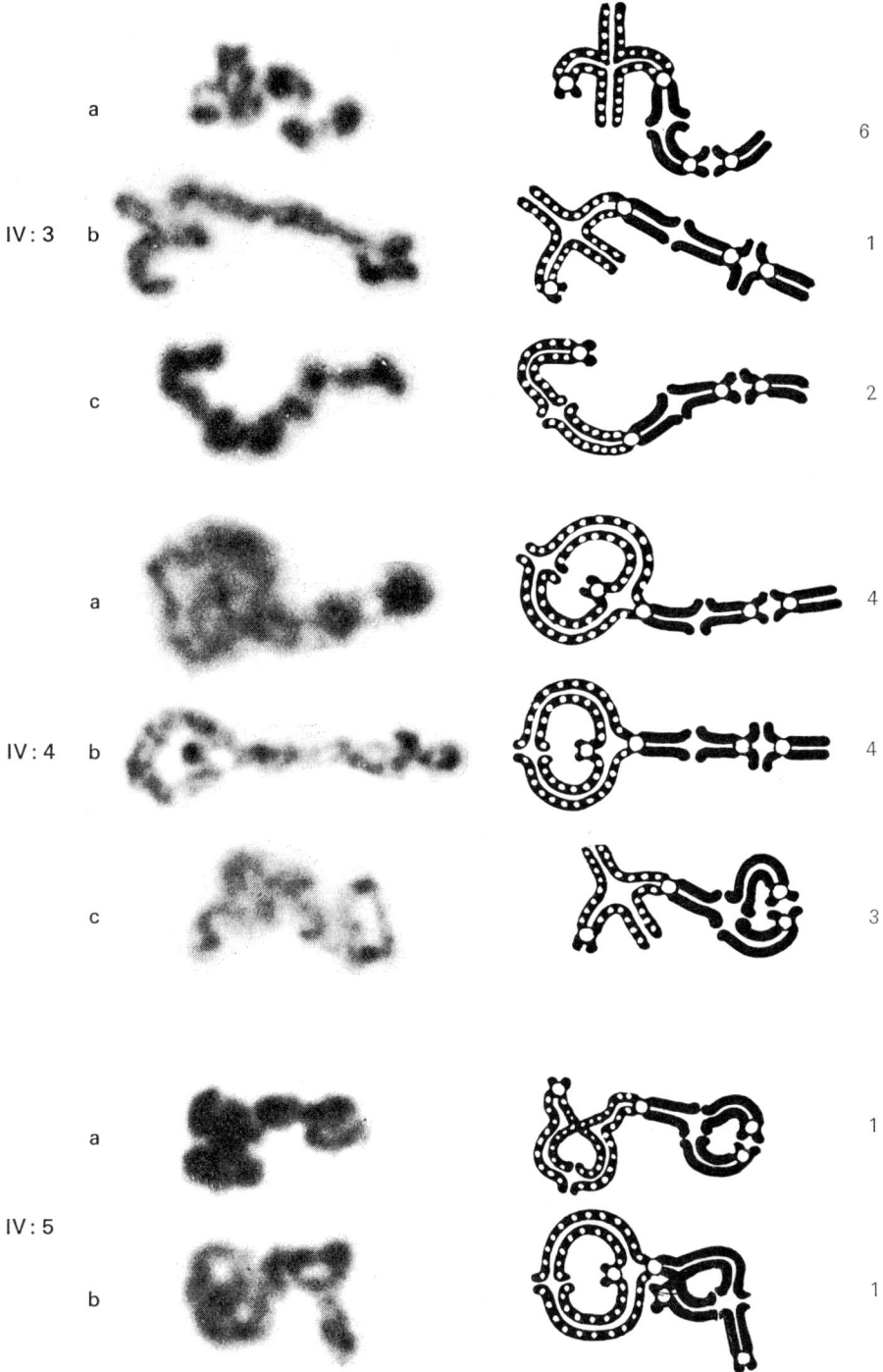

46, XY, D−, t(Dq Gq) +

FIGURE 12

Cut-out quadrivalents (IV) from first spermatocytes from the D/G translocation mongol A.Å. The quadrivalents have been divided into three main groups according to the number of chiasmata (IV : 3, IV : 4 and IV : 5) illustrated in the drawings to the right of the photographs. The number of observed quadrivalents of a specific type is given to the right of the drawings

TABLE 6. Chiasma frequency for the autosomal bivalents

Subject	Chromosome constitution	Total number of chiasmata per cell	
		range	mean± S.D.
C.M.	47,XY,21+	46–64	54·0± 3·7
B.K.	45,XY,D–,G–,t(DqGq)+	42–59	51·1± 3·4
J.K.	45,XY,D–,G–,t(DqGq)+	41–52	48·4± 3·6
A.Å.	46,XY,D–,t(DqGq)+	48–62	53·6± 3·7
L.K.	46,XY,D–,t(DqGq)+	50–57	53·5± 2·9
H.A.	45,XY,D–,D–,t(DqDq)+	42–59	49·2± 4·9
Å.S.	45,XY,D–,D–,t(DqDq)+	47–61	52·6± 3·3
Control 23 years	46,XY	39–54	45·9± 3·4
Control 46 years	46,XY	43–58	50·4± 3·8
Control 52 years	46,XY	42–59	51·1± 3·6
Control 72 years	46,XY	41–63	53·4± 4·5

The estimated total number of chiasmata is given in Table 6. There was a slight, continuous increase of the mean number of chiasmata with age in the controls. Since the number of terminalised chiasmata was fairly constant the terminalisation coefficient decreased with age. The highest number of chiasmata was found in the trisomic mongol and the next highest in the translocation mongols. The significance of these preliminary observations is difficult to evaluate.

The number of single-chiasma bivalents for different groups of chromosomes in each cell was also estimated (Table 7). Since one bivalent in the E group sometimes had only one chiasma, and the size of this bivalent was not very different from the G bivalents, the E—G groups of bivalents were pooled. As far as the trisomic mongol is concerned the following tentative conclusions can be drawn. Among cells with two G bivalents plus a univalent, 25 per cent of the cells had one single-chiasma bivalent and 65 per cent had two such bivalents, while the percentages were 81 and 19, respectively, in the cells with a trivalent. Thus, the G bivalents are the main source of single-chiasma bivalents in the groups considered. Cells with none or one single-chiasma bivalent in the E—G groups obviously have two and one two-chiasma bivalents, respectively. Since the single-chiasma bivalents were mainly G bivalents, the frequency of G bivalents with two

Terminalisation coefficient		Number of cells analysed
range	mean	
0·52–0·80	0·62	40
0·54–0·76	0·64	42
0·61–0·80	0·71	8
0·53–0·72	0·61	14
0·50–0·60	0·57	4
0·52–0·71	0·65	13
0·50–0·80	0·62	31
0·52–0·95	0·69	39
0·53–0·89	0·68	37
0·44–0·79	0·65	50
0·40–0·83	0·59	50

chiasmata can tentatively be estimated. The mean number of two chiasmata G bivalents in the four controls was 29/2 × 176 (8 per cent). This figure is lower than the finding of 22 per cent of cells with a trivalent with two chiasmata in the trisomic mongol. Thus, trivalents may be formed with a frequency higher than that of normal G bivalents with two chiasmata. Three main factors affect the efficiency of this estimate. Firstly, the material is scarce and there was a relatively large number of cells (28 per cent) in which the modality of pairing could not be evaluated. Secondly, the two G bivalents might normally form two chiasmata with different frequencies. Thirdly, the trisomic condition might influence the relative numbers of single-chiasma bivalents in the E and G groups.

The observation of a relatively high proportion of cells with trivalents (22 per cent) in the trisomic mongol shows that at least two chiasmata can be formed between homologous G chromosomes. The existence of trivalents of type III:2:c and III:3 in the trisomic mongol (Fig. 6) indicates that both chiasmata might be on the long arm. However, types III:2:a and b indicate that the short arm is involved quite often in chiasma formation. This is supported by the absence of two chiasmata on the long arm material of the G chromosomes in the trivalents of the D/G translocation carriers,

TABLE 7. Number of single-chiasma bivalents in the D and E – G groups

Subject	Chromosome constitution	Number of cells with single-chiasma bivalents							
		Number of single-chiasma bivalents: *							
		larger than a D bivalent					with the size of a D biv.		
		0	1	2	3	4	0	1	2
C.M.	47,XY,21+								
	II+II+I	19(95)	1(5)	–	–	–	10(50)	5(25)	4
	III	14(88)	2(13)	–	–	–	11(69)	2(13)	2
	III? or II+II?	4(100)	–	–	–	–	3(75)	1(25)	–
B.K.	45,XY,D–, G–,t(DqGq)+	40(95)	2(5)	–	–	–	17(40)	18(43)	7
J.K.	45,XY, D–, G–,t(DqGq)+	21(91)	2(9)	–	–	–	3(13)	12(52)	8
A.Å.	46,XY,D–, t(DqGq)+	13(93)	1(7)	–	–	–	9(64)	5(63)	–
L.K.	46,XY,D–, t(DqGq)+	4(100)	–	–	–	–	–	3(75)	1
H.A.	45,XY,D–, D–,t(DqDq)+	13(100)	–	–	–	–	8(62)	3(23)	2
Å.S.	45,XY,D–, D–,t(DqDq)+	29(94)	2(6)	–	–	–	12(39)	11(36)	6
Control 23 years	46,XY	31(79)	7(18)	1(3)	–	–	5(13)	18(46)	10
Control 46 years	46,XY	35(95)	2(5)	–	–	–	17(46)	13(35)	6
Control 52 years	46,XY	47(94)	3(6)	–	–	–	20(40)	20(40)	4
Control 72 years	46,XY	49(98)	1(2)	–	–	–	18(36)	21(42)	6

* Figures within brackets indicate per cent.   † Both cells had an unequal D/DG bivalent

where the short arm of the G chromosome would be missing when the translocation is of the centric fusion type. It should, however, be mentioned that four presumptive trivalents with two chiasmata between the G chromosome material were observed in one of the D/G translocation mongols. It is difficult to evaluate whether there are localised chiasmata on the G chromosomes. Chiasma formation on the D chromosomes will be discussed below.

The finding in the trisomic mongol of apparently normal cells with two ordinary G bivalents but without a univalent or an obvious trivalent indicates that an extra G chromosome can be eliminated during meiosis, provided that the patient is not a mosaic. Because of the difficulties in

	with the size of E – G bivalents					Total number of cells analysed
4	0	1	2	3	4	
1(5)	–	5(25)	13(65)	2(10)	–	20
–	–	13(81)	3(19)	–	–	16
–	1(25)	–	3(75)	–	–	4
–	9(21)	30(71)	3(7)	–	–	42
–	3(13)	15(65)	5(22)	–	–	23
–	–	12(86)	2†(14)	–	–	14
–	–	3(75)	1(25)	–	–	4
–	–	1(8)	10(77)	2(15)	–	13
–	–	1(3)	24(77)	3(10)	3(10)	31
1(3)	–	6(15)	21(54)	12(31)	–	39
–	–	7(19)	28(76)	2(5)	–	37
–	–	6(12)	40(80)	3(6)	1(2)	50
1(2)	–	10(20)	36(72)	4(8)	–	50

analysing the second spermatocytes we cannot estimate the degree of such an elimination, nor its mechanism.

Separation of the X and Y chromosomes in first spermatocytes is well known since the study of the human male meiotic chromosomes by Ford and Hamerton [13]. These authors found separation in 14 per cent of the cells, Kodani [14] in 40 per cent, Sasaki and Makino [15] in 27 per cent, Kjessler [16] in 4 per cent only, Eberle [17] in 12 per cent, and Luciani [2] in 6 per cent only. Of these authors only Sasaki and Makino [15] used the air-drying technique while the others used squash preparations. Separation of the X and Y was observed in 16/103 (23 years), 32/100 (46 years), 18/100 (52 years), and 24/100 (72 years) cells from our four control subjects. The number of cells with separated X and Y chromosomes was low in the D/G translocation carriers and mongols (8/188 cells). The D/G carrier and his mongol brother showed no separation at all. Whether this is a technical matter, a family trait, an effect of the translocation or some other factor, remains to be shown. The second alternative seems less likely since the father showed separation in 32/100 cells. Hamerton [10] and Mikkelsen [11] found in two D/G translocation carriers separation in 11/42 and 7/90 cells, respectively, which speaks against an interchromosomal effect of the translocation.

MEIOSIS IN D/D TRANSLOCATION CARRIERS

Meiotic studies have so far only been reported in one D/D translocation carrier [16]. Three types of trivalents with different numbers of chiasmata can be seen in the cells reproduced by Kjessler [16], demonstrating that non-homologous chromosomes were involved in the translocation. The patient was childless, had 3–10 million sperm per ml of semen, 47–70 per cent of which were morphologically abnormal. Testis histology revealed partial spermatogenic arrest at the spermatid level.

D/D translocation between non-homologous chromosomes. One 81-year-old healthy man with a D/D translocation was detected when used as a control for a study of chromosome breakage in somatic cells. Autoradiography indicated that the translocation was between non-homologous chromosomes even though it was not possible to differentiate between a 14/15 and a 13/14 translocation. The relatives have not been studied as most of them are dead. He was not able to deliver a semen specimen. His testicular histology showed a decrease in all stages of spermiogenesis which, however, was considered apparently normal for his age.

Seventeen of the 20 spermatogonial metaphases analysed had 45 chromosomes (Table 1, Fig. 3).

A trivalent could be identified in almost all of the available first spermatocytes (Table 5). There were three main types of trivalents with 2, 3 and 4 chiasmata, respectively (Figs. 13, 14). The translocation figures were easy to classify due to their size and to the position of chiasmata.

FIGURE 13 (facing)
Two first spermatocytes from the non-homologous D/D translocation carrier H.A. Twenty-two structures including one trivalent. (*a*) Trivalent with two chiasmata between two of the long arms of the D chromosomes and one on the others. (*b*) Trivalent with two chiasmata between the long arms of the D chromosomes on both sides

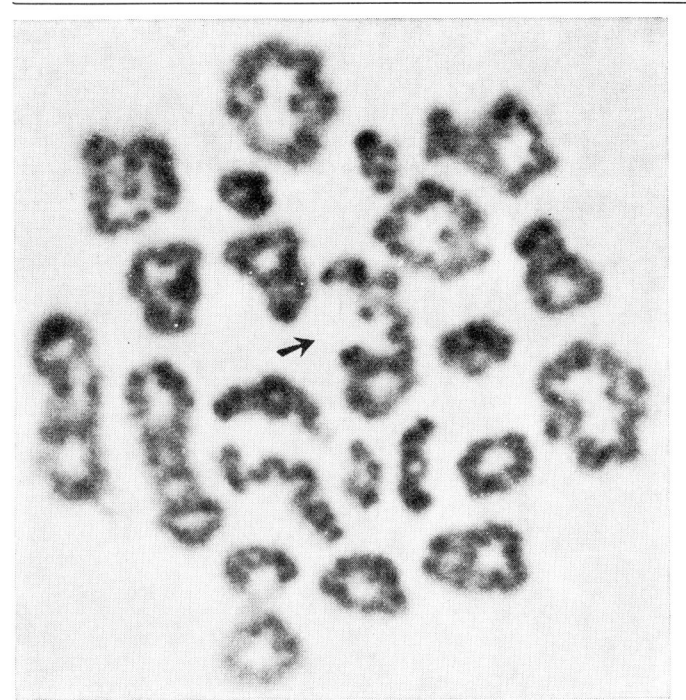

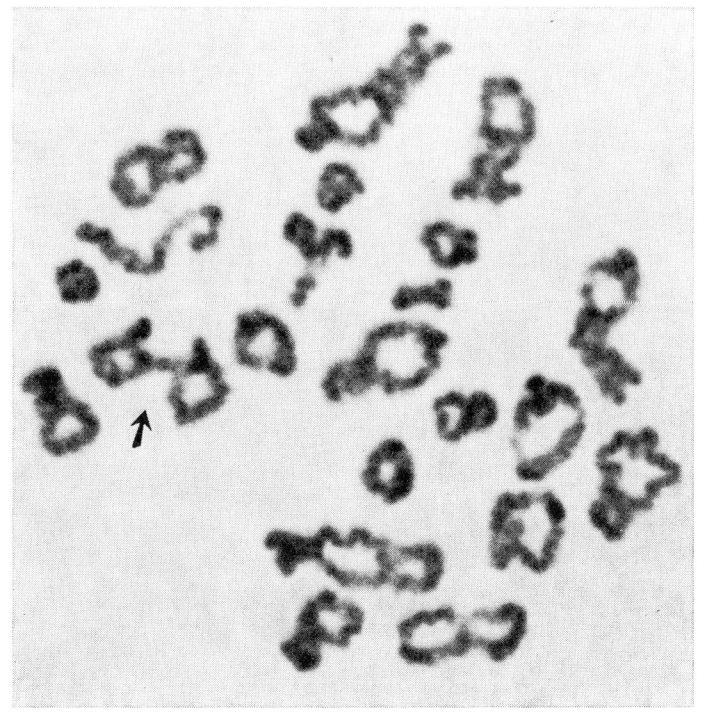

b FIGURE 13

MALE MEIOSIS

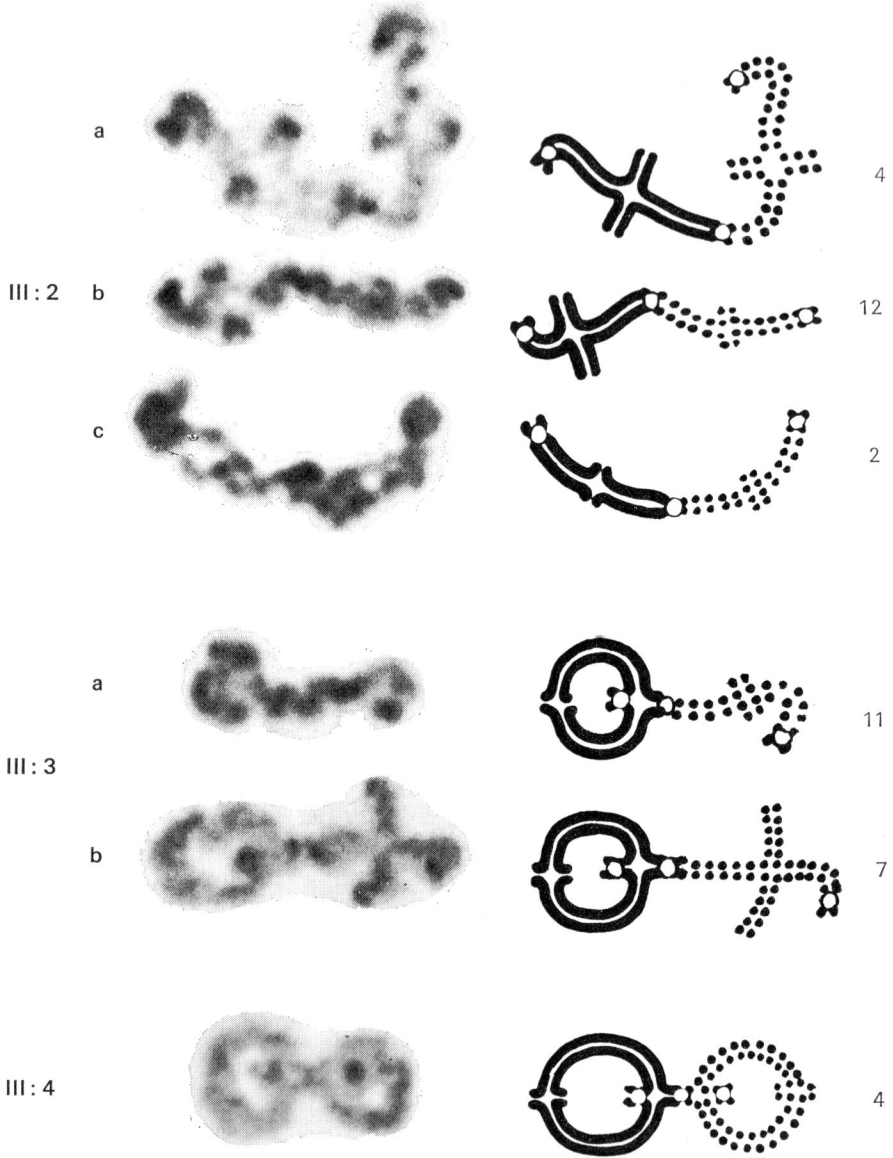

45, XY, D−, D−, t(DqDq) +

FIGURE 14

Cut-out trivalents (III) from first spermatocytes from the non-homologous D/D translocation carrier H.A. The trivalents have been divided into three main groups according to the number of chiasmata (III : 2, III : 3 and III : 4) illustrated in the drawings to the right of the photographs. The number of observed trivalents of a specific type is given to the right of the drawings

D/D translocation between homologous chromosomes. One 35-year-old healthy man, whose two wives had produced a total of eight spontaneous abortions turned out to be a D/D translocation carrier without evidence of mosaicism in cells from lymphocyte and skin cultures. Autoradiographic studies of the DNA replication of the D chromosomes have so far failed to give a decisive answer to which chromosome pair is involved in the translocation. The spermiogram and testicular histology were entirely normal (100–149 million sperm per ml, 31–33 per cent abnormal).

The spermatogonial metaphases were rather easy to count and were generally not of the type with very contracted chromosomes. Forty-five chromosomes and three chromosomes with the size of a no. 3 could be identified in the majority of cells (Table 1).

Twenty-two autosomal structures were consistently found in 365 out of 390 first spermatocytes (Table 5). Eighteen cells had 21 autosomal structures, six had 20 and one had 19 structures. None of these latter cells had an obvious trivalent. The D/D translocation, i.e. the metacentric univalent, could not be identified with certainty in any of the cells (Fig. 15).

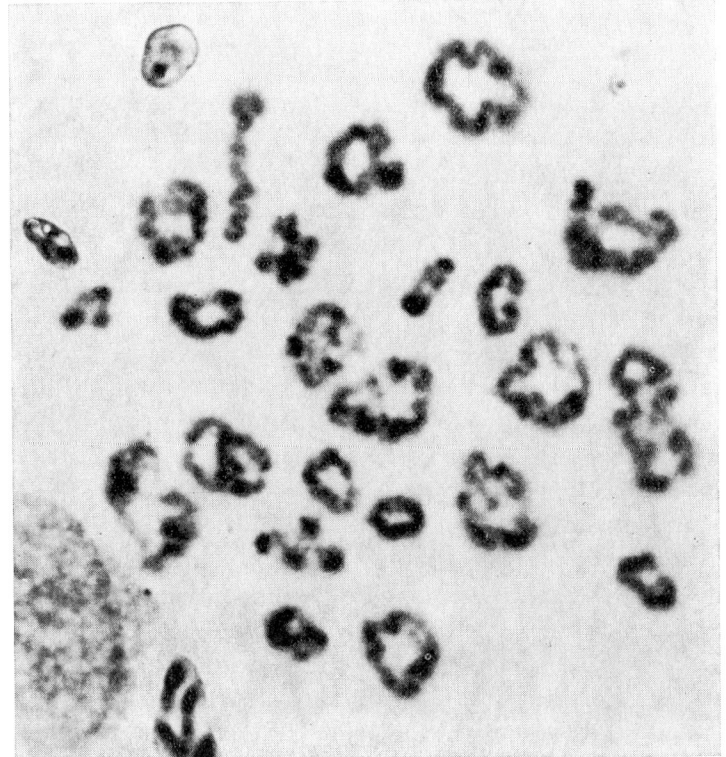

FIGURE 15
Two first spermatocytes from the carrier of an homologous D/D translocation (Å.S.). Twenty-three structures and no obvious metacentric univalent (see also p. 48)

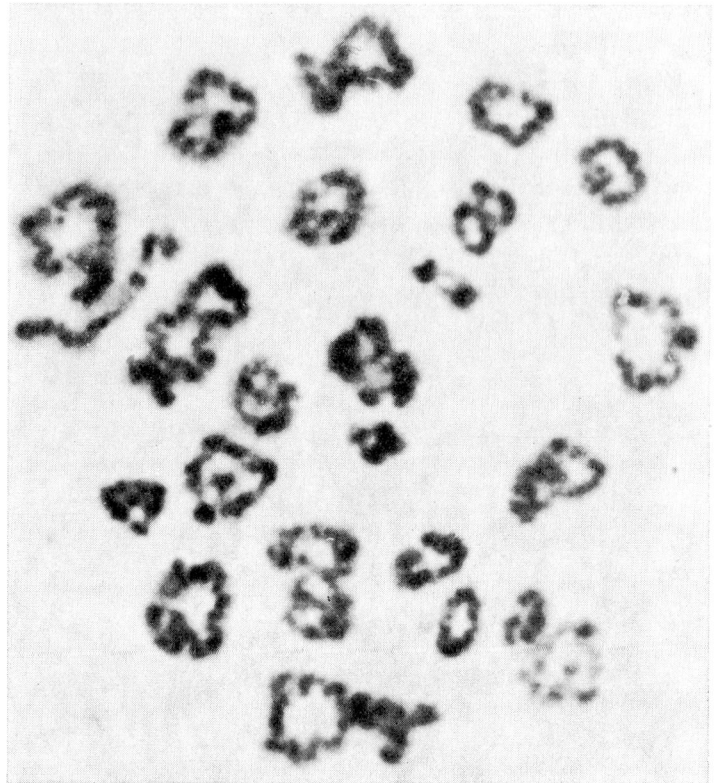

FIGURE 15b (caption p. 47)

Obviously chiasma formation has taken place between the two arms in such a way that the shape of the chromosome does not differ from that of the ordinary D bivalents. This also means that it was counted as having two chiasmata while in fact it had only one. Therefore, the total chiasma count given in Table 6 is probably one chiasma too high.

The second metaphases were as usual very difficult to karyotype. There were, however, occasional cells with apparently two chromosomes the size of chromosome no. 3 (Fig. 16).

One obvious explanation of the finding of apparently normal first spermatocytes would be that the patient is a mosaic and has mainly normal cells in his testes. The finding of spermatogonial metaphases with only 45 chromosomes and of second spermatocytes with two no. 3-like chromosomes speaks, however, against this interpretation. Furthermore, during the course of this study the patient's wife became pregnant. A therapeutic abortion was carried out and an empty amniotic sac was found. Chromosome analysis of this material showed male cells with 46 chromosomes, five D chromosomes and the D/D translocation.

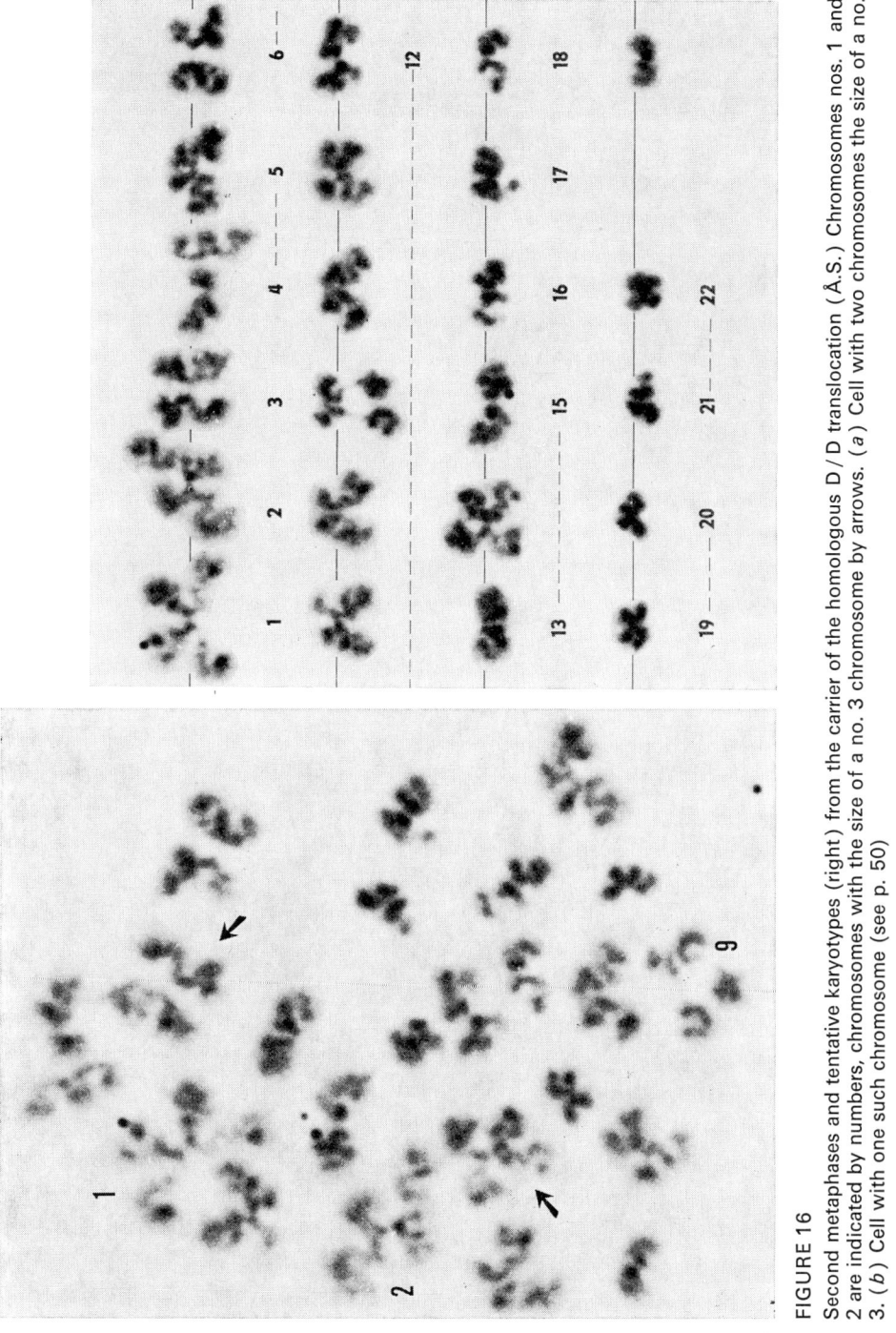

FIGURE 16

Second metaphases and tentative karyotypes (right) from the carrier of the homologous D/D translocation (Å.S.) Chromosomes nos. 1 and 2 are indicated by numbers, chromosomes with the size of a no. 3 chromosome by arrows. (*a*) Cell with two chromosomes the size of a no. 3. (*b*) Cell with one such chromosome (see p. 50)

MALE MEIOSIS

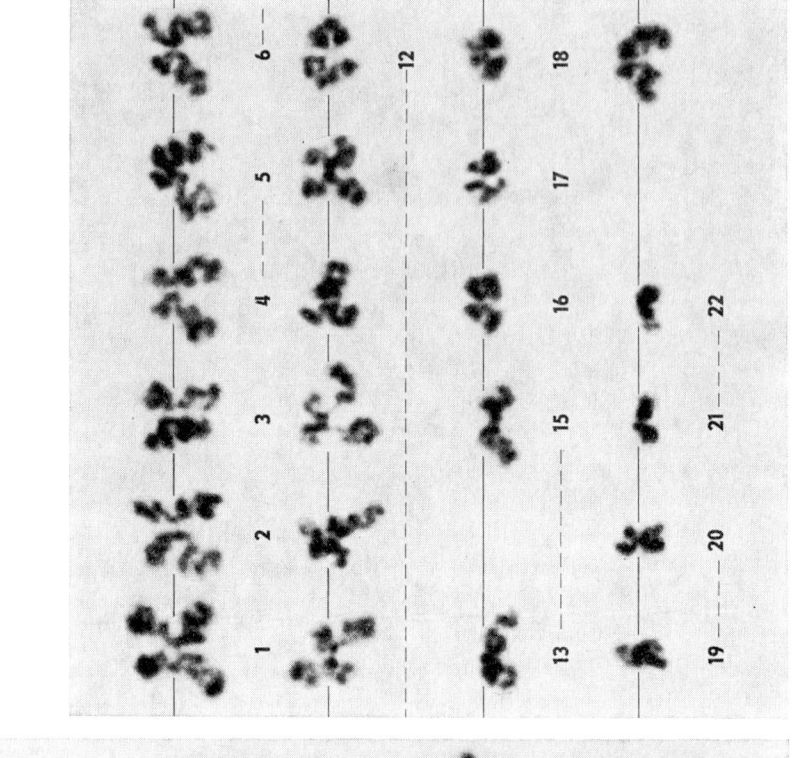

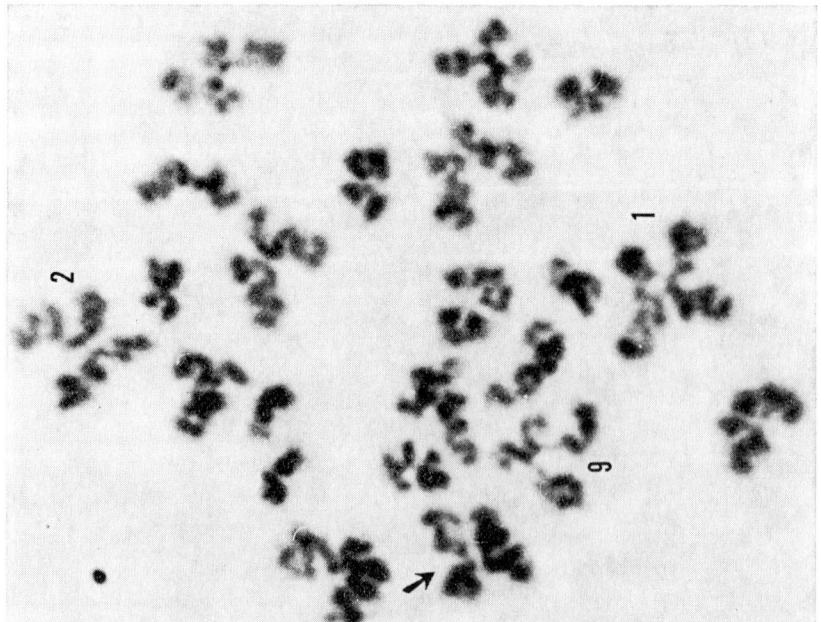

FIGURE 16b (caption p. 49)

Comments. D/D translocation carriers may be found among men with oligoteratospermia [16]. Our finding of a normal spermiogram in an homologous D/D translocation carrier who can only form chromosomally unbalanced gametes shows that a normal spermiogram does not exclude a translocation. Thus, chromosomally abnormal sperm can be morphologically normal.

The location of chiasmata on the D chromosomes involved in the D/G and D/D translocations is mainly of two types. In the first type there is one chiasma on the middle of the long arm material, in spite of the fact that the chiasma on the long arm material of the G chromosomes in the D/G translocation was completely terminalised. In the other type there are two chiasmata, one close to the centromere which apparently does not terminalise, and the other at the distal end of the long arm material. There were very few transitional stages. These observations indicate that there might be chiasmata localised to specific regions of the D chromosomes. Whether chiasmata might be formed on the short arm of the D chromosomes cannot be evaluated.

The number of cells with different numbers of single-chiasma bivalents in the D group is shown in Table 7. It can be calculated that in the controls 178 out of 528 (34 per cent) D bivalents in a total of 176 cells had one single chiasma. In the D/G translocation carriers 62 of the 99 cells (63 per cent) which could be evaluated had a trivalent with one chiasma on the long arm material of the D chromosomes (Table 3). The corresponding value for the D/G mongols was 45 per cent (Table 4, Fig. 12) and for the D/D non-homologous translocation carrier 68 per cent (41 cells) (Table 5, Fig. 14). These figures are higher than that found in the controls, which might indicate that single-chiasma bivalents are formed more often because part of the D chromosome is missing, or that different D bivalents form one chiasma with different frequencies. The relatively high frequency of single-chiasma bivalents in the D/D homologous translocation carrier is noteworthy (Table 7). If the different D bivalents normally form single-chiasma bivalents in similar proportions, and if there is no interchromosomal effect of the translocation on chiasma formation, one would have expected a decrease in single-chiasma bivalents in this case.

THE BEHAVIOUR OF EXTRA UNIDENTIFIED SMALL CHROMOSOMES DURING MEIOSIS

A large number of cases with an extra, small centric fragment have been described in apparently normal individuals and also in patients with varying pathological phenotypes. The nature and origin of these small chromosomes is generally not known. We have studied meiosis in two males with such an abnormal chromosome.

MALE MEIOSIS

The first one is a 19-year-old man (T.E.) previously reported by Hultén et al. [18]. He is 182 cm tall, slightly mentally retarded, and has irregular and defective spermatogenesis. A semen specimen could not be obtained. An extra, minute chromosome was found in his somatic cells (Fig. 17). He was the last of seven pregnancies, three of which terminated in spontaneous abortions. The parents and two of the sibs had normal chromosomes, other relatives have not been karyotyped. The chromosome could be identified in spermatogonial metaphases as well as in first (in 33 out of 55 cells), and second spermatocytes. Cells with separated X and Y chromosomes plus the extra small chromosome were rather frequent (13 out of 33 cells with the extra chromosome). No obvious pairing or association was observed in diakinesis – first metaphase.

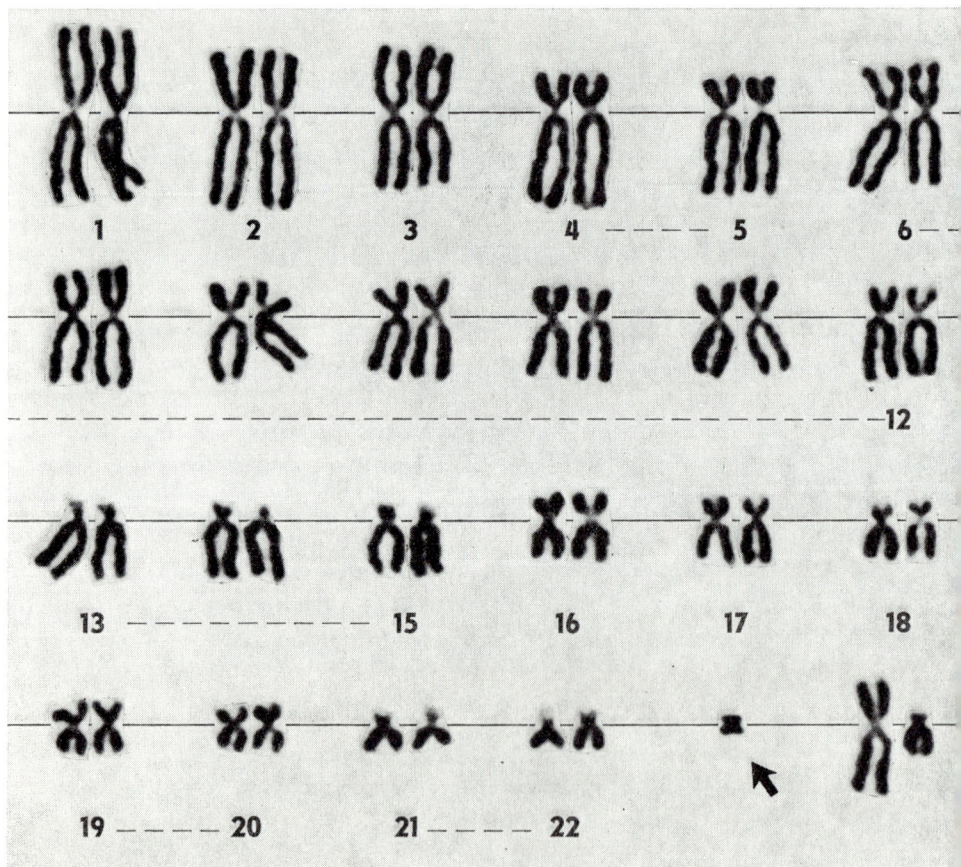

FIGURE 17
Mitotic karyotype from T.E. Note the small extra metacentric chromosome

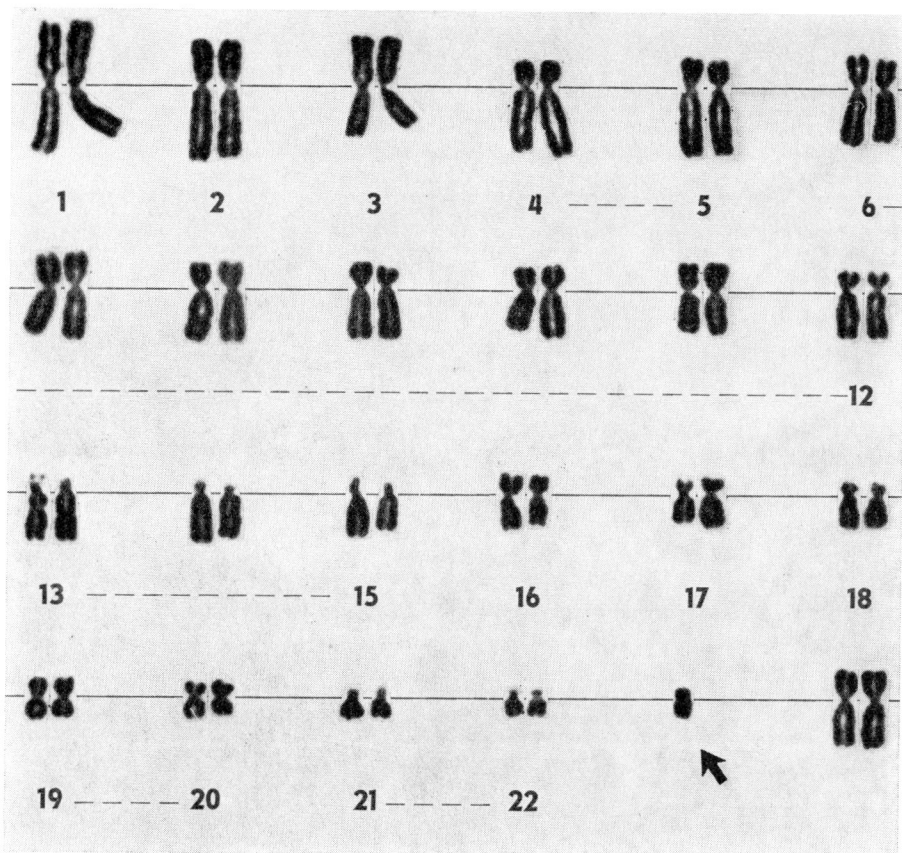

FIGURE 18

Mitotic karyotype from subject III : 7 in Figure 19. Note the extra chromosome of the same size as the G chromosomes (arrow)

In the second case, the abnormal chromosome (Fig. 18) segregated in a family (Fig. 19) ascertained by a grossly malformed stillborn girl who, however, had a normal female karyotype. So far we have not been able to demonstrate any deleterious phenotypic effect of this chromosome. It is obvious already from the pedigree that the chromosome can pass all stages of meiosis but it was also confirmed in one of the males (S.O., II: 1 in Fig. 20). The extra chromosome was observed as a univalent in the majority of cells (40 out of 43 first spermatocytes). He had ankylosing spondylitis for which he had received roentgen therapy (15 R testicular dose). His testicular histology revealed peritubular fibrosis and considerable reduction of all stages of spermiogenesis. A semen specimen could not be obtained.

MALE MEIOSIS

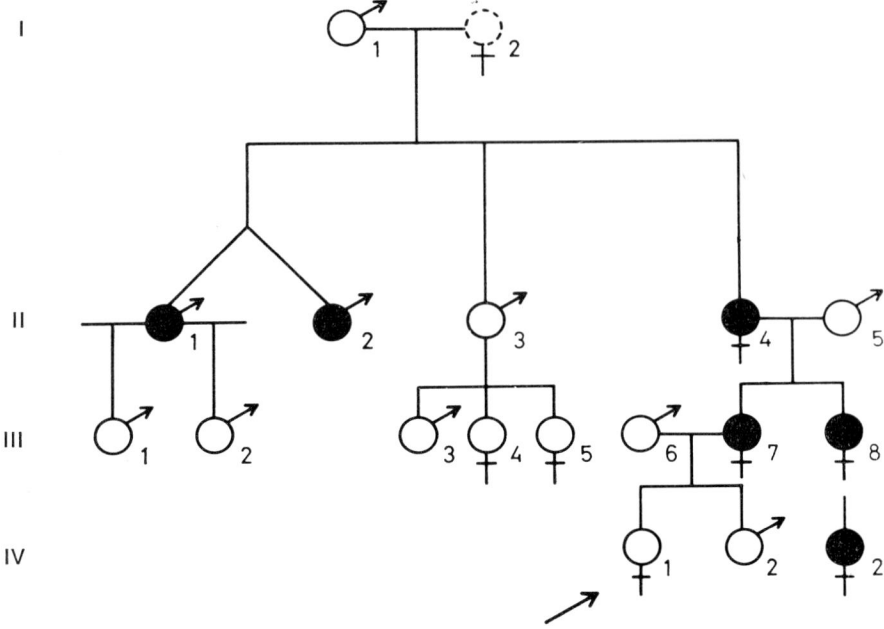

FIGURE 19
Pedigree of family H. An extra small chromosome (Figure 18) was found in several apparently healthy family members (filled symbols)

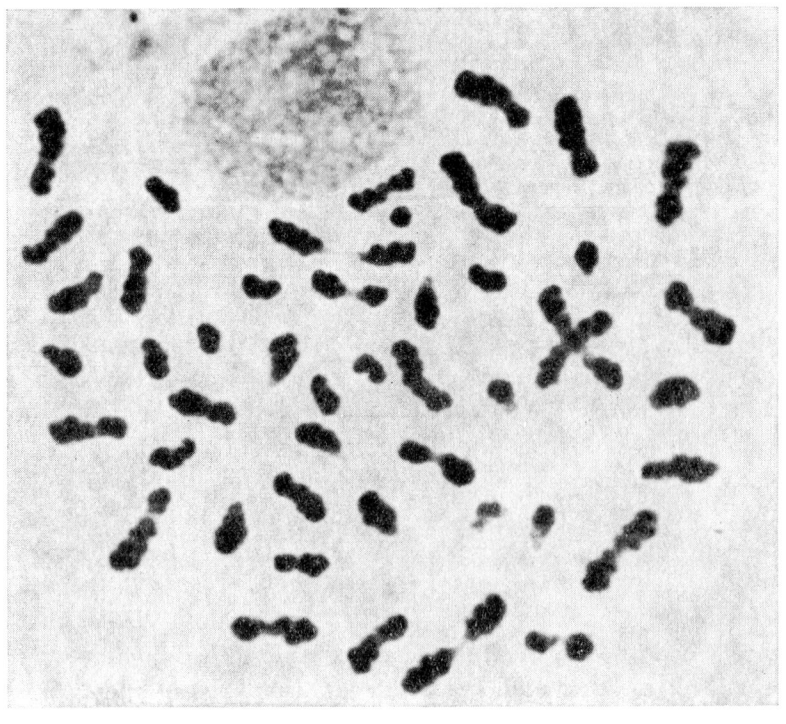

FIG. 20a

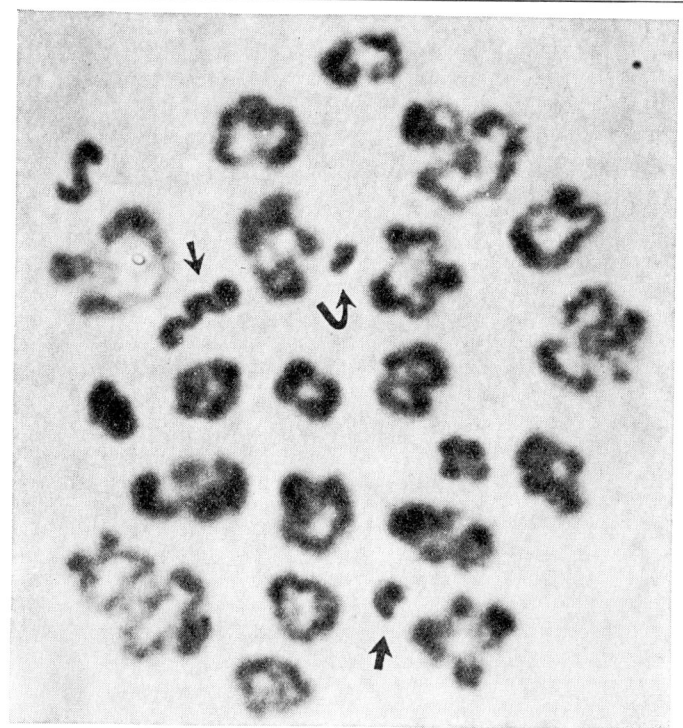

FIGURE 20
Subject S.O. (II : 1 in Figure 19). Spermatogonial metaphase (*a*) and first (*b*) and second (*c*) (see p. 56) spermatocytes with the extra small chromosome (Figure 18). This chromosome is not associated with any of the bivalents in the first spermatocyte where the X and Y are separated

Comments. The meiotic studies in these cases did not help very much in solving the problem of the nature of the small extra chromosomes. If they are structurally altered normal chromosomes, one would expect pairing, at least occasionally, judging from the results on the trisomic mongol. Pairing might of course have taken place in T.E. where the chromosome was obvious only in a proportion of the cells, but hardly so in S.O. The lack of phenotypic effects in the second family indicates that the chromosome is genetically relatively inert. Failure of pairing could be explained by the fact that the tiny chromosome consists mainly of the centromere, and of the paracentromeric regions in which the probability of chiasma formation may be low. Furthermore, if there are localised chiasmata on the human chromosomes, and the regions for such chiasmata are absent, one would also expect the finding of a univalent.

On reconsidering the case T.E. we noticed that he has some of the features found in XYY males, e.g. tall stature and slight mental retardation. Thus, the extra chromosome could be an isochromosome for the short arm of

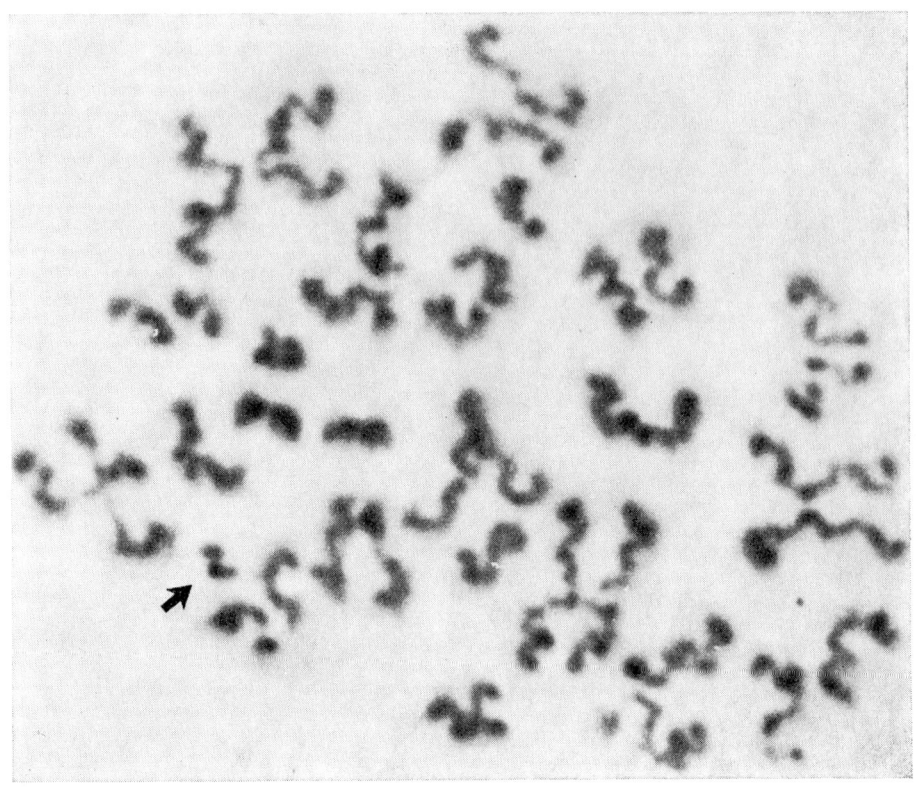

FIGURE 20c (caption p. 55)

the Y or a deletion of the long arm. In this case we may conclude that this structurally abnormal Y passes meiosis in contrast to what has been reported in one XYY male [19]. We have studied also two XYY males and have found it extremely difficult to decide whether the extra Y passes meiosis. However, a few spermatogonial metaphases seemed to have 47 'true' chromosomes, and the appearance of many 'XY bivalents' could be suggestive of the presence of two Y chromosomes. However, an extra Y univalent was not found.

The fact that the presumptively abnormal Y of T.E. is unpaired is, if anything, in favour of an association between the long arm of the Y and the X, in agreement with our earlier findings [4] but in disagreement with Court Brown [20]. Court Brown based his opinion on the finding of a low frequency of association of the X and a dicentric Y with short arm material between the centromeres. However, the mere fact that the X and Y did associate speaks in favour of the long arm being associated with the X.

IDENTIFICATION OF TRANSLOCATIONS BY MEANS OF MEIOTIC STUDIES

Reciprocal translocations are generally ascertained through phenotypically abnormal offspring of translocation carriers. In most cases, both chromosomes involved can be identified in the mitotic karyotype. In some instances the interpretation of an aberration as a translocation is purely presumptive. A study of the meiotic chromosomes is, in those cases, a prerequisite for a correct diagnosis, and thus for a better evaluation of the risk of an unbalanced chromosome constitution to future offspring. Such a case involving a 2/C translocation has been published by Hultén et al. [21]. Another example is given here.

Two brothers with an acrocephalo-syndactyly-like syndrome (Apert's syndrome) and their healthy father turned out to have a G chromosome with an unusually long long arm (Figs. 21, 22). Meiotic studies were performed on the father whose testicular histology and repeated spermiograms

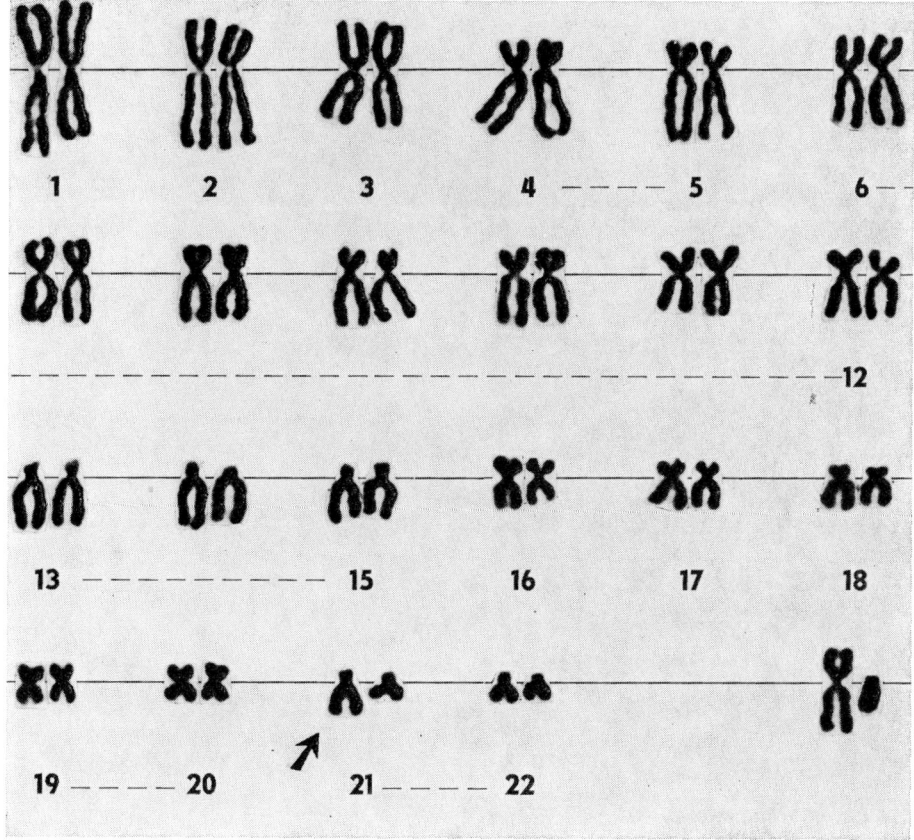

FIGURE 21

Mitotic karyotype from subject III : 18 in Figure 22. Note the G chromosome with an unusually long long arm (arrow)

MALE MEIOSIS

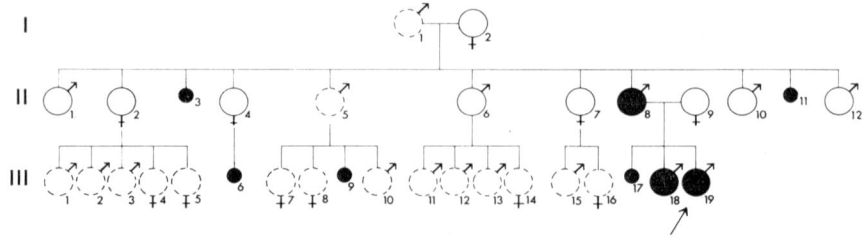

FIGURE 22
Pedigree of family K. with a reciprocal A/G translocation. Filled symbols indicate subjects with the unusually long G chromosome shown in Figure 21

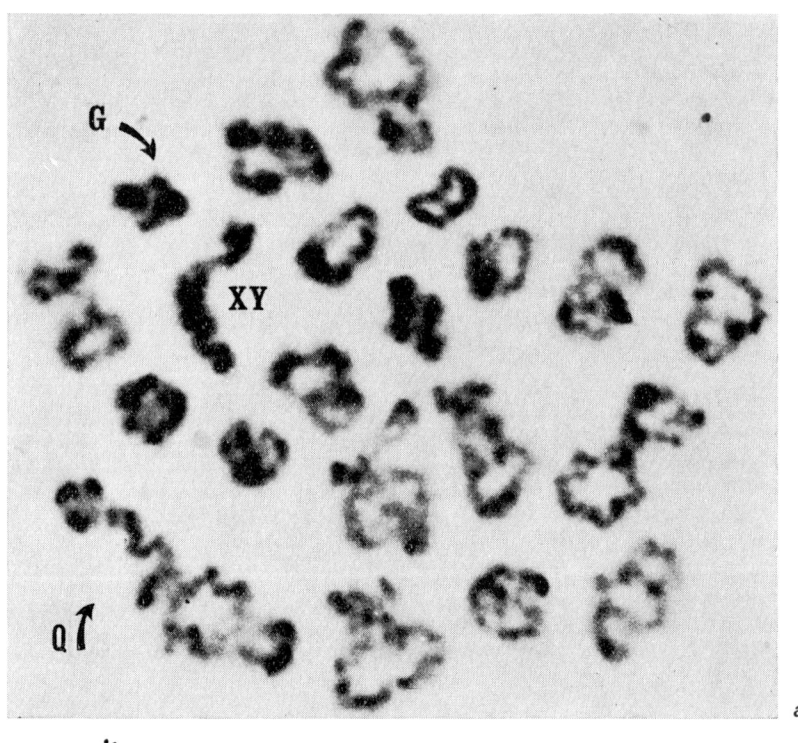

FIGURE 23
First spermatocytes from subject II : 8 in Figure 22 with an unusually long G chromosome (Figure 21). Twenty-two structures including one quadrivalent. The size and shape of the quadrivalent indicate that an autosomal bivalent of group A is involved in the translocation. (*a*) Quadrivalent with no chiasma between two of the adjacent chromosomes as indicated in the drawing below. (*b*) Quadrivalent with chiasmata between all adjacent chromosomes as indicated in the drawing below (see p. 59)

were normal (140–148 million sperm per ml, 31–36 per cent abnormal) [22].

Fifty-six out of 62 first spermatocytes analysed had an obvious quadrivalent involving the G and one of the largest autosomal bivalents in the complement. Two main types of quadrivalents were seen. One had apparently no chiasma between two of the adjacent chromosomes (29/56 cells), and therefore only one chiasma on the long arm material of the G chromosomes. The other type had chiasmata on all four chromosomes and two on the G chromosome material (27/56 cells) (Fig. 23). One cell had a trivalent plus a univalent, the remaining five cells could not be fully evaluated.

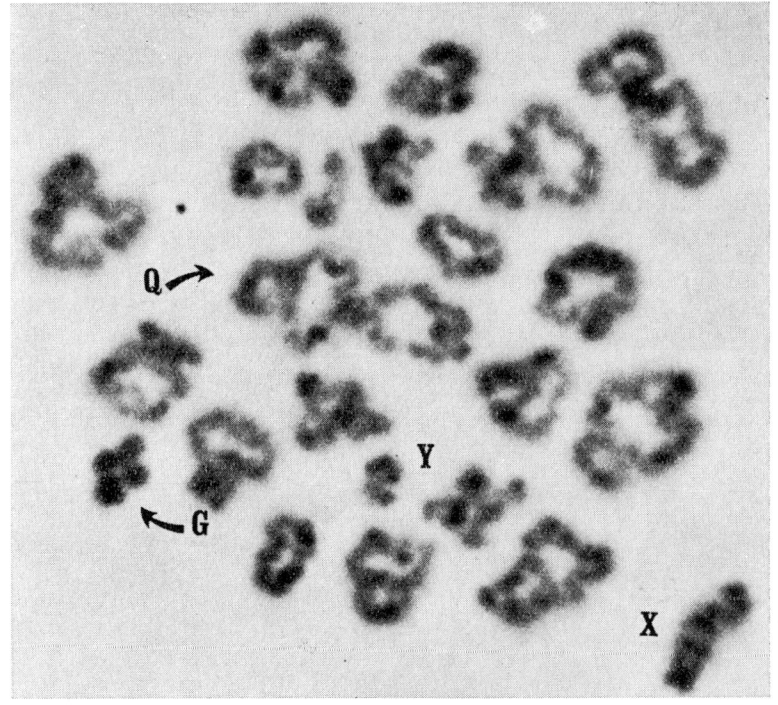

FIGURE 23*b* (caption facing)

Comments. In this family only meiotic studies revealed the presence of a reciprocal translocation. If possible such studies should therefore be applied before more sophisticated or speculative explanations are given. The family also demonstrates that a chromosome abnormality may be found occasionally in a clinical condition not generally considered to be related to such an abnormality. As in the case of the D/D translocation, it can be concluded that translocation carriers can have completely normal testicular histology and spermiograms. The appearance of the quadrivalents can be interpreted in several ways depending on where the chiasmata are considered to have been formed. The finding of about 50 per cent of quadrivalents with two chiasmata on the long arm material of the G chromosomes is high in comparison with our estimates in the normal males and the trisomic mongol. A detailed discussion of these findings will be made elsewhere [22].

ACKNOWLEDGEMENTS

The expenses for the present study were covered by a grant from the Swedish Medical Research Council. The karyotypic analysis on skin and foetal cells and the autoradiographic studies were made in collaboration with Drs M. Fraccaro and L. Tiepolo, Pavia, Italy. We would like to thank Drs B. Fredriksson, Stockholm, and K.-H. Gustavson, Uppsala, who kindly allowed us to study subjects J.K. and A.Å., respectively, and Dr R. Eliasson, Stockholm, who performed the semen analyses. The skilful technical assistance of Miss Kerstin Hansson and Miss Anita Tillberg is gratefully acknowledged.

REFERENCES

[1] Fraccaro, M., Hultén, M. and Lindsten, J., *Ann. N.Y. Acad. Sci.*, **155**, 667, 1968
[2] Luciani, J.-M., *Recherches sur les chromosomes meiotiques de l'homme*. Dissertation, Marseille, 1968.
[3] Hultén, M., Lindsten, J., Lidberg, L. and Ekelund, H., *Annls Génét.*, **11**, 201, 1968.
[4] Hultén, M., Lindsten, J., Pen-Ming., L. M. and Fraccaro, M., *Ann. hum. Genet.*, **30**, 119, 1966.
[5] Evans, E. P., Breckon, G. and Ford, C. E., *Cytogenetics*, **3**, 289, 1964.
[6] Mittwoch, U., *Ann. Eugen.*, **17**, 37, 1952.
[7] Miller, O. J., Mittwoch, U. and Penrose, L. S., *Heredity, Lond.*, **14**, 456, 1960.
[8] Sasaki, M., *Chromosoma*, **16**, 652, 1965.
[9] Finch, R. A., Böök, J. A., Finley, W. H., Finley, S. C. and Tucker, C. C., *Ala. J. med. Sci.*, **3**, 117, 1966.
[10] Hamerton, J. L., Cowie, V. A., Giannelli, F., Briggs, S. M. and Polani, P. E., *Lancet*, **2**, 956, 1961.
[11] Mikkelsen, M., *Ann. hum. Genet.*, **30**, 147, 1966.
[12] Penrose, L. S. In *Advances in Teratology*, vol. I. Ed. D. H. M. Wollam. New York/London : Academic Press, 1966.
[13] Ford, C. E. and Hamerton, J. L., *Nature, Lond.*, **178**, 1020, 1956.
[14] Kodani, M., *Am. J. hum. Genet.*, **10**, 125, 1958.
[15] Sasaki, M. and Makino, S., *Chromosoma*, **16**, 637, 1965.
[16] Kjessler, B., *Karyotype, Meiosis and Spermatogenesis in a Sample of Men attending an Infertility Clinic*. Monographs in Human Genetics 2. Ed. S. Karger. Basel, 1966.
[17] Eberle, P., *Die Chromosomenstruktur des Menschen in Mitosis und Meiosis*. Stuttgart: Gustav Fischer Verlag, 1966.
[18] Hultén, M., Lindsten, J., Fraccaro, M., Mannini, A., and Tiepolo, L., *Lancet*, **2**, 22, 1966.
[19] Thompson, H., Melnyk, J. and Hecht, F., *Lancet*, **2**, 831, 1967.
[20] Court Brown, W. M. *Human population cytogenetics*. Amsterdam : North-Holland Publ. Co., 1967.
[21] Hultén, M., Lindsten, J., Pen-Ming., L. M., Fraccaro, M., Mannini, A., Tiepolo, L., Robson, E. B., Heiken, A. and Tillinger, K.-G., *Nature, Lond.*, **211**, 1067, 1966.
[22] Hultén, M., Bodegård, G., Eriksson, B., Fraccaro, M., Heiken, A. and Lindsten, J. (in preparation, 1969).

Robertsonian Translocations
Evidence on Segregation from Family Studies

JOHN L. HAMERTON

Paediatric Research Unit*
Guy's Hospital Medical School, London

* Present address: Department of Genetics, Children's Hospital of Winnipeg, 685 Bannatyne Avenue, Winnipeg, Manitoba, Canada.

¶ HAMERTON [1, 2] has summarised the available data on the cytogenetics and segregation of the three Robertsonian translocations which have been reported in human populations. In the present paper, segregation data on forty-seven t(DqGq), forty-two t(DqDq) and six t(21q22q) [3] families are considered. These include the families which were previously discussed [1, 2].

MATERIAL AND METHODS

Family data. The available family data are summarised in Tables 1 and 2.

TABLE 1. Family data on three Robertsonian translocations

	t(DqGq)	t(DqDq)	t(21q22q)
Total kindreds	47	42	6
Heterozygotes			
Female	96	67	12
Male	46	58	6
Unknown sex	14	15	2
Total	156	139	20

TABLE 2. Progeny in three Robertsonian translocations

	t(DqGq)	t(DqDq)	t(21q22q)
Total (uncorrected)	623	610	72
Total (corrected)	526	573	58
Number of probands	97 *	38 †	14 *

* Probands and key heterozygotes (see text)
† Probands only. Four kindreds ascertained through founder heterozygote

These 95 families (Appendixes) have largely been collected from the literature and by personal communication with the authors concerned. With the exception of two families, all the t(DqGq) and t(21q22q) families were ascertained through probands with Down's syndrome. Two exceptional families may involve different chromosomes and will be considered separately. Ascertainment of the t(DqDq) families was much more varied and is summarised in Table 4. Twenty-five of these families were ascertained through probands with various abnormal phenotypes, not obviously associated with the translocation, while thirteen families were ascertained by chance in various surveys of different population groups. Three families were ascertained because of a proband with the clinical features of Patau's syndrome (D-trisomy).

TABLE 3. Ascertainment of 56 t(DqGq) and t(21q22q) families

	Kindreds
t(DqGq)	
Down's syndrome	47
Prison survey (No. D.S.)	1
Infertility	1
t(21q22q)	
Down's syndrome	6

TABLE 4. Ascertainment of 41 t(DqDq) families

Reason for ascertainment		No. of kindreds
Clinical		
Down's syndrome		11
Patau's syndrome		3
Mental and/or developmental retardation		4
Infertility and abortion		3
Ovarian dysgenesis		2
Cerebral palsy		1
Nystagmus & mental retardation		1
Polydysspondyly		1
Chronic myeloid leukaemia		1
	Total	27
Not known		2
Surveys		
Cancer radiotherapy		3
Ankylosing spondylitis		2
Atomic energy workers		1
Penal institutes		3
Mental hospital		1
Newborn		2
Chromosome abnormality in spouse		1
	Total	13
	Total	42

Correction for ascertainment. Hamerton [1, 2] has discussed the necessity for the correction of any numerical bias resulting from the way in which these families were ascertained, and concluded [2] that the following corrections were necessary:

(1) t(DqGq) and t(21q22q). All probands were excluded from the calculation of segregation ratios; because of the finding of an aberrant sex ratio among the t(DqGq) heterozygotes, all the heterozygous parents of probands were also excluded (key heterozygotes).

(2) t(DqDq). This translocation has been largely ascertained by chance and there is no evidence of any sex ratio distortion among the heterozygotes; for this reason only the probands were excluded. The same corrections have been applied to the present data, but it is recognised that they may be far from ideal and may result in an over-correction against unbalanced progeny in the t(DqGq) families.

Calculation of gametic frequencies. The gametes and resultant zygotes which may be produced by a t(DqGq) heterozygote are summarised in Table 5. Six classes of gamete should result from concordant orientation of the trivalent (discordant orientation will not be considered). Two of these classes have never been detected; these result from the second type of adjacent segregation which may not occur. The third class (G –) has never been observed but must be formed at meiosis as it is complementary to the D –, t(DqGq) + class. For all practical purposes therefore, three classes of gametes are formed and might be expected to be observed in equal numbers. The same arguments apply to the t(DqDq) and the t(21q22q) translocations and will form the theoretical basis of the discussion about the observed progeny frequencies.

RESULTS

(a) *t(DqGq) Down's syndrome ascertained* (47 families)

Examination of Tables 6 to 10 gives the following information:

(1) *Sibship size and spontaneous abortion* (Table 6)

The mean sibship size for heterozygous fathers was found not to differ significantly from that for heterozygous mothers. Female heterozygotes show a slight but insignificant increase in the frequency of spontaneous abortion compared to males ($0.2 > P > 0.1$). The frequency of spontaneous abortion is not, however, increased over the level of 15 to 20 per cent of live births which is generally accepted as the incidence for the population at large.

(2) *Progeny frequencies.* These are summarised in Table 7 and the main point to note is the deficiency of unbalanced progeny (6.25 per cent). Comparison of the frequency of subjects with normal chromosomes and

TABLE 5

Translocation	Chromosome complement of balanced heterozygote	Gametes			
		Alternate segregation		Adjacent segregation *	
t(DqGq)	45,XX or D−,G−,t(DqGq)+ 45,XY	23,X or 23,Y	22,X or D−,t(DqGq)+ 22,Y	23,X or D−,t(DqGq)+ 23,Y	22,X or G− 22,Y
Zygotes (assuming the above gametes are fertilized by normal 23,X or 23,Y gametes)	Chromosome complement	46,XX or 46,XY	45,XX or D−,G−,t(DqGq)+ 45,XY	46,XX or D−,t(DqGq)+ 46,XY	45,XX or G− 45,XY
	Genome	Normal	Balanced heterozygous	Unbalanced Gq duplication	Unbalanced G monosomic
	Phenotype		Normal	Down's syndrome	Not observed

* A second type of adjacent segregation is possible giving 23,X or 23,Y, G−, t(DqGq)+ and 22,X or Y, D−. These are not included. There is no evidence that they are found

TABLE 6. t(DqGq): Sibship size and spontaneous abortions

Sex of heterozygote	Male		Female	
No. of sibships	46		96	
Mean size of sibship ± SD	3·26 ± 2·05		3·42 ± 2·26	
Number of abortions	12		42	
Abortions as per cent total progeny*	7·50		11·60	
Abortions per sibship	0·26		0·44	
	$\chi^2 = 1.81$	$0.2 > P > 0.1$		

* Corrected totals

[67

TABLE 7. t(DqGq): Liveborn progeny of balanced heterozygotes by chromosome consititution

Heterozygotes		Segregation*				
		Alternate		Adjacent		
Sex	No.	46,XX or XY	45,XX or XY, D−,G−,t(DqGq)+	46,XX or XY, D−,t(DqGq)+	45,XX or XY, G−	Total
Female	96	82	88 (100)	16 (82)	—	186 (264)
Male	46	45	57 (65)	4 (10)	—	106 (120)
Unknown	14	14	44 (49)	1 (1)	—	59 (64)
Totals		141	189 (214)	21 (93)	—	351 (448)
Per cent		40·17	53·85	5·98	—	100·00

*Uncorrected totals in ( )

TABLE 8. t(DqGq): Comparison of the frequency of balanced heterozygotes and normal progeny

	Observed			Expected hetero-zygotes
	Normal	Balanced hetero-zygote	Total	
Female	82	88	170	90·63
Male	45	57	102	54·38
Totals	127	145	272	145·01

$\chi^2_{1:1} = 1·06$ $0·3 > P > 0·2$

TABLE 9. t(DqGq): Frequency of Down's syndrome per sibship (uncorrected data)

Heterozygote	No.	No. of D.S.	D.S./sibship
Female	96	82 (31·1%)	0·85
Male	46	10 (8·3%)	0·22

$\chi^2 = 17·79$ $P > 0·0005$

balanced translocation heterozygotes indicates an excess of the latter. However, if the children of heterozygous parents of unknown sex are not considered, this excess is no longer significant ($0.3 > P > 0.2$, Table 8). The reason for the finding of such an excess of heterozygous children in these sibships is almost certainly selection bias. The heterozygous parents in this category are invariably deceased or untested and can only be classed as a heterozygote if more than one of their children carries the translocation. This inevitably must lead to an excess of balanced heterozygotes among their progeny.

Tables 9 and 10 summarise the data on children affected with Down's syndrome in these families. Table 9 gives the uncorrected data; this shows a highly significant excess of affected children in female sibships (0.9 per sibship) compared to male (0.2 per sibship) ($P < 0.0005$). After correction of the data for ascertainment (Table 10) this difference is no longer significant ($0.2 > P > 0.1$). This will be discussed below.

(b) $t(DqDq)$ (42 families). No significant difference in mean sibship size or in the frequency of spontaneous abortion can be found in these 41 families (Table 11).

TABLE 10. t(DqGq) : Frequency of unbalanced progeny (corrected data)

Heterozygote	Observed			Expected D.S.
	Translocation D.S.	Phenotypically normal	Total	
Female	16 (8.60%)	170	186	12.74
Male	4 (3.77%)	102	106	7.26
Totals	20	272	292	

$\chi^2 = 2.30 \qquad 0.2 > P > 0.1$

TABLE 11. t(DqDq) : Sibship size and spontaneous abortions

Sex of heterozygote	Male	Female
No. of sibships	58	67
Mean size of sibship	2.97 ± 2.01	3.41 ± 2.49
Number of abortions	36	67
Abortions as per cent total progeny*	17.2	22.6
Abortions per sibship	0.62	1.00

$\chi^2 = 1.65 \qquad 0.2 > P > 0.1$

* Corrected totals

TABLE 12. t(DqDq): Liveborn progeny of balanced heterozygotes by chromosome constitution

Heterozygotes			Segregation*				
			Alternate		Adjacent		
Sex	No.	46,XX or XY	45,XX or XY, D−,D−,t(DqDq)+	46,XX or XY, D−,t(DqDq)+	45,XX or XY, D−	Total
Female	67	67 (71)	81 (100)	1 (3)	—	149 (174)
Male	58	46 (46)	74 (84)	− (1)	—	120 (131)
Unknown	15	17 (17)	38 (40)	—	—	55 (57)
Totals		130 (134)	193 (224)	1 (4)	—	324 (362)
Per cent		40·12	59·57	0·31	—	100·00

*Uncorrected totals in ( )

TABLE 13. t(DqDq): Comparison of the frequency of balanced heterozygotes and normal progeny

	Observed			Expected heterozygotes	χ^2
	Normal	Balanced heterozygote	Total		
Female	67	81	148	85·60	0·25
Male	46	74	120	69·41	0·31
Totals	113	155	268	155·01	0·56

$\chi^2_{1:1} = 6\cdot 27$ $0\cdot 025 > P > 0\cdot 01$

TABLE 14. t(21q22q): Sibship size and spontaneous abortions

Sex of heterozygote	Male	Female
No. of sibships	6	12
Mean size of sibship	4·00	3·50
No. of abortions	1	12
Abortions as per cent total*	5·26	34·3
Abortions per sibship	0·16	1·00

*Corrected totals

Progeny frequencies. These are summarised in Table 12. Two points to note are the almost complete absence of children with unbalanced chromosome complements (0·3 per cent), and an apparent excess of balanced heterozygotes. If the children of heterozygous parents of unknown sex are not included for the reasons discussed above, an excess of heterozygotes still remains (0·025 > P > 0·01, Table 13). This excess applies equally to male and female sibships which no longer differ significantly from each other (P = 0·7) as originally reported [2].

(c) t(21q22q). The number of families known carrying this translocation is still too low for any firm conclusion to be reached about the frequency of the different progeny classes. The available results are summarised in Tables 14 and 15.

DISCUSSION

The results reported here differ in two essential respects from those given in an earlier paper [2], in which it was concluded that there was an excess of heterozygotes in both t(DqGq) and t(DqDq) families, and that this comprised progeny of male and unknown sex heterozygotes only. It has been shown above that the progeny of unknown sex heterozygotes are subject to selection bias and should not have been included. Removal of this class of progeny reduces the excess of heterozygotes in t(DqGq) families to an insignificant level (Table 8); with the increased amount of data no difference can be observed between the frequency of heterozygotes in male and female sibships. The t(DqDq) families continue to show an excess of balanced heterozygotes over normal progeny (0·025 > P > 0·01), but a comparison of male and female sibships shows no significant heterogeneity. Thus, with the increased data which are now available, these translocations no longer provide any evidence for gametic selection [2]. There is however still the suggestion that there may be an excess of heterozygotes present especially in t(DqDq) families. This suggests that these may have a selective advantage and serve to maintain these translocations in the population. The second essential departure from earlier results concerns the level of affected children in the t(DqGq) families. The uncorrected data show a large excess of these children among the progeny of female compared to male heterozygotes (Table 9) which is removed when the data are corrected for ascertainment. Inspection of the pedigrees [1] reveals that among the families reported earlier many were ascertained because of the presence of two or more affected children, while fewer of the families reported later [2 and present paper], were ascertained because of multiple affected births. This suggests that the correction applied to these data, involving as it does, the almost complete elimination of the affected progeny, is too severe. It is impossible to arrive at a more satisfactory

TABLE 15. t(21q22q) : Liveborn progeny of balanced heterozygotes by chromosome constitution

Heterozygotes			Segregation*				
			Alternate		Adjacent		
Sex	No.	46,XX or XY	45,XX or XY, 21−,22−,t(21q22q)+	46,XX or XY, 22−,t(21q22q)+	45,XX or XY, 21−		Total
Female	12	11	6 (7)	2 (8)	—		19 (26)
Male	6	11	6 (10)	− (1)	—		17 (22)
Unknown	2	—	5 (6)	—	—		5 (6)
Totals		22	17 (23)	2 (9)	—		41 (54)
Per cent		53.7	41.5	4.9	—		100.1

*Uncorrected totals in ( )

TABLE 16. Comparison of the frequency of unbalanced progeny in t(DqGq) and t(DqDq) families

	Observed		Total	Expected Unbalanced	χ^2
	Unbalanced	Balanced			
t(DqGq)	20	272	292	10.92	7.55
t(DqDq)	1	268	269	10.06	8.16
Totals	21	540	561	20.98	15.71
				P > 0.0005	

method of data correction because of heterogeneity and incomplete method of family ascertainment. The data do, however, suggest that the risk of heterozygous mothers having an affected child lies between 10 and 15 per cent, while the risk to a heterozygous father is probably about 5 per cent. The reason for this difference may be the result of selection against unbalanced sperm, or the pattern of segregation may be different in female meiosis, resulting in a higher level of adjacent segregation in oogenesis compared to spermatogenesis. Finally, differential lethality of unbalanced zygotes is possible. This would imply a maternal foetal incompatability when the zygote is unbalanced and the mother normal, not when the mother is heterozygous; this seems an unlikely possibility. A comparison of the t(DqGq) and the t(DqDq) families (Table 16) indicates a highly significant excess of unbalanced progeny in the t(DqGq) families. Hamerton [1, 2] concluded that a higher level of adjacent segregation was to be expected from an asymmetrical trivalent such as the t(DqGq) which is formed from chromosomes of greatly disparate length, rather than from a symmetrical trivalent, found in the t(DqDq) in which the translocation is between chromosomes of similar length.

The mean size of sibship does not differ significantly between heterozygous mothers or fathers, which indicates a similar fertility for the heterozygotes of both sexes. The frequency of spontaneous abortion in both the t(DqGq) and the t(DqDq) families does not differ significantly from the 15 to 20 per cent expected in the general population. In both the t(DqGq) and t(DqDq) families, heterozygous mothers have a slightly increased frequency of abortion compared with sibships in which the father was heterozygous, but in neither group is this significant. This could be a reflection of a higher level of adjacent segregation in females, which would give an increased frequency of chromosomally unbalanced zygotes, and therefore the possibility of increased foetal loss. It could also be a reflection of more accurate data collection from mothers than from fathers. The level of spontaneous abortion does not differ significantly from the expected 15 to 20 per cent found in the population at large for either group of families. This is not in agreement with earlier data [4] in which levels of 35 and 43 per cent for male and female heterozygotes respectively were recorded for 20 t(DqDq) sibships in which a nearly complete record of conception was available. It is well known how difficult it is to obtain such complete records, and in doing so, families with a particularly high level of spontaneous abortion may inadvertently have been selected. Alternatively, the present data may include a large number of sibships with incomplete conceptual histories, so reducing the observed levels of abortion.

The results reported here differ from those given earlier [1, 2] and this reflects the difficulties of handling this type of material. In particular the problems resulting from ascertainment bias in such incompletely ascertained families are almost insuperable. The most satisfactory data are undoubtedly those families in which the proband was ascertained at random in surveys of different population groups, and these families should be studied fully in respect of their reproductive histories. At present about thirteen such families are known among the forty-one t(DqDq) families, and accumulation of this extremely valuable material clearly requires an extensive international collaborative study, in which the standards of ascertainment and follow up are established in advance.

REFERENCES

[1] Hamerton, J.L. In *Chromosomes Today*, p. 237. Eds. C.D. Darlington and K.R. Lewis. Edinburgh : Oliver and Boyd, 1966.
[2] Hamerton, J.L., *Cytogenetics,* **7,** 260, 1968.
[3] See Appendices for full references and data.
[4] Court Brown, W.M., *Human Population Cytogenetics.* Amsterdam : North-Holland Publ. Co., 1967.
[5] Macintyre, M.N., Staples, W.I., Steinberg, A.G. and Hempel, J.M., *Am. J. hum. Genet.,* **14,** 335, 1962.
[6] Sergovich, F., Soltan, H. and Carr, D., *Can. med. Ass. J.,* **87,** 852, 1962.
[7] Hamerton, J.L., Cowie, V.A., Giannelli, F. and Briggs, S.M., *Lancet,* **2,** 956, 1961.
[8] Hayman, D., personal communication, 1964, quoted by Hamerton, 1966.
[9] Penrose, L.S., Ellis, J.R. and Delhanty, J.D.A., *Lancet,* **2,** 409, 1960.
[10] Carter, C.O., Hamerton, J.L., Polani, P.E., Gunalp, A. and Weller, S.D.V., *Lancet,* **2,** 678, 1960.
[11] Giannelli, F., Hamerton, J.L. and Carter, C.O., *Cytogenetics,* **4,** 186, 1965.
[12] German, J.L., Demayo, A.P. and Bearn, A.G., *Am. J. hum. Genet.,* **14,** 31, 1962.
[13] Forssman, H. and Lehmann, O., *Acta Paediat., Stockh.,* **51,** 180, 1962.
[14] Van der Hagen, C.B., personal communication, 1963, quoted by Hamerton, 1966.
[15] Buckton, K.E., Harnden, D.G., Baikie, A.G. and Woodcock, G.E., *Lancet,* **1,** 171, 1961.
[16] Warkany, J., Soukup, S.W. and Weinstein, E.D., personal communication, 1963, quoted by Hamerton, 1966.
[17] Breg, W.R., Miller, O.J. and Schmickel, R.D., *New Engl. J. Med.,* **266,** 845, 1962.
[18] Atkins, L., O'Sullivan, M.A. and Pryles, C.V., *New Engl. J. Med.,* **266,** 631, 1962.
[19] Edwards, J.H. and Clarke, C., personal communication, 1961, quoted by Hamerton, 1966.
[20] Mellman, W.J., personal communication, 1962, quoted by Hamerton, 1966.
[21] Jacobs, P.A., personal communication, 1969, M.R.C. Clinical and Population Cytogenetics Research Unit. Kindred Registry Nos. K1 (see Ref. 15), K11, K2, K22, K6, K33, K38, K32, K29, K56, K7, K18, K16, K49, K13, K44, K41, K37, K52, K46, and 92/68.
[22] Macintyre, M.N., personal communication, 1966, quoted by Hamerton, 1968.
[23] Soudek, D., personal communication, 1967, quoted by Hamerton, 1968.
[24] Wright, S.W., personal communication, 1966, quoted by Hamerton, 1968.
[25] Valdmanis, A., Mann, J.D. and Johns, D., *Hum. Chromos. Newsl.,* **18,** 25, 1966.

[26] Hustinx, T. W. J., *Cytogenitisch onderzoek bij enige families,* Drukkerij-Uitgeverij, Brakkenstein, 1966.
[27] Mikkelsen, M., *Ann. hum. Genet.,* **30,** 147, 1966.
[28] Paediatric Research Unit, unpublished data, 1969, P4403/H6740, P1827/H4000, P3328/H5477 and P4399/H6744.
[29] Pearson, P. L., Clarke, G., Davison, B. C. C., Lennox, I. G. and Pritchard, P. M. M., *J. med. Genet.,* 1969 (in press).
[30] Shaw, M. W., *Cytogenetics,* **1,** 141, 1962.
[31] Ferguson-Smith, M., personal communication, 1969.
[32] Uchida, I., personal communication, 1969.
[33] Walker, S. and Harris, R., *Ann. hum. Genet.,* **26,** 151, 1962.
[34] Jagiello, G. M., *New Engl. J. Med.,* **269,** 66, 1963.
[35] Hamerton, J. L., Giannelli, F. and Carter, C. O., *Cytogenetics,* **2,** 194, 1963.
[36] Oikawa, K., Gromultz, J. M. and Hirschhorn, K., personal communication, 1962, quoted by Hamerton, 1966.
[37] Wahrman, J., personal communication, 1964, quoted by Hamerton, 1966.
[38] Zergollern, L., Hoefnagel, D., Benirschke, K. and Corcoran, P. A., *Cytogenetics,* **3,** 148, 1964.
[39] Dill, F. J. and Miller, J. R., personal communication, 1963, quoted by Hamerton, 1966.
[40] Hamerton, J. L., Giannelli, F. and Carter, C. O., unpublished data, 1964.
[41] Gerald, P. S., personal communication, 1965, quoted by Hamerton, 1968.
[42] Wilson, J. A., M.D. Thesis, University of Liverpool, 1967.
[43] Engel, E., McGee, B. J., Hartman, R. C. and Engel-de-Montmollin, M., *Cytogenetics,* **4,** 157, 1965.
[44] Yunis, J. J., Hook, E. B. and Mayer, M., *Am. J. hum. Genet.,* **17,** 191, 1965.
[45] Marsden, H. B., Mackay, R. I., Murray, A. and Ward, H. E., *J. med. Genet.,* **3,** 56, 1966.
[46] De Grouchy, J., Mlynarski, J.-C., Maroteaux, P., Lamy, M., Deshaies, G., Benichou, C. and Salmon, C., *C.R. Acad. Sci., Paris,* **256,** 1614, 1963.
[47] Dekaban, A. S., *Am. J. hum. Genet.,* **18,** 288, 1966.
[48] Nuzzo, F., Marini, A., Flauto, U. and Santachiara-Benerecetti, A. S., *Atti Ass. genet. Ital.,* **11,** 411, 1966.
[49] Richards, B. W., Stewart, A., Sylvester, P. E. and Jasiewicz, V., *J. ment. Defic. Res.,* **9,** 245, 1965.
[50] Palmer, C., *J. med. Genet.,* 1969 (in press).
[51] Krompotic, E., Ramanathan, K. and Grossman, A., *J. med. Genet.,* **5,** 205, 1968.
[52] Walzer, S., Breau, G. and Gerald, P. S., *J. Pediat.,* **74,** 438, 1969.
[53] Atkins, L. Bartsolas, C. S. and Porter, P. J., *J. med. Genet.,* **5,** 314, 1968.
[54] Pfeiffer, R. A. *Lancet,* **1,** 1163, 1963.
[55] Jackson, J. F. and Ashford, W. P., *J. Am. med. Assoc.,* **200,** 722, 1967.

APPENDIX I. t(DqGq) Complete Family Data

Ref. No.	Ascertainment and family identification	Progeny of male heterozygotes							Progeny of female heterozygotes							Progeny of heterozygotes of unknown sex						
		Number	Normal chromosomes	Balanced heterozygotes	Unbalanced heterozygotes	Not tested	Spontaneous abortions		Number	Normal chromosomes	Balanced heterozygotes	Unbalanced heterozygotes	Not tested	Spontaneous abortions		Number	Normal chromosomes	Balanced heterozygotes	Unbalanced heterozygotes	Not tested	Spontaneous abortions
5	Down's Syndrome	3	1	3(1)	—	—	—		5	4	3	4(1)	—	10		1	—	4	—	1	—
6	"	4	9	5(1)	1(1)	—	—		1	2	—	—	—	—		1	—	5	—	1	—
7	"	2	—	10(1)	—	3	—		1	—	3	1(1)	—	1		—	—	—	—	—	—
8	"	1	—	—	2	—	—		3	1	1	2(2)	4	1		1	4	5(2)	—	—	—
9	"	—	—	—	—	—	—		2	—	2(1)	2(2)	—	2		—	—	—	—	—	—
10	"	—	—	—	—	—	—		3	1	3(1)	3(2)	4	—		—	—	—	—	—	—
11	"	—	—	—	—	—	—		3	3	3(1)	2(1)	1	1		—	—	—	—	—	—
12	"	—	—	—	—	—	—		1	1	1	1(1)	2	2		—	—	—	—	—	—
13	"	1	—	1(1)	1	—	—		—	—	—	3(3)	—	1		—	—	—	—	—	—
13	"	—	—	—	—	—	—		1	1	1	2(2)	—	—		1	—	3	—	1	—
14	"	2	2	2(1)	1(1)	—	2		1	—	—	—	—	—		—	—	—	1	—	—
15	" (K1)	—	—	—	—	—	—		2	—	2(1)	3(3)	1	1		1	—	3	—	2	—
16	"	—	—	—	—	—	—		1	—	—	2(2)	—	—		—	—	—	—	—	—
17	"	1	—	—	—	—	—		1	—	—	1(1)	1	—		—	—	—	—	—	—
18	"	—	—	—	—	—	—		1	—	—	3(3)	—	—		—	—	—	—	—	—
19	"	—	—	—	—	—	—		1	—	—	1(1)	1	—		—	—	—	—	—	—
20	"	—	—	—	—	—	—		1	—	1	1(1)	1	—		—	—	—	—	—	—
21	" (K11)	1	—	1	1(1)	—	—		2	—	4(1)	1(1)	—	1		—	—	—	—	—	—
21	" (K2)	—	—	—	—	—	—		1	1	—	—	—	—		—	—	—	—	—	—
21	" (K22)	1	—	—	—	—	—		—	—	—	1(1)	—	—		—	—	—	—	—	—
21	" (K6)	—	—	—	—	6	2		3	3	3(1)	2(2)	3	—		1	2	2	—	8	—
21	" (K33)	3	2	—	—	3	2		6	7	4	2(1)	2	2		1	4	8	—	2	—

26	"	(FAM B)	6	11	13	1(1)	–	5	1	–	4	–	1	1	–	3(1)	–	6	–
26	"	(FAM C)	2	5	11(1)	1	–	2	6	6	10	5(1)	17	2	1	–	2	–	4	–
27	"	(7)	–	–	–	–	–	–	1	1	–	2(2)	–	1	–	–	–	–	–	–
27	"	(4)	–	–	–	–	–	–	1	1	–	2(2)	1	1	–	–	–	–	–	–
27	"	(J)	1	1	–	1(1)	–	1	–	–	–	–	–	–	–	–	–	–	–	–
27	"	(L)	–	–	–	–	–	–	2	–	1(1)	1(1)	–	–	1	–	2	5(1)	–	1	–
27	"	(1)	1	–	1	–	–	–	6	6	8	4(2)	3	–	1	1	2	–	1	–
27	"	(16)	2	–	3(1)	–	1	–	3	–	3	2(2)	–	4	1	–	2	–	2	–
28	"	(P4403)	–	–	–	–	–	–	1	2	1	1(1)	2	–	–	–	–	3	–	1	–
29	"	(AS/60)	–	–	–	–	–	–	6	3	6(1)	2(2)	2	2	1	–	–	3	–	–	–
29	"	(CD/290)	2	2	1	–	1	–	3	4	3(1)	2(2)	2	2	–	–	–	–	–	–	–
29	"	(CDH/164)	–	–	–	–	–	–	1	2	–	2(2)	2	–	–	–	–	–	–	–	–
30	"	(K)	5	5	4	–	9	–	6	10	10(1)	6(3)	4	–	1	1	–	3	–	2	–
31	"	(CE 68230)	–	–	–	–	–	–	1	1	1	1(1)	–	–	–	–	–	–	–	–	–
31	"	(SN 67263)	–	–	–	–	–	–	1	1	–	1(1)	–	1	–	–	–	–	–	–	–
31	"	(McD650151)	–	–	–	–	–	–	1	2	3	2(2)	–	–	–	–	–	–	–	–	–
31	"	(DN680090)	1	–	–	1(1)	–	–	–	–	–	–	–	–	–	–	–	–	–	–	–
32	"	(1)	1	1	3	–	1	–	3	5	3(1)	2(2)	–	–	1	–	–	2(1)	–	4	1
32	"	(2)	–	–	–	–	–	–	1	–	–	3(2)	–	–	–	–	–	–	–	–	–
32	"	(3)	–	–	–	–	–	–	1	3	1	1(1)	1	–	–	–	–	–	–	–	–
32	"	(4)	–	–	–	–	–	–	1	–	1	1(1)	–	1	–	–	–	–	–	–	–
TOTALS			46	45	65(8)	10(6)	28	12	96	82	100(12)	82(66)	56	42	14	14	49(5)	1	36	1
CORRECTED			46	45	57	4	28	12	96	82	88	16	56	42	14	14	44	1	36	1
(Families not ascertained through Down's Syndrome)																				
21	Prison Survey (K38)	1	4	1	–	–	–	3	5	8(1)	–	8	–	–	–	–	–	–	–
21	Infertility (92/68)	–	–	–	–	–	–	1	1	–	–	–	–	–	–	–	–	–	–

APPENDIX II. t(DqDq) Complete Family Data

Ref. No.	Ascertainment and family identification	Progeny of male heterozygotes						Progeny of female heterozygotes						Progeny of heterozygotes of unknown sex				
		Number	Normal chromosomes	Balanced heterozygotes	Unbalanced heterozygotes	Not tested	Spontaneous abortions	Number	Normal chromosomes	Balanced heterozygotes	Unbalanced heterozygotes	Not tested	Spontaneous abortions	Number	Normal chromosomes	Balanced heterozygotes	Unbalanced heterozygotes	Not tested	Spontaneous abortions
33	Primary amenorrhoea	—	—	—	—	—	—	3	3	7(1)	—	4	4	1	2	4	—	—	—
34	Cerebral palsy	1	1	3	—	1	—	1	1	1(1)	—	1	—	1	2	3	—	4	—
35	Down's syndrome	—	—	—	—	—	—	2	5(1)	—	—	—	—	1	1	2	—	7	—
26	Down's syndrome (Family D)	1	1	—	—	1	—	3	7	6(1)	—	—	1	—	—	—	—	—	—
26	Turner's syndrome (Family E)	—	—	—	—	—	—	1	—	1(1)	—	10	—	—	—	—	—	—	—
36	D trisomy	1	1	—	—	—	—	2	2	3	2(1)	—	—	—	—	—	—	—	—
8	Mental retardation	1	2	3	—	3	—	3	2	2(1)	—	—	—	—	—	—	—	—	—
37	Nystagmus	—	—	—	—	—	—	1	—	4(1)	—	—	—	—	—	—	—	—	—
38	Down's syndrome	3	1	5(1)	—	2	—	—	—	—	—	—	—	—	—	—	—	—	—
39	D trisomy	1	—	—	1(1)	1	—	—	—	—	—	—	—	—	—	—	—	—	—
40	Down's syndrome	3	1	5(1)	—	2	—	—	—	—	—	—	—	—	—	—	—	—	—
41	Down's syndrome (65)	—	—	—	—	—	—	2	2(1)	2	—	3	7	—	—	—	—	—	—
41	Down's syndrome (74)	—	—	—	—	—	—	1	1	1(1)	—	—	1	—	—	—	—	—	—
41	D trisomy (327)	1	2	2	—	—	2	—	—	—	1(1)	—	—	—	—	—	—	—	—
28	Ovarian dysgenesis (P1827)	2	2	3(1)	—	—	—	1	1	1	—	—	—	1	2	3	—	2	—
42	Habitual abortion	3	2	4(1)	—	3	—	1	1	—	—	—	3	1	—	3	—	2	—
43	Chronic myeloid leukaemia	1	2	1	—	—	1	1	—	1	—	2	—	—	—	—	—	—	*
44	Down's syndrome	4	1	5	—	—	5	2	2(1)	1	—	—	2	1	—	1	—	—	—

49	Down's syndrome	–	–	–	–	–	3	5(1)	–	8	–	2	2	3	–	14	–
50	Down's syndrome	2	4	3	–	–	2	5(1)	–	1	14	–	–	–	–	–	–
51	Devl. retardation & failure to thrive	–	–	–	–	3	–	–	–	–	–	–	–	–	–	–	–
52	Newborn survey (A157)	–	–	–	–	–	2	2(1)	–	1	–	–	–	–	–	–	–
52	Newborn survey (B2040)	1	–	1(1)	–	–	1	1(1)	–	–	1	–	–	–	–	–	–
52	Down's syndrome	–	–	–	5	–	–	1(1)	–	–	1	–	–	–	–	–	–
53	Down's syndrome	1	2	2(1)	–	–	–	–	–	–	1	–	–	–	–	–	–
31	Not known (MY68234)	1	–	–	6	–	1	2	–	–	–	1	–	2	–	4	–
21	Spouse with chromosome abnormalities (K32)	4	3	9(1)	–	–	–	–	–	–	–	–	–	–	–	–	–
21	Radiotherapy patient (K29)	10	8	15	–	3	6	7	–	5	10	1	1	4(1)	–	6	–
21	Infertility (K56)	–	–	–	–	5	3	3	–	4	–	–	–	–	–	–	*
21	Recurrent abortion (K7)	1	–	–	–	2	1	2(1)	–	–	–	–	–	–	–	–	–
21	Survey, penal institution (K18)	–	–	–	–	–	3	3(1)	–	–	9	–	–	–	–	–	–
21	Survey, atomic workers (K16)	–	–	–	–	–	–	–	–	–	–	–	–	–	–	–	–
21	Survey, penal institution (K49)	1	–	2	1	–	1	1(1)	–	–	–	1	–	2(1)	–	1	–
21	Ankylosing spondylitis (K13)	–	–	–	–	–	1	2(1)	–	–	1	–	–	–	–	–	–
21	Survey, mental hospital (K44)	3	4	3	–	4	1	3(1)	–	–	1	–	–	–	–	–	–
21	Survey, penal institution (K41)	2	–	1(1)	–	3	1	1	–	1	4	1	3	2	–	–	–
21	Brother of pat. with ankylosing spondylitis (K37)	6	5	8(1)	–	3	9	17	–	12	6	1	1	6	–	5	–
21		1	1	2	–	–	–	–	–	–	–	–	–	–	–	–	*
21	Cancer patient, radiotherapy survey (K52)	–	–	–	–	–	2	4	–	–	3	–	–	–	–	–	*
21	Cancer patient, radiotherapy survey (K46)	1	2	–	–	1	2	4(1)	–	1	–	1	2	2	–	2	–
	Total	58	46	84(10)	1(1)	42	36	67	71(4)	100(19)	3(2)	56	67	15	17	40(2)	–	47	–
	Corrected	58	46	74	–	42	36	67	67	81	1	56	67	15	17	38	–	47	–

* Proband is founder heterozygote

APPENDIX III. t(21q22q) Complete Family Data

Ref. No.	Ascertainment and family identification	Progeny of male heterozygotes						Progeny of female heterozygotes						Progeny of heterozygotes of unknown sex					
		Number	Normal chromosomes	Balanced heterozygotes	Unbalanced heterozygotes	Not tested	Spontaneous abortions	Number	Normal chromosomes	Balanced heterozygotes	Unbalanced heterozygotes	Not tested	Spontaneous abortions	Number	Normal chromosomes	Balanced heterozygotes	Unbalanced heterozygotes	Not tested	Spontaneous abortions
30	Down's Syndrome	2	7	5(2)	1(1)	—	—	1	—	3	1(1)	—	—	—	—	—	—	—	—
54	,,	1	—	1	—	—	1	3	2	2	1(1)	3	3	1	—	3(1)	—	4	—
55	,,	1	2	2(1)	—	1	—	2	3	1	2(1)	—	4	—	—	—	—	—	—
44	,,	—	—	—	—	—	—	2	1	—	2(1)	—	2	—	—	—	—	—	—
28	,, (P.3328)	1	2	—	—	—	—	3	5	1(1)	1(1)	1	3	1	—	3	—	—	—
28	,, (P.4399)	1	—	2(1)	—	1	—	1	—	—	1(1)	—	—	—	—	—	—	—	—
Totals		6	11	10(4)	1(1)	2	1	12	11	7(1)	8(6)	4	12	2	—	6(1)	—	4	—
Corrected		6	11	6	—	2	1	12	11	6	2	4	12	2	—	5	—	4	—

Reciprocal Translocations in Human Populations
A Preliminary Analysis

J. LEJEUNE, B. DUTRILLAUX *and*
J. DE GROUCHY

Institut de Progénèse, Paris

RECIPROCAL TRANSLOCATIONS

¶ THE GENETIC equilibrium of reciprocal translocations in man is difficult to assess because of the fundamental biases in the detection of carriers.

The great majority of recorded reciprocal translocations have been found because one child received an unbalanced genome through malsegregation of the translocation, and only 3 of the reported families [1, 6, 43] have been ascertained randomly. In two others [25, 37], the proband was affected by a sex chromosome aberration. Hence any attempt to analyse the genetic fitness of balanced translocations must take into account the biases of ascertainment, a rather difficult task considering the scarcity of data. Many different translocations have been found but the data on any particular translocation is insufficient to assess its genetic fitness. We therefore decided to pool all published data [1–48] together with our personal observations [49–50], deliberately putting together different types of translocations in an attempt to get an over all impression. Only families where a translocation carrier has more than one child have been considered. Centric fusions between acrocentrics have not been included in this study.

At least one example of a partial trisomy has been recorded for each group of chromosomes, except the F group. Propositi with the partial monosomies, 5p –, 18p –, 18q – and Dq – have also been found.

In order to estimate the segregation ratio, two approaches can be made with this material:
(i) an estimation of the risk of malsegregation in the progeny of translocation carriers,
(ii) an estimation of the ratio of normal karyotypes to translocation carriers amongst the phenotypically normal descendants of carriers.

(i) *Risk of malsegregation*
(A) *Children born from a carrier mother.* The overall sample includes 80 sibships containing 69 affected persons with an unbalanced karyotype in a total of 268 children. The net frequency of affected individuals is thus 0·26. Three main biases affect this figure:
(1) All the families include at least one affected child (truncated distribution).
(2) The probability of ascertainment is related to the actual number of affected individuals (multiple selection).
(3) Malsegregations resulting in non-viable zygotes are not included in this survey (bias impossible to assess).

The overall sample can be split up into subsamples in order to apply statistical corrections.
(a) *Sibships of probands born from a carrier mother.* There are 136 children including 58 affected persons in a total of 43 sibships.

The multiple selection equation is

$$p_1 = \frac{R-N}{T-N} \pm \sqrt{\frac{(T-R)(R-N)}{(T-N)^3}}$$

where N = number of sibships
R = number of affected children
T = total number of children
p_1 = estimated proportion of affected children.

This gives $p_1 = 0.161 \pm 0.038$.

This value does not agree with the value obtained by application of the truncated binomial according to the Haldane formula [52] for which

$p_2 = 0.2587 \pm 0.0527$

This indicates that bias no. 1 (at least one child affected) is not the only one operating but that multiple selection is also occurring (bias no. 2).

(b) *Other sibships excluding the sibships of the probands.* There are 132 children, of whom eleven are affected:

$p_3 = 0.083 \pm 0.032$

We can split the sample into two parts. In the sibships born from a carrier grandmother or great-grandmother of a proband, we find 4 affected out of 72 children (i.e. $p_4 = 0.056 \pm 0.028$). This estimate is too low because of the necessary inclusion of the carrier (the parent of the proband).

The second subsample consists of sibships born from carrier aunts, great-aunts or female cousins of a proband. There are 7 affected out of 60 children, i.e. $p_5 = 0.117 \pm 0.041$. This is likely to be the least biased estimate for the whole sample.

It must be remarked that $p_5 = 0.117 \pm 0.041$ does not differ significantly from the estimate (after the correction for multiple selection) for the sibships of the probands: $p_1 = 0.161 \pm 0.038$.

(B) *Children born from a carrier father.* The whole sample consists of 34 sibships giving a total of 96 children, of which 21 are affected (i.e. $p = 0.22$). The same analysis can be performed:

(a) *Sibships of probands born from a carrier father.* The multiple selection equation gives a value of $p_1 = 0.091 \pm 0.062$.

The truncated binomial would give $p_2 = 0.380 \pm 0.037$ and, as noted for the carrier mothers, this discrepancy indicates that multiple selection is playing the major role.

(b) *Other sibships excluding the sibships of probands.* If the sibships of the probands are excluded then the rest of the sample is limited to the sibships born from a carrier uncle, great-uncle or cousin of a proband, and

$$p_3 = \frac{2}{24} = 0.083 \pm 0.053$$

RECIPROCAL TRANSLOCATIONS

Empirical risk of malsegregation
From the data it can be surmised that the risk to the progeny of carrier mothers is of the order of 0·20 to 0·10. The risk to the progeny of carrier fathers is probably less. This difference, however, seems less pronounced than in the case of centric fusions [51]. This difference between the sexes is also indicated by the fact that the carrier parent is the mother of the proband in 43 cases, and the father in 13 cases.

(ii) *Segregation ratio in phenotypically normal persons*
The maintenance of a balanced translocation in a given population can be due either to mutational pressure or to a selective advantage by the carriers. Cases arising from non-carrier parents prove abundantly that mutation pressure is strong, but it is difficult to measure. A selective advantage could result from two kinds of effects in the progeny of heterozygotes:
(1) carriers have *more children* than normal people.
(2) the *translocation is more often transmitted* than the normal karyotype.

The first hypothesis is difficult to assess because all the families recorded contain at least two children: i.e. childless families and one-child families are not analysed. However, the segregation of the translocation among phenotypically normal persons can be directly analysed.

The whole sample of children born from a carrier parent includes 145 carriers and 101 normals. This segregation differs significantly from the expected 1 : 1 ratio ($\chi^2 = 7\cdot86$ for $v = 1$). However, here also ascertainment is heavily biased, and statistical corrections are necessary and must be adapted to each subsample.

Sibships of the carrier parent of a proband. Of 54 persons born to a carrier *female* ancestor, there are 37 carriers and 17 normals. Of 27 persons born to a carrier *male* ancestor there are 20 carriers and 7 normals.

Application of the truncated binomial (since the carrier parent is included by necessity), gives an estimate higher than $p = 0\cdot60$ which is beyond the range of the published tables. This could indicate a preferential transmission of the translocation, but it can be assumed that, in this subsample, the bias is large not only because of the inclusion of at least one carrier in each sibship, but also because the likelihood of recording the family is effectively much greater if there are many carriers than if there are few.

Hence an application of the multiple selection equation is valid and gives $p = 0\cdot528 \pm 0\cdot083$ for the progeny of a female carrier ancestor and $p = 0\cdot588 \pm 0\cdot119$ for the progeny of a male carrier ancestor. These two estimates do not differ significantly from each other and are fully compatible with an expected 1 : 1 segregation among phenotypically normal persons.

In other sibships in which the ascertainment is very close to random, as far as the translocation in phenotypically normal persons is considered, we find:

Progeny of a carrier mother, including phenotypically normal children in the sibships of probands and the sibships of aunts, great-aunts and female cousins of the probands:

$$p = \frac{52}{106} = 0.49 \pm 0.047$$

Progeny of a carrier father, including phenotypically normal children in the sibships of probands and in the sibships of uncles, great-uncles and male cousins of probands:

$$p = \frac{20}{39} = 0.51 \pm 0.08$$

These two values are in agreement with a 1 : 1 segregation ratio and this is the most bias-free sample we can utilise.

A final remark concerns the distribution of sexes. Among the carrier children there are 54 males and 74 females but among the normals, there are 48 males and 47 females. This apparently curious sex ratio does not depart significantly from homogeneity ($\chi^2 = 1.52$ for $\nu = 1$). However, the multiple ascertainment bias and other possible mechanisms which could explain this tendency require further investigation.

SUMMARY

1. The general risk of malsegregation resulting in a live birth is of the order of ten to twenty per cent in the progeny of carrier mothers. The risk is possibly half in the progeny of carrier fathers.

2. The segregation ratio between the balanced translocation and the normal complement among phenotypically normal children is apparently not different from 1 : 1 in the progeny of carrier parents. If ascertainment is properly taken into account, the less biased the sample, the closer is the fit of the data to a normal segregation ratio.

There is no evidence that balanced reciprocal translocations have a strong selective advantage in the present conditions. It must be stressed, however, that with the available data, a real but small selective advantage would not be detected.

REFERENCES

[1] Buchanan, J.G., Scott, P.J., McLachlan, E.M., Smith, F., Richmond, D.E. and North, J.D.K., *Am. J. Med.,* **42,** 1003, 1967.
[2] Capoa, A. de., Warburton, D., Breg, W.R., Miller, D.A. and Miller, O.J., *Am. J. hum. Genet.,* **19,** 586, 1967.
[3] Clarke, G., Stevenson, A.C., Davies, P. and Williams, C.E., *J. med. Genet.,* **1,** 27, 1964.
[4] Cohen, M.M. and Lockwood, M.A., *Pediat. Res. (Basel),* **1,** 104, 1967.
[5] Cooke, P., *J. med. Genet.,* **5,** 200, 1968.
[6] Court Brown, W.M., Mantle, D.J., Buckton, K.E. and Tough, I.M., *J. med. Genet.,* **1,** 35, 1964.
[7] Day, R.W. and Miles, C.P., *J. Pediat.,* **67,** 399, 1965.
[8] Edwards, J.H., Fraccaro, M., Davies, P. and Young, R.B., *Ann. hum. Genet.,* **26,** 163, 1962.
[9] Falek, A., Schmidt, R. and Jervis, G., *Pediatrics,* **37,** 92, 1966.
[10] Gray, J.E., Dartnall, J.A. and Macnamara, B.G.P., *J. med. Genet.,* **3,** 62, 1966.
[11] Gropp, A., Marsch, W. and Brodehl, J., *Nature, Lond.,* **207,** 374, 1965.
[12] Grouchy, J. de and Canet, J., *Annls Génét.,* **8,** 16, 1965.
[13] Grouchy, J. de, Roy, C., Lachance, R., Frézal, J. and Lamy, M., *Archs fr. Pédiat.,* **24,** 849, 1967.
[14] Grouchy, J. de, Thieffry, S., Aicardi, J., Chevrie, J.J. and Zucher, G., *Archs fr. Pédiat.,* **24,** 859, 1967.
[15] Grouchy, J. de, Émerit, I. and Aicardi, J., *Annls Génét.* (in press, 1969).
[16] Hauschteck, E., Murset, G., Prader, A. and Bühler, E., *Cytogenetics,* **5,** 281, 1966.
[17] Jacobsen, P. and Mikkelsen, M., *J. ment. Defic. Res.,* **12,** 144, 1968.
[18] Kaplan, M., Grumbach, R., Fischgrund, A. and Ferragu, O., *Sém. Hôp. Paris* (Ann. Pédiat.), **42,** 2329, 1966.
[19] Kontras, S.B., Currier, G.J., Cooper, R.F. and Ambuel, J.P., *J. Pediat.,* **69,** 635, 1966.
[20] Koulischer, L., Petit, P. and Hayez-Delatte, F., *Annls Génét.,* **10,** 150, 1967.
[21] Laurent, C. and Robert, J.M., *Annls Génét.,* **9,** 113, 1966.
[22] Laurent, C. and Robert, J.M., *Annls Génét.,* **9,** 134, 1966.
[23] Laurent, C. and Robert, J.M., *Annls Génét.,* **11,** 28, 1968.
[24] Lee, C.S.N., Bowen, P., Rosenblum, H. and Linsao, L., *New Engl. J. Med.,* **271,** 12, 1964.
[25] Lejeune, J., Lafourcade, J., Berger, R., Haynes, M. and Turpin, R., *Annls Génét.,* **6,** 3, 1963.
[26] Lejeune, J., Lafourcade, J., Berger, R., and Réthoré, M.O., *Annls Génét.,* **8,** 11, 1965.
[27] Lejeune, J. and Berger, R., *Annls Génét.,* **8,** 21, 1965.
[28] Lejeune, J., Berger, R., Réthoré, M.O., Salmon, C. and Kaplan, M., *Annls Génét.,* **9,** 12, 1966.
[29] Lejeune, J., Réthoré, M.O., Berger, R., Abonyi, D., Dutrillaux, B. and See, G., *Annls Génét.,* **11,** 171, 1968.
[30] Lindsten, J., Fraccaro, M., Klinger, H.P. and Zetterqvist, P., *Cytogenetics,* **4,** 45, 1965.
[31] Lord, P.M., Casey, M.D. and Laurence, B.M., *J. med. Genet.,* **4,** 169, 1967.
[32] Melnyk, J., personal communication.
[33] Migeno, N.R. and Young, W.J., *Bull. Johns Hopkins Hosp.,* **115,** 379, 1964.

[34] Mikkelsen, M., Mortensen, E., Skakkekack, N. E. and Yssing, M., *Acta genet. Basel,* **18,** 241, 1968.
[35] Miller, J. R. and Dill, F. J., *Proc. Am. Soc. hum. Genet.,* **37,** 1963.
[36] Noel, B., Quack, B. and Thiriet, M., *Annls Génét.,* **11,** 247, 1968.
[37] Ockey, C. H. and de La Chapelle, A., *Cytogenetics,* **6,** 178, 1967.
[38] Pfeiffer, R. A., Laerman, J. and Heidtmann, H. L., *Helv. Paediat. Acta,* **22,** 558, 1967.
[39] Pitt, D. D., Webb, G. C., Wong, J., Robson, M. K. and Ferguson, J., *J. med. Genet.,* **4,** 171, 1967.
[40] Punnett, H. H., Pinsky, L., Di George, A. M., and Gorlin, R. S., *Am. J. hum. Genet.,* **18,** 572, 1966.
[41] Rhode, R. A. and Cate, B., *Lancet,* **2,** 838, 1964.
[42] Shaw, M., Cohen, M. M. and Hildebrandt, H. M., *Am. J. hum. Genet.,* **17,** 54, 1965.
[43] Stalder, G. R., Buhler, E. M., Gadola, G., Widmer, R. and Freuleur, F., *Humangenetik,* **1,** 197, 1964.
[44] Summitt, R. L., *Am. J. hum. Genet.,* **18,** 172, 1966.
[45] Uchida, I. A., Wang, H. C., Laxdal, O. E., Zaleski, W. A. and Duncan, B. P., *Cytogenetics,* **3,** 81, 1964.
[46] Valdmanis, A., Pearson, G., Siegel, A. E., Hoeksema, R. H. and Mann, J. D., *Annls Génét.,* **4,** 159, 1967.
[47] Walzer, S., Favara, B., Ming, L. and Gerald, P. S., *New Engl. J. Med.,* **275,** 290, 1966.
[48] Zaremba, J., Zajaczkowska, K., Abramovicz, T. and Wald, I., *Polski Tygod lek.,* **20,** 811, 1965.
[49] Grouchy, J. de, unpublished observation No. 3760.
[50] Lejeune, J., unpublished observations Nos I.P. 3039, 3430, 3481, 3831, 3927, 4336, 4461 and 4685.
[51] Dutrillaux, B. and Lejeune, J., *Annls Génét.* (in press, 1969).
[52] Lejeune, J., *Biometrics,* **14,** 513, 1958.

The Inheritance of Randomly Ascertained Chromosome Abnormalities

PATRICIA A. JACOBS

MRC, Clinical and Population Cytogenetics Research Unit, Western General Hospital, Edinburgh

¶ INDIVIDUALS in the population who are carrying chromosome abnormalities which are inherited are frequently ascertained through a clinically abnormal propositus who on cytogenetic investigation is found to have an unbalanced form of a structural rearrangement. The parents and relatives of such individuals are often found to be carrying the balanced form of the chromosome rearrangement. Many families ascertained in this way have a poor reproductive history with a high frequency of abortions, stillbirths and children with multiple congenital abnormalities, and these are presumed, or are actually shown by cytogenetic examination, to be due to the unbalanced form of the structural rearrangement [1, 2]. Until recently the great majority of structural rearrangements in man have been ascertained in this way (with the exception of D/D translocations of the centric fusion type) and it has been assumed that the majority of such rearrangements are associated with a lowered reproductive fitness.

The advent of the blood culture technique and all its various simplifications has, in recent years, made it technically possible to do chromosome surveys on fairly large numbers of people. As a result it is becoming increasingly obvious that in man structural rearrangements of the chromosomes are not uncommon. Court Brown and Smith considered that about three per 1,000 of the population were carrying a structural abnormality of chromosomes which was detectable in mitotic cells [3].

During the course of cytogenetic studies on a variety of different subgroups of the population we have found a number of people who have a structural abnormality of their chromosomes which on investigation of their relatives has been found to be familial. These abnormalities consist of Robertsonian translocations of the D/D and D/G types, reciprocal translocations, pericentric inversions and supernumerary chromosomes. It is the purpose of this paper to describe the pattern of inheritance of the last three types of abnormality. Robertsonian translocations will not be mentioned further as their pattern of inheritance has been dealt with by Dr Hamerton in a previous paper.

Method of ascertainment of the chromosome abnormalities. Before dealing with the data in detail it is necessary to consider the method of ascertainment of the abnormalities which will be described. Broadly these fall into four categories. First there are chromosome abnormalities found in individuals who can be considered to be representative of the population at large. These include unselected newborn babies, individuals chosen at random from the lists of general practitioners, and spouses of members of families in which a chromosome abnormality is known to be segregating (Table 1). Secondly there are the abnormalities found in individuals whose chromosomes were examined because they were in penal or corrective training institutions. These include surveys of prisons, borstals and approved schools (Table 2). Thirdly there are the abnormalities ascertained through the study of selected groups of individuals in which there is, on present evidence, no reason to suspect a higher incidence of structural abnormalities than in the general population. These include surveys of patients with a malignant disease, patients in general hospitals and parents of individuals known to have a sex chromosome abnormality (Table 3). Lastly there are those people found in surveys of individuals where it is known that there is an increased frequency of structural rearrangements, or where such an

TABLE 1. Chromosome abnormality by method of ascertainment. Category 1

Chromosome abnormality	Method of ascertainment of index case
46,t(Bp−;Eq+)	Spouse of individual in family with chromosome abnormality
46,inv(Yp+q−)	Survey of patients in a general practice.

TABLE 2. Chromosome abnormality by method of ascertainment. Category 2

Chromsome abnormality	Method of ascertainment of index case
46,t(Bp+ ;Eq−)	Approved School Survey
46,t(1 ?+ ;Cq−)	Borstal Survey
46,t(1 ?−;16p+)	Survey of Tall Prisoners
46,inv(Cp+q−)	Borstal Survey
46,inv(Yp+q−)	Approved School Survey
46,inv(Yp+q−)	Approved School Survey

TABLE 3. Chromosome abnormality by method of ascertainment. Category 3

Chromosome abnormality	Method of ascertainment of index case
46,t(1 ?−;Cq+)	Cancer of the breast
46,inv(Cp+q−)	Secondary amenorrhoea
46,inv(Cp+q−)	Child with low hair line
46,inv(Yp+q−)	Hospital patient used as a control to obtain 'normal' marrow
46,inv(Yp+q−)	Cancer of the mouth
47,mar+	Mother of a 45,X child

increase could be expected on theoretical grounds, but where the chromosome abnormality found appears to be co-incidental to the reason for which the chromosomes were examined. These include surveys of patients in mental subnormality hospitals, surveys of babies with congenital malformations and surveys of individuals attending a subfertility clinic (Table 4).

TABLE 4. Chromosome abnormality by method of ascertainment. Category 4

Chromosome abnormality	Method of ascertainment of index case
46,t(1 ?−;Eq+)	Survey of Mental Sub-normality Hospital
46,t(Cq−;Cq+)	Survey of Mental Sub-normality Hospital
46,t(Cq−;Cq+)	Survey of Mental Sub-normality Hospital
46,t(Bp−;Eq+)	Survey of babies with congenital malformations
46,t(Dp−;Gp+)	Survey of babies with congenital malformations
46,t(Cp+ ;Dq−)	Sub-fertility Clinic
46,Xinv(Yp+q−)	Sub-fertility Clinic
47,mar+	Sub-fertility Clinic

It is only in the first category of individuals that the chromosome abnormality can be said with certainty to be randomly ascertained as it is only these people who were selected without reference to their physical or mental state or to their behaviour. In the remaining three categories the chromosome abnormality appears to be co-incidental to the reason for which the patients' chromosomes were examined. However at present it is better to be cautious and to regard these abnormalities as 'presumptively random' since it is possible, particularly in the fourth category, that the condition of the patient may be related in some as yet undetermined way to the chromosome abnormality.

Methods of Collection of the data. The pedigree data was obtained in the first instance by interviewing various members of the family, and for Scottish families the data were checked with, and augmented by, information in the central registers of births, deaths and marriages held by the Registrar General for Scotland, and on the basis of this, complete pedigrees were obtained. Unfortunately for the small number of English families which will be described, the pedigrees are based on the data obtained directly from the members of the family. The reasons why the pedigrees of the English families are not constructed with the same degree of accuracy as those of the Scottish families are, first that there is much more limited access to the data in the custody of the Registrar General for England and Wales than for that in the custody of the Registrar General for Scotland, and secondly, and more importantly, the data on English birth, marriage and death certificates is very much more limited than that on the Scottish counterparts.

In addition to the data obtained in the above way, the General Practitioners of every individual with a chromosome abnormality and all their first degree relatives are contacted once a year and all additional information obtained in this way is recorded on the pedigree. This is done for all families, both English and Scottish and it is hoped that in this way the majority of the pedigrees are complete up to the end of 1968.

The conceptional histories of the chromosomally abnormal individuals and the controls, who are sibs of the same sex who are known to have a normal chromosome constitution, are considered accurate for live births if the data was obtained from the records of the Registrar General for Scottish families, and from the mother herself for English families. However, as no records of stillbirths and abortions can be obtained from the Registrar General's data these are only considered accurate for both Scottish and English families if they were obtained directly from the mother, or from the general practitioner or from the maternal obstetric case notes obtained from the hospital in which the mother had her last confinement. Where data on abortions and stillbirths could not be obtained from one or more of these sources the relevant conceptional history is considered accurate only for livebirths. Where information on abortions and stillbirths was obtained only from the husband or another close relative it is recorded but it is recognised that the conceptional history is not necessarily complete.

Results. In all, 22 individuals have been ascertained who have a chromosome abnormality which is present in one or more of their relatives and in whom the method of ascertainment can be considered 'random' or 'presumptively random'. The abnormalities themselves are of four types, namely, reciprocal translocations, pericentric inversions of an autosome, pericentric inversions of the Y chromosome and supernumerary chromosomes. As the theoretical consequences of these four types of abnormality are very different they will be considered separately.

Reciprocal translocations. Eleven propositi were ascertained who had an abnormality which was presumed to be a balanced reciprocal translocation although in no instance has it been possible to confirm this by meiotic studies. In each family the individuals examined were found to have either the balanced form of the translocation or to have a normal chromosome constitution and no individual was found to have an unbalanced form of the translocation in any family.

An attempt was made to examine the segregation ratio for each family separately by study of those sibships where one parent had been shown to carry the abnormality. There was no evidence of any marked departure from a ratio of 0·5 within an individual pedigree (nor was there found to be any significant difference in the sex ratio of the abnormals to the normals) but the data on individual families was rather sparse. The eleven pedigrees have therefore been considered together, even although the translocations involve different chromosomes. This may subsequently prove to have been an invalid combination but, on present evidence, there would seem to be no differential effects on the carrier of the various translocations considered. Table 5 shows the combined results of the segregation of the translocations in matings where one parent is known to carry the abnormality. By testing the segregation ratio in all sibships where at least one individual was known to carry the translocation an estimate was obtained of $0.48 \pm$ s.e. 0.05.

No individual with a reciprocal translocation has any phenotypic abnormality which appears attributable to the abnormal chromosomes. It is noteworthy that there is no evidence among the live born children of any of these families of the presence of either unbalanced form of the translocation, in spite of the fact that gametes with the unbalanced form of the abnormality would be expected to be formed with a frequency equal to the combined frequencies of normal gametes and gametes containing the balanced form of the translocation. The absence of individuals with the unbalanced forms of the translocation in these families could be due either to the failure of production of the unbalanced forms of the gametes, or to selection against such gametes effecting fertilisation, or to embryos carrying the unbalanced forms of the translocation being so severely affected that they are non-viable. If the latter is the explanation for the absence of people with unbalanced translocations the foetus could die during the very early part of pregnancy when the pregnancy would be unrecognised, but the affected individuals might have, on average, fewer children than the non-affected or a longer average birth interval. Alternatively the affected foetus could die in a later stage of pregnancy when it would be recognised as an abortion or a stillbirth. The data on the reproductive histories of all individuals known to have a translocation was examined to see whether they had an increased frequency of abortions, or stillbirths, fewer children, more childless marriages or a different birth interval than their sibs who were known to have a normal chromosome constitution. It is recognised that the unbalanced forms of one translocation might behave quite differently from the unbalanced forms of any other translocation. The data on each translocation family have been examined separately but as there was no obvious difference between them the data from all eleven were

TABLE 5. Segregation of abnormal chromosomes in sibships where one parent is known to have the reciprocal translocation

No. of sibships	Chromosomes of parents		Offspring			
			Abnormal		Normal	
	Father	Mother	M	F	M	F
9	A	N	8(+ 3P)	11	8	4
5	A	NK	0(+ 2P)	3	3	0
13	N	A	9(+ 2P)	4(+ 2P)	5	12
6	NK	A	3	4(+ P)	5	4
33			20(+ 7P)	22(+ 3P)	21	20

TABLE 6. Conceptional histories. Families with a translocation

Parents			Live births				Deaths <5 Years	Miscarriages	No Pregnancies
			Chromosomes			Total			
Sex	Ch.	No.	Normal	Balanced abnormal	Not examined				
F	A	26	25	23(+ 4P)	15	63(+ 4P)	5	11	4
F	N	18	–	–	–	50	2	5	2
M	A	18	16	19(+ 6P)	6	41(+ 6P)	2	12	6
M	N	16	–	–	–	40	2	3	3

combined and the results are shown in Table 6. Because of a possible bias towards greater accuracy in the conceptional history of affected individuals the degree of reliability with which the conceptional history was obtained was compared between the normal and affected females, and the normal and affected males. The conceptional histories of the affected and normal females were all regarded as satisfactory but there were a number of conceptional histories in both the normal and affected males in which it is likely that there has been an under-reporting of abortions and stillbirths. However as the proportion of incomplete conceptional histories is about the same in both groups it is unlikely that these have produced a serious bias.

The total number of pregnancies of the females who had the translocation was compared with the total number of pregnancies of their female sibs who were known to have a normal chromosome constitution, and there was no significant difference. This was also found to be the case when the proportions of pregnancies in each group which terminated as abortions and stillbirths were compared. Similarly there were no significant differences between the total number of pregnancies fathered by males with a translocation as compared with their normal sibs, nor in the number of abortions and stillbirths in the two groups. However it is worth noting that while the differences in the proportions of pregnancies which terminated as abortions and stillbirths is not significantly increased in either the affected males or the affected females, in both groups there are more abortions than among their normal sibs.

The mean birth interval between live born children of affected females was compared with that for control females, and a similar comparison was made for affected and control males. The mean birth interval for affected females was 38·3 months and for control females 32·2 months, while for affected males it was 30·8 months and for normal males 33·7 months. Neither of these differences was significant.

It therefore appears that in these eleven families carrying reciprocal translocations there is no marked evidence of foetal wastage nor early childhood deaths to explain the unexpected absence of individuals with the unbalanced forms of the translocation.

The absence of any evidence of the unbalanced form of the translocation is in contrast to the majority of published data on reciprocal translocations. The difference is presumably due almost entirely to the method of ascertainment of the translocation. In the 129 families with a translocation, summarised by Ford and Clegg, 118 were ascertained through the presence of a child with multiple congenital malformations, six because of a history of spontaneous abortions and only five in a random or presumptively random way [4]. In the 129 families there were 161 matings where one parent was known to be heterozygous for a translocation and Ford and Clegg showed that the ratio of normal and balanced translocations to that of zygotes presumed to be carrying the unbalanced form of the translocation was about 1·4 to 1·0. They concluded that the deviation from the expected 1 : 1 ratio was due either to the failure of some karyotypically abnormal zygotes to implant and give clinical signs of pregnancy, or to the under-reporting of abortions and stillbirths, or to an excess of the type of meiotic disjunction which gives rise to the normal and balanced form of the translocation.

The absence of evidence of the unbalanced forms in the families reported here cannot be due to under-reporting of abortions and stillbirths as the data on these was collected in the same way for both affected and control individuals, nor is it likely to be entirely due to the loss of the unbalanced forms before they are recognised as a pregnancy, as there is no decrease in family size nor increase in birth interval in the affected individuals. It is therefore likely to be due either to the failure of production of unbalanced gametes, or to selection against such gametes effecting fertilisation.

It is of interest to see whether the chromosomes involved in translocations ascertained randomly are different from those where the ascertainment is through an unbalanced form, as this might go some way to explaining the different types of segregation between the two groups. As can be seen from Table 7 the data on the twenty-two breaks involved in the eleven translocations described here are too few to give any evidence on whether certain chromosomes are involved in translocations more often than would be expected, where the expected number of breaks is assumed to be proportional to the length of the chromosome group. Ford and Clegg found a significant excess of break points in groups B, D, E and G and too few break points in the chromosomes in groups A, C, and F.

TABLE 7. Chromosomes involved in reciprocal translocations

	Chromosome number or group									
	1	2	3	B	C	D	16	E	F	G
Breaks observed	4	0	0	3	7	2	1	4	0	1
Expected	1·85	1·73	1·46	2·62	8·43	2·21	0·68	1·24	1·02	0·74

Pericentric inversions. Three families were found who had a structurally abnormal chromosome replacing one of the group C chromosomes. The most reasonable interpretation of the abnormal chromosome was that it was the result of a pericentric inversion. One family was ascertained through two propositi—one was a male identified during a survey of a borstal population, while the other was a child whose chromosomes were looked at because he was noted at birth to have a number of congenital malformations including micrognathism, club feet and poorly formed ears. It was only during the construction of the pedigrees of both individuals from the Registrar General's data that it was realised that the father of the borstal boy was the missing paternal grandfather of the congenitally abnormal baby. As data on these three families have already been published [5] they will only be dealt with in summary. The chromosomes appeared identical in all the families and the segregation of the abnormal chromosome was tested using the same two methods as were used for the reciprocal translocations. Table 8 shows the combined results of the segregation of the inverted chromosome in matings where one parent is known to carry the abnormality. An estimate of the segregation ratio by the sibship method was 0·53 ± S.E.0·08.

With the exception of one of the propositi whose chromosomes were examined because he had a number of congenital malformations no individual with the abnormal chromosome was phenotypically abnormal. The conception histories of the males and females who were known to have a pericentric inversion were compared with their normal sibs and the results are shown in Table 9. Again there seemed to be no difference between affected and controls in terms of total number of pregnancies nor

TABLE 8. Segregation of abnormal chromosomes in sibships where one parent is known to have the pericentric inversion

No. of sibships	Chromosomes of parents		Offspring			
			Abnormal		Normal	
	Father	Mother	M	F	M	F
6	A	N	5(+P)	4(+P)	1	6
5	A	NK	3	2(+P)	1	3
7	N	A	3(+P)	5(+2P)	5	2
2	NK	A	0	2	2	0
20			11(+2P)	13(+4P)	9	11

TABLE 9. Conceptional histories. Families with a pericentric inversion

Parents			Live Births				Deaths <5 Years	Miscarriages	No Pregnancies
			Chromosomes						
Sex	Ch.	No.	Normal	Balanced abnormal	Not examined	Total			
F	A	10	9	10(+3P)	0	19(+3P)	0	2	1
F	N	8	–	–	–	19	1	4	2
M	A	8	10	14(+4P)	7	31(+4P)	5	2	0
M	N	4	–	–	–	8	0	0	2

TABLE 10. Liveborn offspring of males with a 46,Xinv(Yp+q−) chromosome constitution

Sibship no.	No. of males	Offspring	
		Males	Females
K8*	6	7(+P)	9
K25	5	6(+P)	3
K48	20	35(+P)	41
K53	3	5(+P)	8
K54	1	8(+P)	0
K58*	4	6(+P)	2
	39	67(+6P)	63

* May be related

was there a difference in the proportion which terminated in abortions or stillbirths nor in the mean birth interval. The absence of any evidence of the genetically unbalanced gamete with a duplication/deficiency, which would be expected if a chiasma was formed in the inversion loop, may be due to the fact that cross-overs do not occur within the inversion loop, or to the fact that selection operates against the gametes carrying the duplication deficiency.

Pericentric inversions of the Y. There is a special class of pericentric inversion in man, namely that involving the Y chromosome which, by its nature, has to be considered separately. We have ascertained six individuals in whom the Y chromosome is replaced by a metacentric chromosome similar to the members of group F. This abnormal chromosome was interpreted as being due to a pericentric inversion of the Y and, as expected, it is present in all the male relatives of the propositi and absent from all the female relatives.

It is almost certain that two of the families are related as they have the same name. Through the records of the Registrar General we have traced both families back to the eighteenth century. Unfortunately however we have not as yet been able to establish a definite ancestral link.

The conceptional histories of the males in the families was obtained from the men themselves or from the records of the Registrar General. They are therefore considered accurate only for live births, and Table 10 shows the live births by sex for the males in the six families. As can be seen there is no significant departure from the expected in the sex ratio. It is worth noting that family K54 consists at present of a single sibship of nine, all of whom are males. The father has the abnormal Y chromosome but it has not yet been possible to examine his sibs. From the pedigree data we have obtained, there does not appear to be an unusual sex ratio in any of the other sibships in this family.

There is no evidence in any of the families in which the abnormal Y is segregating for the presence of individuals with a sex chromosome aneuploidy. Such individuals might occur more often than expected in these families if the structural abnormality of the Y in any way increases the likelihood of its non-disjunction at meiosis. On the data collected to date there is no evidence to suggest that this form of the Y chromosome behaves in any way differently from the normal acrocentric Y chromosome.

Supernumeraries. Two people were found to have 47 chromosomes, the additional chromosome being a structurally abnormal chromosome which was smaller in size than the group G autosome. The abnormal chromosome was found in several members of the families of both propositi. In one family the supernumerary chromosome was only slightly smaller than the group G autosomes and appeared to be acrocentric with non-satellited

INHERITANCE

short arms and satellites or distinctly heterochromatic regions at the distal part of the long arms. In the other family the supernumerary chromosome was a very small almost metacentric chromosome with a pale heterochromatic region at the distal end of the slightly shorter arms. In both families the individuals examined were either found to have a normal chromosome constitution or to have the supernumerary chromosome present in every cell examined. A minimum of thirty cells were counted and analysed from peripheral blood cultures of every member of the family from whom a sample was obtained but there was no evidence of mosaicism. Segregation of the abnormal chromosome in matings where one parent is known to have the chromosome is shown for both families combined in Table 11. As can be seen the abnormal chromosome has only been transmitted by females, but no significance can be placed on this observation in view of the very small number of carriers. The sibship method gave an estimated segregation ration of $0.51 \pm $ S.E.0.14.

Table 12 gives the conceptional histories of the females with a supernumerary chromosome by comparison with their normal sibs. Within the limits of the very small number of people studied there does not appear to be any difference between the two groups.

TABLE 11. Segregation of abnormal chromosomes in sibships where one parent is known to have the supernumerary

| No. of sibships | Chromosomes of parents | | Offspring | | | |
| | Father | Mother | Abnormal | | Normal | |
			M	F	M	F
0	A	N	0	0	0	0
0	A	NK	0	0	0	0
1	N	A	0	0	2	1
3	NK	A	2	4(+2P)	3	1
4			2	4(+2P)	5	2

TABLE 12. Conceptional histories. Families with a supernumerary

| Parents | | | Live births | | | | Deaths <5 Years | Miscarriages | No Pregnancies |
| Sex | Ch. | No. | Chromosomes | | | Total | | | |
			Normal	Balanced abnormal	Not examined				
F	A	6*	7	6(+2P)	1	14(+2P)	0	1	2*
F	N	3	—	—	—	7	0	0	0
M	A	0	0	0	0	0	0	0	0
M	N	6	—	—	—	12	1	0	1

*1 proposita ascertained through sub-fertility clinic

The presence of individuals with a supernumerary chromosome, and by these is meant a chromosome which is additional to the normal complement of 46, which is morphologically distinguishable from the other chromosomes and which appears to be without phenotypic effect, seems to be a not uncommon abnormality in man. Walzer and his colleagues found three among 2,400 phenotypically normal newborn babies [6]. Although the supernumerary chromosomes which are described here and by other authors appear to be largely heterochromatic at least in mitotic metaphases, they also appear to have a euchromatic segment, and it is perhaps surprising that this euchromatic region has no obvious effect on the phenotype. Supernumerary chromosomes, at least in the two families described here, appear to be mitotically very stable. In spite of the fact that relatively large numbers of cells were scored from each member of the family there was no evidence of mosaicism. At present the origin and role of these supernumerary chromosomes in man is obscure.

SUMMARY

In summary 22 families have been described with a structural chromosome abnormality which was ascertained in a random or 'presumptively random' way. These abnormalities include reciprocal translocations, pericentric inverisons and supernumerary chromosomes. In no family was there any demonstrable effect of the chromosome abnormality in terms of the phenotype or the reproductive history. The proportion of structural abnormalities in man which fall into this apparently 'harmless' category is not known, but it is likely to be substantial. The true proportion can only be found by the study of large unbiased samples of the population and the subsequent investigation of the families of individuals in this population who are found to have a structural rearrangement of the chromosomes.

ACKNOWLEDGEMENTS

I would like to thank all those members of the Medical Research Council's Clinical and Population Cytogenics Research Unit who helped in the study of the families who are described. I am particularly grateful to Mrs Anna Frackiewicz for her work in constructing the pedigrees, to Mrs Pamela Law for her help in abstracting the data and to Dr James Aitken and Dr Marjorie Newton for collecting the blood and much of the raw data.

REFERENCES

[1] Walzer, S., Favara, B., Pen-Ming, L.M. and Gerald, P., *New Engl. J. Med.,* **275,** 290, 1966.
[2] Falek, A., *Ped.,* **37,** 92, 1966.
[3] Court Brown, W.M. and Smith, P.G., *Br. med. Bull.,* **25,** 74, 1969.
[4] Ford, C.E. and Clegg, H.M., *Br. med. Bull.,* **25,** 110, 1969.
[5] Jacobs, P.A., Cruickshank, G., Faed, M.J.W., Frackiewicz, A., Robson, E.B., Harris, H. and Sutherland, I., *Ann. hum. Genet.,* **31,** 219, 1967.
[6] Walzer, S., Breau, G. and Gerald, P., *J. Pediat.,* **74,** 439, 1969.

Chromosome Abnormalities
and Spontaneous Abortions

D. H. CARR

Department of Anatomy, McMaster University
Hamilton, Canada

¶ALL INVESTIGATORS agree that the prevalence of chromosome anomalies in spontaneous abortions is much higher than in liveborn infants. There is, however, a wide variation in the percentage of chromosomally abnormal abortuses from one study to another. Even among the three series which have 100 or more specimens karyotyped, this varied from 8 to 38 per cent (Table 1). There are a number of factors which may account for these differences and they will be discussed below. The factors may be divided into those primarily involving the progenitor* and those mainly relevant to the conceptus.

TABLE 1. Heteroploidy in published studies of more than 100 abortions

Reference	No. studied	Trisomy	Triploid	Tetraploid	45,X	Other	% Abnormal
1	132	27	10	1	8	4	38
2	227	26	9	2	12	1	22
3	101	1	3	0	2	2	8
Total	460	54	22	3	22	7	

In arriving at a working figure for the prevalence of chromosome anomalies among spontaneous abortions it must be qualified. There seems little doubt that anomalies of cell division in the early zygote are responsible for the loss of many early conceptuses. Hertig and co-workers [4] found that 4 of 8 human morulas or blastocysts were grossly abnormal and showed unequal size of blastomeres and cellular degeneration. It is reasonable to suppose that the chromosomal content of these zygotes was abnormal and that they would probably have been lost before implantation. These early losses are important when one attempts to estimate zygotic loss but they are really in a different category from clinically definable abortions. In order to arrive at an estimate for chromosome anomalies in abortions, no account will be taken of these very early losses. In arriving at a figure for the prevalence of heteroploidy in abortions I shall rely heavily on our own data, only because of the greater availability of details.

FACTORS RELATED TO THE PROGENITOR

1. *Geographical.* A variety of environmental or local factors could affect the prevalence of chromosome anomalies in abortions. There is evidence which suggests that chromosome anomalies at term show geographic variation. Subray and Prabhaker [5] failed to find any chromatin-positive males among 2,058 neonates. In Western countries about 1 in 500 newborn

---

* The word 'parent' is so regularly associated with a living child that 'progenitor' will be used to describe those individuals responsible for the production of an abortus. They may or may not be 'parents'.

males are chromatin-positive [6]. The fact that the frequency of Down's syndrome in India was also found to be low [7] suggests that chromosome anomalies at term are less common in that country. On the other hand, there appears to be an increased frequency of Down's syndrome in certain countries, which is independent of maternal age [7].

There is further evidence of regional variation in the occurrence of chromosome anomalies. Robinson and Puck [8] noted a seasonal variation in the incidence of sex chromosome anomalies which did not apply to Down's syndrome. In addition to the seasonal variation, low socio-economic status appeared to predispose individuals to produce offspring with gonosomal anomalies. Two studies from Europe, based on month of birth, failed to show a seasonal variation in the incidence of sex chromatin-positive males, but socio-economic factors were not considered [9, 10].

A variation in prevalence of Down's syndrome has been found in urban and rural areas, but in British Columbia the difference did not quite reach statistical significance [11]. In Scotland, there were some remarkable regional variations in the prevalence of Down's syndrome. However, they did not appear to be between urban and rural communities and remain unexplained [12].

Any agent which alters the prevalence of chromosome anomalies at term may, presumably, apply equally to spontaneous abortions. One factor, which might apply specifically to ascertainment of spontaneous abortions, is hospital admission patterns. The majority of specimens in published series come from women who have been admitted to hospital. The tendency to seek admission, as well as hospital policy regarding admission, may vary considerably from place to place. Thus, in some areas, women aborting early may avoid hospital treatment and unless arrangements were made to collect specimens from the home, a considerable bias would be introduced.

2. *Induced abortions.* It is obvious that the presence of a number of unsuspected induced abortions within a series of supposed spontaneous abortions will obviously decrease the prevalence of chromosome anomalies in the population studied. The exclusion of illegal abortions is not easy as most patients are, naturally, anxious to conceal the facts from the physician. Similarly, the physician is reluctant to reveal privileged information on the patient's hospital record.

In one of the larger cytogenetic studies of spontaneous abortions, only 8 per cent had chromosome anomalies [3]. In a personal communication, Stenchever suggests that this may have been partly due to a relatively high proportion of induced abortions in the area. This is certainly a factor which could vary greatly from region to region.

3. *Age of female.* It has been known for some time that Down's syndrome is much commoner among the offspring of older mothers, especially those

over 40 years of age [13]. The same is true for other chromosome anomalies compatible with livebirth such as 13-trisomy, 18-trisomy and some cases of Klinefelter's syndrome [14–16]. However, the 45,X anomaly is not age dependent [17]. The same situation seems to pertain to chromosomally abnormal abortuses. The 45,X anomaly and polyploidy are not age dependent, while certain trisomies are commoner in abortions from older women [18]. There is one notable exception, trisomy 16, which is not associated with increased maternal age [19].

It is possible that reproductive habits of different populations could affect the prevalence of chromosome anomalies among abortuses. The relative rarity of chromosomal aneuploidy in India has already been noted and this may be related to maternal age, though other factors are probably involved [7]. In countries in which women continue to become pregnant into late reproductive life, chromosome anomalies may be increased among abortions. In addition to an increase in certain aneuploidies in the progeny of older women, these women have an increased risk of abortion [20]. Thus, they make a relatively larger contribution to a cohort of abortions, as well as contributing a greater proportion of certain types of chromosome anomaly.

FACTORS RELATED TO THE CONCEPTUS

1. *Tissue selected.* As Inhorn [21] has pointed out, if cultures are made only from embryos and foetuses and not from the placenta, a considerable selection factor enters into the data because foetal material is frequently absent from abortuses. However, we should be clear what is meant by the word 'placenta'. Placenta means a cake and there is no 'cake' in the early conceptus. All or most of the chorion is covered with villi and is readily separated from the overlying decidua either during abortion, or by the investigator. The chorion is clearly distinguishable from decidua with a little experience and there is virtually no chance of growing maternal tissue if the chorionic membrane is used for culture instead of the tips of the villi. In spontaneous abortions, amnion and chorion are usually fused, a long-recognised anomaly of development not found in the normal early conceptus.

If the investigator cultures a mass of tissue cut from the true placenta of later gestational age, there is the possibility of contamination with maternal cells. Again, this can be avoided by culturing chorionic membrane from the foetal surface, as well as umbilical cord and foetal tissue itself if it is viable.

2. *Phenotype.* Thiede and Metcalfe [22] found chromosome anomalies in cells cultured from 88 per cent of 'blighted ova'. This included specimens consisting of an intact sac with or without a stunted embryonic remnant.

On the other hand, only 5 per cent of phenotypically normal embryos had a chromosome anomaly. It is apparent, therefore, that selection could arise if more 'blighted ova' were submitted for chromosome analysis. This could happen without the knowledge of the investigator if specimens were submitted for analysis rather than consisting of a consecutive series. Such a situation should be suspected if the percentage of abnormal specimens rises significantly as an investigation progresses.

TABLE 2. Percentage of heteroploid specimens during collection of material

Reference	No. specimens	% Abnormal
Carr [23]	35	23
[24]	100	21
difference	65	20
[25]	200	22
difference	100	23
Boué et al. [26]	40	30
[27]	86	32
difference	46	35
[1]	132	37
difference	46	48

Table 2 indicates the frequency of chromosomally abnormal specimens at different stages in the collection of two large series of abortions. It shows that in the studies by Boué and co-workers, chromosome abnormalities were more than 50 per cent commoner among the last 46 conceptuses than the first 40 abortions examined. Boué (personal communication) believes that this is due, at least in part, to the elimination of induced abortions in the later material.

3. *Age of the conceptus.* As Thiede and Metcalfe [22] have noted, the menstrual age, as recorded on hospital charts, is a highly unreliable method of dating a conceptus. However, it is the only means by which the ages of abortuses may be compared. Morphological staging is a biased method of comparison because most chromosomally abnormal conceptuses show gross retardation in development and would be included as first trimester abortions regardless of gestational age. The menstrual age is valid, as long as it is interpreted with reason. This will be discussed in detail later.

The gestational age is about 16 days less than the menstrual age, in women with a 30-day cycle, and the majority of chromosome anomalies are found in abortuses under 11 weeks. By selecting for early abortions, Szulman found chromosome anomalies in 16 of 25 specimens [28, 29]. Even for a selected series the figure seems remarkably high and some factor other than early gestational age may have been involved.

On the other hand, there is ample evidence that chromosome anomalies are rare in mid-trimester abortions. Wingate [30] studied 50 such specimens and found normal chromosomes in all but one, which had G-trisomy

Stenchever (personal communication), feels that the low percentage of chromosome anomalies in their series [3] was due, in part, to an excessive number of mid-trimester abortions (61 per cent).

The proportion of early conceptuses among a collection of abortuses could vary from one population to another. Hospital admission practices have already been discussed. In studying the data from different published series of abortions, the proportion of early conceptuses depends on the source of the material. In the first three series listed in Table 3 there was at least partial ascertainment for abortions which occurred at home [31-33]. It will be seen that this made a great difference to the frequency of early abortions. Even in those instances where there is a special reason or inducement to report very early terminations, ascertainment is probably incomplete. Estimates based on foetal loss during successive months suggest that abortion in early pregnancy is much higher than is indicated from collection data [34, 35].

TABLE 3. Comparison of incidence of abortion by menstrual age among four populations

	Total abortions	End of week (postmenstrual)			
		8	13	17	21
Javert [31]	1117	27	52	13	8
Stevenson *et al.* [32]	1073	20	61	14	5
Shapiro *et al.* [33]	802	19	58	16	7
(Mean for 31, 32, 33)	997	22	57	14	7
Carr [2]	186	4	41	32	23

RE-EXAMINATION OF 'LONDON' DATA

How do these observations relate to our own data? London is regarded by the manufacturing industry as a 'typical' Ontario city and is frequently used to assess sales promotion response. However, it is actually a rather affluent, predominantly commercial city which is rurally located and lacks heavy industry and pollution. The water supply at the time of study was from wells and was not artificially fluoridated. The city is located 100 miles due north of Cleveland, Ohio, the site of another large abortion survey [3].

As far as is known there were no unusual disease epidemics during or immediately prior to the survey. Oral contraceptives were not in extensive use during the period January 1963 to June 1965. This is based on actual questioning of patients. During the period of survey, abortuses conceived in the winter months (October to March) had a higher incidence of chromosome abnormality than those conceived in the summer months. However, the difference did not approach statistical significance.

The problem of detecting illegal abortions has already been mentioned. In studying the hospital records, certain signs were taken as evidence that interference was suspected by the attending physician. These signs were: high fever on hospital admission with or without leucocytosis and the administration of high doses of antibiotics. Of course, some women may have come to hospital because of excessive bleeding following illegal abortion. Without evidence of infection, the detection of these cases is difficult. The attending physician was contacted in doubtful situations. It is felt that spontaneity was judged in a realistic way and that induced abortions contributed little to the series.

Nine of the abortuses (4 per cent) came from women over 40 years of age. Three of these specimens had a chromosome anomaly (33·3 per cent). For the years covered by the abortion survey, 3 per cent of live births in Ontario were from women over 40 years of age [36]. Nineteen per cent of the 152 specimens from women under 30 years of age had a chromosome anomaly. For 69 abortuses from women over 30, the figure was 30·4 per cent. However, as seen in Figure 1, there is an early peak in the frequency of chromosome anomalies among conceptuses from women under 20 years of age.

The mean maternal age for the chromosomally normal and abnormal conceptuses did not differ significantly. However, the trisomic specimens alone came from older women and the difference was significant [25]. As noted already, even the trisomic abortuses do not form a homogeneous group in this respect. The mean maternal age of progenitors of abortuses with trisomy 16 was 26·3 years while for those trisomic for D and G group chromosomes the figure was 33·3 and 33·5 years respectively [18]. On the other hand, polyploidy and the 45, X anomaly were not age dependent.

The differing results from one centre to another are most likely the result of differences related to the conceptus and our own procedure for the collection of abortuses will be discussed in detail. One of our staff called personally every morning at Victoria Hospital and collected all specimens from three locations, emergency, gynaecology ward and operating floor. We were notified when specimens were to be picked up from the other hospital. It is probable that the abortuses from the latter institution represented only a percentage of the total, though there is no reason to suspect selection. In addition, there were 37 specimens received from cities in Southern Ontario, other than London. Only four of these abortuses had chromosome abnormalities, a figure considerably lower than for the series

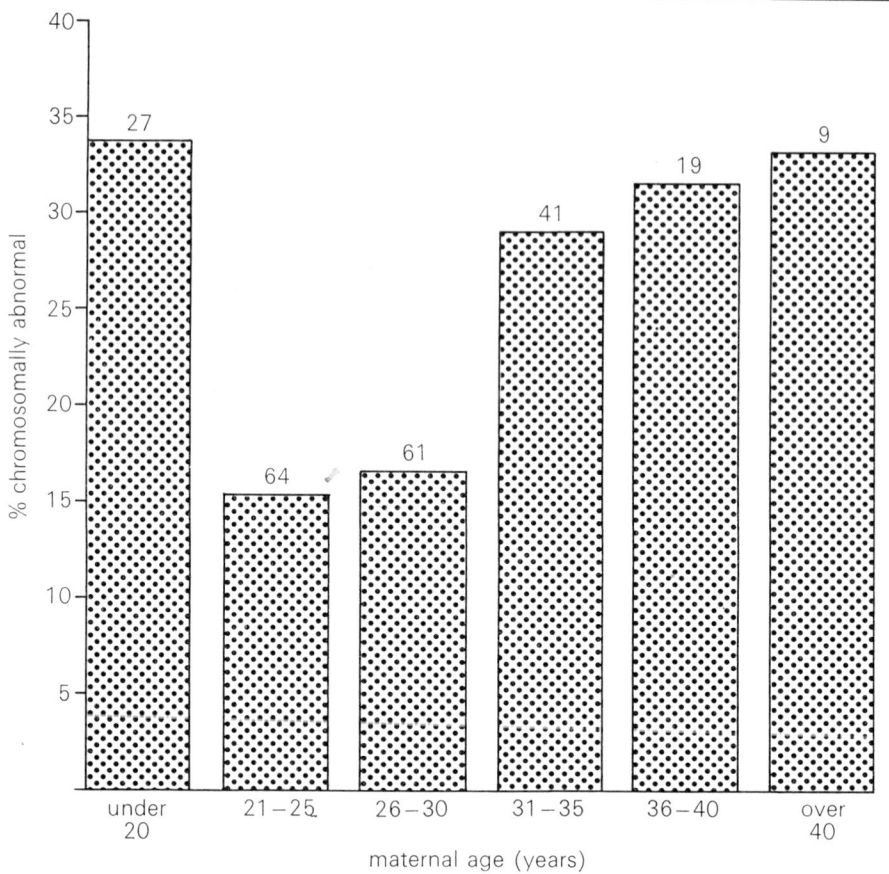

FIGURE 1
Frequency of heteroploidy in 221 unselected spontaneous abortions by maternal age

as a whole. It is clear that some factor was acting against random selection for these specimens. This was perhaps due to increased vigour of normal as against abnormal tissue which had been shipped by mail, as well as selection of later foetuses for analysis. However, removal of these from the series has little effect on the total result.

About two in five specimens failed to grow in culture and the effect this may have had on our results has been discussed in detail elsewhere [37]. As far as could be judged on the basis of gestational age of the specimens and age and obstetrical histories of the progenitors, the failure of growth of about 40 per cent of the specimens did not influence the results for the series as a whole. The high failure rate was partly due to deliberate efforts to discover just what specimens were viable. No selection of material was made in any way. Culture was attempted on every specimen which contained an embryo, foetus, umbilical cord, amnion or chorion. Just for interest, attempts were made to grow decidua and, with the technique which we used [38], these invariably failed to yield usable cells. This is not to say that it is impossible to grow maternal cells but I do not believe that this happened in our material for the following reasons : 1. A true membrane, chorionic or amniotic, was always selected for culture from early specimens as well as umbilical cord and embryonic material when available. 2. An admixture of male and female cells was never found. One would have expected this to occur occasionally if male villi and female decidua both grew. Similarly, admixtures of abnormal (for example, triploid) and normal female cells were never encountered. 3. As stated above, deliberate attempts to grow decidua consistently failed to produce cells in division.

We were particularly concerned that pre-selection might occur as information spread to local physicians about the types of abortuses most likely to yield chromosome anomalies. The percentage of heteroploid specimens remained constant throughout the $2\frac{1}{2}$-year period of the study which seems to exclude this particular bias (Table 2).

It remains now to consider the ages of the abortuses we received. Notwithstanding the inaccuracy of menstrual ages, this method of dating was retained for reasons already stated. The crude data, on the basis of menstrual age are shown in Figure 2. These data were then subjected to more detailed study. Singh [39] noted that several foetuses were far more advanced in their development than was suggested by their menstrual age. In order to correct for these obvious mistakes in dating, embryos and foetuses whose morphological ages were advanced more than one week beyond their menstrual ages, were removed from the study. No less than five of these specimens were among conceptuses with a menstrual age of less than 60 days. This left only eight abortuses in this group, three of which were abnormal (Table 3). In order to raise the number of conceptuses in this category, a series of specimens have been collected during the last 18 months, the only selection criterion being a menstrual age 60 days or less. Their developmental ages were compatible with the menstrual history in each case. A total of 18 specimens fell in this category, eight of which were abnormal. These were added to the eight abortuses under 60 days from the 'London' series giving a total of 26 specimens, 11 or 42 per cent being chromosomally abnormal. The corrected frequency for heteroploid specimens by menstrual age is shown in Figure 3. This corrected data shows that the incidence of chromosomal abnormalities in our material is not significantly different throughout the first 90 days (11 weeks of gestational age). The percentage then drops from 40 to about 25 between 11 and 15 weeks of gestation. After that there is a sharp drop to little over 3 per cent, a figure similar to that found by Wingate for mid-trimester abortions [30].

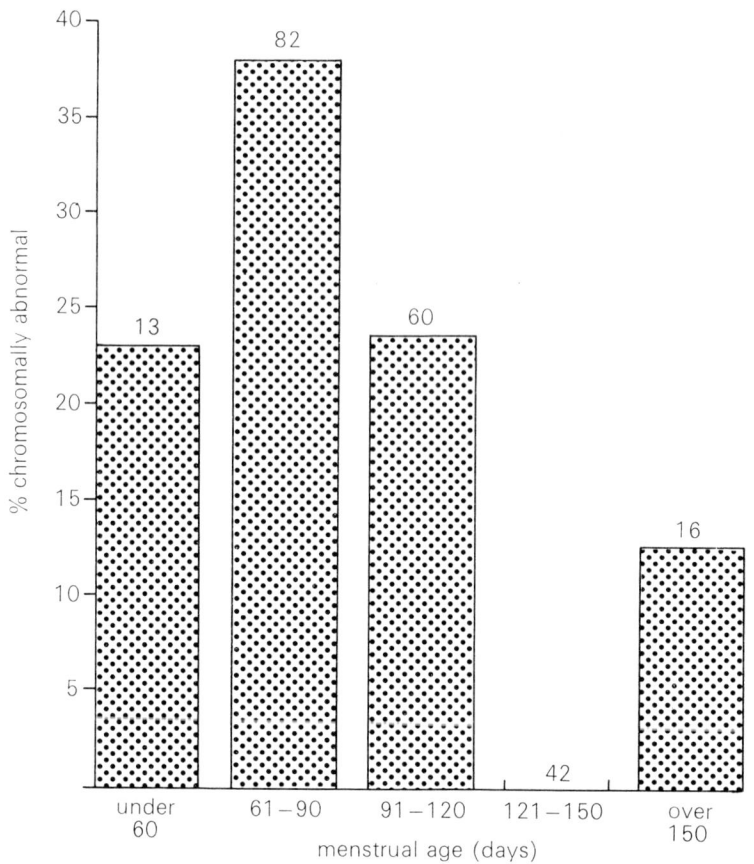

FIGURE 2
Uncorrected data showing percentage of heteroploid abortuses by menstrual age

The gestational age varies somewhat according to the type of abnormality. For the 50 chromosomally abnormal specimens in the unselected series [2], the mean age at the time of abortion for trisomic, polyploid and 45,X specimens was 80, 87·6 and 97 days respectively. Since that time some twenty additional triploid and five tetraploid specimens have been collected. Because of selection factors, the ages of the triploid specimens may not be strictly comparable to the unselected series but the additional data allows the triploid and tetraploid specimens to be separated. The mean menstrual age of 29 triploid specimens was 97·4 days and of seven tetraploid specimens, 77 days. The latter anomaly therefore appears to be much more lethal than the former. The results are summarised in Table 4.

The various types of chromosome anomaly in our series each contributed about the same to the total as those in the study by Boué and colleagues [1].

The latter investigators have continued to collect large numbers of chromosomally abnormal conceptuses by selection. The only categories in which the proportion of different heteroploidies differ from our own are C-trisomy and X-monosomy (Boué, personal communication).

TABLE 4. Mean menstrual ages at time of abortion of heteroploid specimens (days)

	Unselected series (2)	Selected study
Trisomies	80	—
Triploid		97.4
Polyploid	87.6	
Tetraploid		77
Monosomy X(45,X)	97	—

The classification of 'blighted ovum' is not a scientific one but it is understood to mean a conceptus in which the embryo is either absent or is only a stunted remnant. In our series, 50 per cent of blighted ova had a chromosome anomaly [40]. Thiede and Metcalfe [22] found heteroploidy in 88 per cent of these specimens. The reason for this difference is unknown. As far as phenotypically normal embryos and foetuses were concerned, the same investigators found chromosome anomalies in almost the same proportion in each study, 6 and 5 per cent respectively [40, 22].

Estimates of the pre-viable conceptual loss, known as spontaneous abortion, vary considerably according to the method of study. Figures based on patients admitted to hospital are much too low, as many abortions occur at home and often without medical aid. Frequencies based on personal interview suggest that about 15 per cent of pregnancies terminate spontaneously before the foetus reaches viability [20, 41]. However, pregnancy losses estimated by extrapolation suggest that 24 to 27 per cent occur before the end of 20 weeks of gestation [34, 35].

DISCUSSION

The apparent differences in the percentage of chromosomally abnormal abortuses in published studies by different investigators may depend on a number of factors, as just outlined. It really means very little to present a figure for the overall prevalence of heteroploidy in spontaneous abortions unless the population is clearly defined. This should include any conditions which may show large regional variation such as the ascertainment of total spontaneous abortions in the population, the likelihood of undetected induced abortions and mean maternal age.

Geographical and social factors for London, Ontario have been outlined above. Sergovich [42] found that about 1 in 200 newborn infants had a chromosome anomaly. The only one for which reliable regional figures are available, G-trisomy associated with Down's syndrome, seemed to be rather infrequent. The incidence in that series of neonates in 1967 was only 1 in 1,000. As noted already, the anomaly shows considerable regional variation, even between hospitals in the same city [7]. The incidence of Down's syndrome during 1967 in the other general hospital in London was approximately the same, 1 in 1,000 (Sergovich, personal communication).

If the chromosomally normal and abnormal abortuses of the unselected series were considered together, the mean maternal age was 27·3 years. This is slightly less than the mean maternal age for women bearing live born children in Ontario which was 28·2 years. However, the latter figure tends to be elevated because it represents legitimate births only.

Even more important are factors related to the conceptus. Is the series truly consecutive? What, if any, material was excluded from the study? Did local physicians submit specimens of their choice for chromosome analysis? In our own series, the most important selection factor, occurring without our knowledge, was the paucity of early abortuses and, therefore, proportional excess of later specimens. There were in fact so few specimens under 60 days menstrual age that we were unable to draw any definite conclusions for the incidence of chromosome anomalies in this period. The percentage of heteroploid abortuses for the corrected data in this period was 37·5 (3 out of 8). It is interesting that the addition of data from eighteen specimens collected in Hamilton, Ontario has not altered this figure appreciably (Fig. 3). In our experience, therefore, the proportion of chromosomally abnormal abortuses at any stage of gestation barely exceeds 40 per cent.

The age group in which heteroploidy was maximal, under 90 days or 13 weeks of menstrual age, was small when compared with other populations in the literature. The three studies in Table 3 were taken from community as well as hospital surveys. In each instance there was an attempt to assess the frequency of spontaneous abortion in pregnant women who were never admitted to hospital. From these data it appears that about 80 per cent of all pre-viable conceptuses are lost before the end of the eleventh week of gestation (90 days or 13 weeks menstrual age). Although the total zygotic loss is higher when based on projection studies, the proportional loss by weeks is about the same as that indicated for the three studies in Table 3 [34, 35].

From our original study we concluded that 'chromosome anomalies are probably a commoner cause of abortion in man than the 22 per cent incidence indicated by this series' [25]. Now that the percentage of heteroploid specimens under 60 days menstrual age has been more clearly defined we

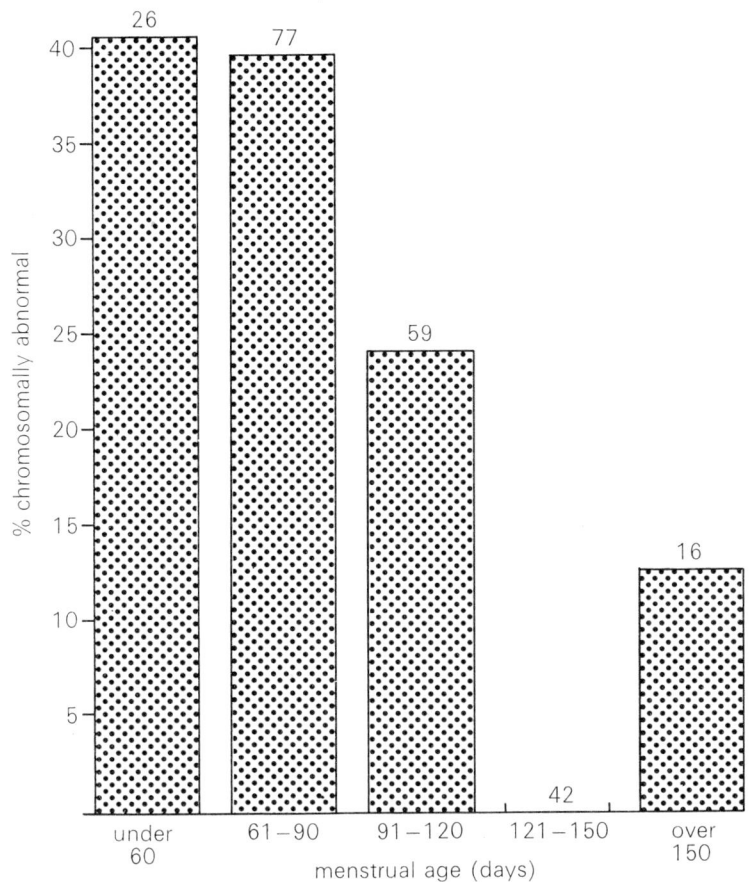

FIGURE 3
Corrected frequency of heteroploidy among abortuses classified by menstrual age

can make some projections. From studies involving results of pregnancy, whether or not a patient is admitted to hospital, we may conclude that 80 per cent of abortions occur before 13 weeks from the first day of the last menstrual period [31–35]. The corresponding figures for the 17 and 21 week periods are 15 and 5 per cent respectively. As noted from Table 3, the proportion of abortuses in these categories was quite different for our hospital population. Now, if we use these figures and consider the proportion of aneuploid abortuses in each category, we get the results illustrated in Table 5.

Forty per cent of the abortions under 13 weeks menstrual age have a chromosome anomaly and about four fifths of pre-viable pregnancy loss occurs in this period. Thus, 32 of 80 abortuses in this age group would be expected to be heteroploid. The equivalent figures for the 14–17 and 18–21 week periods are 3·75 and 0·15 respectively.

If our assumptions are correct, about 36 per cent of all clinically recognisable abortions are associated with a chromosome anomaly. This does not include an estimate of heteroploidy in pre-implantation and early post-implantation losses which are impossible to assess. However, it does include the projected results expected from a study of all clinically recognisable abortions.

TABLE 5. Percentage of heteroploid abortions by groups (menstrual age)

	Under 13 weeks	To 17 weeks	To 21 weeks
1. Unselected series [2]	45	32	23
2. Population studies [31–33]	80	15	5
3. Percentage heteroploid (Fig. 3)	40	25	3
Projected contribution [Based on 2 & 3]	32	3·75	·15
Total heteroploid	\multicolumn{3}{c}{Approx. 36 per cent}		

SUMMARY

1. The prevalence of chromosome anomalies in a population of abortuses is influenced by many factors. These include geographical variation, undetected induced abortions, maternal age and especially the procedure and scope of collection of abortuses.
2. When results of studies are published, these factors must be taken into account, otherwise widely variable figures result.
3. Very early conception loss cannot be realistically considered in an assessment of cytogenetic abnormalities in abortions.
4. Based on our own studies and with the conditions and projections discussed in this paper, the prevalence of chromosome anomalies in spontaneous abortions is about 36 per cent.

ACKNOWLEDGEMENTS

This study was supported by a grant from the Medical Research Council of Canada. The technical assistance of Mrs G. Feleki and Miss H. MacFarlane are gratefully acknowledged. Mrs Herman typed the manuscript with her usual care.

REFERENCES

[1] Boué, J.G., Boué, A. and Lazar, P., *Annls Génét.*, **10**, 179, 1967.
[2] Carr, D.H., *Am. J. Obstet. Gynec.*, **97**, 283, 1967.
[3] Stenchever, M.A., Hempel, J.M. and Macintyre, M.N., *Obstet. Gynec.*, **30**, 683, 1967.
[4] Hertig, A.T., Rock, J. and Adams, E.C., *Am. J. Anat.*, **98**, 435, 1956.
[5] Subray, N. and Prabhaker, S., *Science*, **136**, 1116, 1962.
[6] Taylor, A.I. and Moores, E.C., *J. med. Genet.*, **4**, 258, 1967.
[7] Stevenson, A.C., Johnston, H.A., Stevart, M.I.P. and Golding, D.R., *Bull. Wld Hlth Org. Suppl.*, **34**, 1966.
[8] Robinson, A. and Puck, T.T., *Am. J. hum. Genet.*, **19**, 112, 1967.
[9] Frøland, A., *Lancet*, **2**, 771, 1967.
[10] Tunte, W. and Niermann, H., *Lancet*, **1**, 641, 1968.
[11] Baird, P.A. and Miller, J.R., *Br. J. prev. Soc. Med.*, **22**, 81, 1968.
[12] Ross, H.S., Innes, G. and Kidd, C., *Scot. med. J.*, **12**, 260, 1967.
[13] Penrose, L.S., *Br. med. Bull.*, **17**, 184, 1961.
[14] Magenis, R.E., Hecht, F. and Milham, S., *J. Pediat.*, **73**, 222, 1968.
[15] Polani, P.E., *Br. med. Bull.*, **25**, 81, 1969.
[16] Ferguson-Smith, M.A., Mack, W.S., Ellis, P.M., Dickson, M., Sanger, R. and Race, R.R., *Lancet*, **2**, 46, 1964.
[17] Boyer, S.H., Ferguson-Smith, M.A., Grumbach, M.M., *Ann. hum. Genet.*, **25**, 215, 1961.
[18] Carr, D.H. In *Year Book of Obstetrics and Gynecology*, p. 31. Ed. J.P. Greenhill. Chicago : Year Book Medical Publishers, 1967.
[19] Polani, P.E., *Develop. Med. Child Neurol.*, **8**, 67, 1966.
[20] Warburton, D. and Fraser, F.C., *Am. J. hum. Genet.*, **16**, 1, 1964.
[21] Inhorn, S.L. In *Advances in Teratology*, p. 37. Ed. D.H. Woolam. New York : Academic Press.
[22] Thiede, H.A. and Metcalfe, S., *Am. J. Obstet. Gynec.*, **96**, 1132, 1966.
[23] Carr, D.H., *Lancet*, **2**, 603, 1963.
[24] Carr, D.H. In *Mental Retardation*, p. 35. Ed. G.A. Jervis. Springfield : Charles C. Thomas, 1967.
[25] Carr, D.H., *Obstet. Gynec.*, **26**, 308, 1965.
[26] Boué, J.G. and Boué, A., *C.R. Acad. Sci., Paris*, **263**, 2054, 1966.
[27] Boué, J.G. In *Les Avortements spontanés du premier trimestre*, p. 303. Ed. J. Bret. Paris : Masson et Cie, 1967.
[28] Szulman, A.E., *Fed. Proc.*, **23**, 499, 1964.
[29] Szulman, A.E., *New Engl. J. Med.*, **272**, 811, 1965.
[30] Wingate, M.B., *J. Obstet. Gynec. Br. Commonw.*, **73**, 296, 1966.
[31] Javert, C.T., *Spontaneous and Habitual Abortion*. New York : McGraw-Hill, 1957.
[32] Stevenson, A.C., Dudgeon, M.Y. and McClure, H.I., *Ann. hum. Genet.*, **23**, 395, 1959.
[33] Shapiro, S., Jones, E.W. and Densen, P.M., *Millbank Mem. Fund Quart.*, **40**, 7, 1962.
[34] French, F.E. and Bierman, J.M., *Publ. Hlth Rep. Wash.*, **77**, 835, 1962.
[35] Erhardt, C.L., *Am. J. Publ. Hlth*, **53**, 1337, 1963.
[36] Vital Statistics, *Province of Ontario*. Toronto : Queen's Printer, 1968.
[37] Carr, D.H., Bateman, A.J. and Murray, A.B., *Obstet. Gynec.*, **28**, 611, 1966.

[38] Lejeune, J., Turpin, R. and Gauthier, M., *Rev. fr. Étud. clin. biol.,* **5,** 406, 1960.
[39] Singh, R.P., Ph.D. Thesis., University of Western Ontario. London, 1967.
[40] Singh, R.P. and Carr, D.H., *Obstet. Gynec.,* **29,** 806, 1967.
[41] Roth, D.B., *Int. J. Fert.,* **8,** 431, 1963.
[42] Sergovich, F.R. (Abstr.), *Ann. Mtg. Am. Soc. hum. Genet.,* 1968.

Applications of Quantitative Karyotypy to Chromosome Variation In 4400 Consecutive Newborns

H. A. LUBS *and* F. H. RUDDLE

Yale University, New Haven
Connecticut

¶ HUMAN CYTOGENETICS is now thirteen years old. Like most thirteen-year olds it is just beginning to face the real problems that confront it. Cytogeneticists have studied highly abnormal children and adults and have attempted to make meaningful correlations between their cytogenetic findings and the phenotypes of the subjects. In this fashion, cytogenetics has defined mongolism and a number of other autosomal syndromes and has made a number of fundamental contributions to mammalian biology. The most exciting time for cytogenetics, however, is probably just ahead. To realise its full potential in the scientific practice of medicine, cytogenetics must greatly expand its base of knowledge and improve its technics. It must become a quantitative science, not remain a subjective art. The present conference stems from this realisation, and in a sense represents one of the first signs that cytogenetics is growing beyond its adolescence.

In order to develop a quantitative cytogenetics, automated procedures of chromosome mensuration must be employed. These will permit the sampling of large populations and allow their subsequent statistical evaluation. Such an approach should provide data on the extent of variation within and between variously defined human populations and ultimately contribute to knowledge pertaining to the genetic content of individual chromosomes.

Description of the New Haven newborn study. The first requirement for a precise definition of the range of normal variation is a large representative sample from the human population. The population of the New Haven area is approximately 400,000, and the total number of births each year is about 6,700, of whom about 4,500 are born at the Yale-New Haven Medical Centre. This study is based on all infants born at Yale-New Haven Hospital between October 1967 and October 1968, and these constituted 66 per cent of all infants born in the New Haven area. There is only one other obstetrical service in the area and the two obstetrical services are compared in the Table 1. Yale-New Haven Hospital serves as a combination community hospital and university hospital. The second hospital, St Raphael, is a Catholic community hospital. The major difference between the new born populations at the two hospitals consists of the greater percentage of black infants born at Yale-New Haven Hospital. (The 5% of infants classified as 'other' at Yale-New Haven Hospital in Table 1 were largely Puerto Rican; these newborns were classified as white at St Raphael.)

The microculture technic of Hungerford [1] was employed throughout, and the use of hypnotic KCl during harvesting was particularly important in obtaining excellent metaphase preparations with straight chromosome arms. Two culture bottles were set up for each infant from cord blood, and

TABLE 1. New Haven area births

	Yale-New Haven	St. Raphael
No. births/year	4483 *	2053
Race: White	77%	87%
Black	18%	13%
Other	5%	0
Maternal age: mean	25·7	26·4
<20	12%	10%
>34	8%	10%
Religion: Catholic	47%	67%
Protestant	42%	33%
Other	11%	0
Hospital class: private	74%	82%

* Successful preparations were obtained in approximately 4400 of these infants

six slides generally were made from each bottle. If the culture was not successful or cord blood not obtained, a second preparation was made using blood obtained by heel prick. Only cells of very high quality were selected for inclusion in the study. Cells with overlapped or curved arms were not included, unless necessitated by a poor culture. The adherence to these criteria resulted in exclusion of early metaphase cells, and the majority of metaphase chromosomes were moderately compact.

Karyotype analysis of our overall New Haven sample is being carried out on a number of levels. Two conventional idiograms from each child have been prepared. These cells will also be subjected to quantitative karyotypy by means of a semi-automated procedure utilising an X,Y digitiser [2], and also by means of a Fidac-Fidacsys fully automated methodology [3]. In instances where *major* departures from the norm were encountered, such as aneuploidy and translocation, 30 cells were evaluated. Considerable attention has been given to *minor* deviations from the norm, or so called minor variants. These represent instances where the length of chromosome arms or the size of satellites exceeds certain pre-set values. In such instances 30 cells will also be examined. A large cell sample will also be studied in instances where mosaicism is suspected. Ultimately 30 cells will be studied for each subject in our study. Attention will also be given to the frequency of breaks, gaps, and other abnormalities.

Defining the normal range of variation is a complex process. Clinical and other information about the population must be correlated with the quantitative data extracted from the karyotype of each infant. The input into the present study, therefore, includes a number of different types of information in addition to the cytogenetic data. Information recorded included birth weight, specific anomalies, and course of the baby during the first year; number of abortions and number of relatives with congenital abnormalities or retardation in each infant's family; all known drug, viral, and radiation exposure in the mother; paternal occupation; birth place of the grandparents; religion; and many other items. Placental isozymes were determined for a number of enzymes in the majority of the placentas by Dr Ronald Davidson (U. of Buffalo), and the remaining cord blood was used for globulin determination and viral isolation by Dr Robert McCollum (Yale University). Tests for linkage or gene localisation between the various marker loci and the chromosomal variants detected in the study are also in progress.

Major variants. Determination of the significance of chromosomal variation has been one of the major purposes of the study, and we prefer the terms major and minor variation rather than 'abnormality' and 'normal variant'. The design of the study has permitted a fresh look at the phenotypes associated with both of these types of chromosomal variation.

TABLE 2. 4400 Newborns – New Haven October 1967 – October 1968

Translocations	6
Trisomy D	1
F	1
G	3
47,XYY	3
47,XXY	4
47,XXX	3
45,X	1
Total (1/200)	22

Twenty-two infants had a major variation (Table 2). Half involved the sex chromosomes and half the autosomes. The overall frequency of 1/200 is a minimal estimate. It is likely to be significantly greater when 30 cells per newborn have been analysed and mosaics detected. The frequency of major variants may further increase when the precise cytogenetic nature of certain minor variants (see below) becomes clear. In Table 3 these results are expressed as the number of variants per thousand newborns, without regard to sex, and are generally comparable to other surveys of consecutive newborns by Sergovich et al. [4], and Court Brown et al. [5].

Six infants had definite translocations, and a number of the minor variants may ultimately also be classified as translocations, inversions or other types of rearrangements, when detailed quantitative studies have been completed. Three infants with 47 chromosomes were interpreted as having an XYY karyotype. Since the male–female ratio in the study was 1·02 this represents a frequency of one XYY in 740 male newborns. Four infants had XXY karyotypes, three were XXX, and one XO.

TABLE 3. Number of major chromosal variants/1000 newborns *

	Sergovich et al. (2159)	Court Brown et al. (2006♂)	Lubs and Ruddle (4400)
Grand Total	4·61	(5·50)	5·00
Autosomal Total	2·30	3·0	2·50
Translocations and inversions	0·92 (−)	1·0	1·36
Trisomy D	0 (+)	0	0·23
E	0	0	0·23
G	0·92	2·0	0·69
Deletions	0·46	0	
Sex Total	2·31	−	2·50
47,XYY	1·85	1·00	0·69
47,XXY	0·46	1·00	0·91
47,XXX	0	?	0·69
45,X	0	?	0·23
46,XX (♂)	0	0·50	0

(−) Translocation trisomy included
(+) Translocation trisomy not included
*The incidences at birth are expressed as the No./1000 births without regard to sex. To determine the sex-specific incidence of sex chromosome variants, the rates should be roughly doubled (M/F ratio 1·01)

Trisomy G in the present study was less common than reported in the literature. The mean maternal age of 25·7 years in the present population was about two years lower than that for the United States, England and Australia in recent years [6]. Moreover, there were approximately half as many mothers over 34 in the present study. Although the same size was relatively small for determining frequency differences of uncommon clinical syndromes in different populations, our incidence of 1/1,500 may represent a real difference from the reported frequency of 1/600–700 in the literature. Moreover, in previous years the frequency of Down's syndrome in newborns at Yale-New Haven Hospital (based upon phenotypic diagnosis) has also been lower than 1/700. It is particularly important to record and report the maternal age in population surveys of this type and caution must be exercised in comparing the frequency of major variants in different studies since the maternal age and other factors vary from year to year as well as from population to population.

Robinson et al. [7, 8] observed possible temporal clustering of infants with X chromosome variants and G-trisomy in the Denver population.

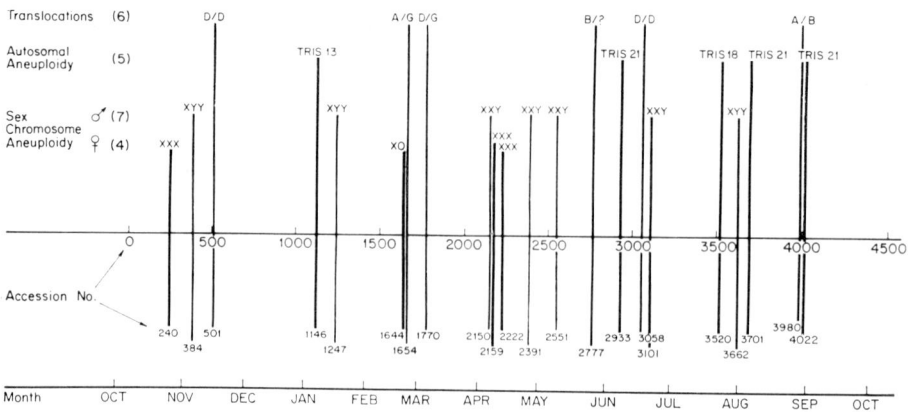

FIGURE 1

The temporal relationships between the 22 infants with major chromosomal variants are shown. On the horizontal axis both the month and the order of birth of the infants (Accession No.) with major variants are shown. Translocations are represented by the highest vertical lines, autosomal aneuploidy by lines of medium height, and sex chromosome aneuploidy by the lowest level of vertical lines. Of the 7 infants with an extra X chromosome, 5 were born in a one month period (Accession Nos. 2150, 2159, 2222, 2391, 2551). The last menstrual period of each of these infants' mothers occurred during the only viral epidemic in the preceding year, a Coxsakie B-5 epidemic in the previous July and August

The temporal occurrence of the 22 major variants from the present study is shown in Figure 1. Five of the seven infants with extra X chromosomes were born in the same one-month period. In the year preceding the study only one major viral epidemic occurred. This was a Coxsakie type B-5 epidemic which occurred in July and August of 1966. The last menstrual period of each of the mothers of these five infants took place during this epidemic. We have collected blood from the parents of children with X chromosome variants as well as control parents with normal children from the same period. Analysis of the antibody titres (which persist for more than a year) is in progress by Dr Robert McCollum (Yale University). This information should enable us to determine whether or not this possible cluster is causally related to the epidemic.

Six translocations have been ascertained (Table 2). Three were new, two were inherited, and the inheritance of the sixth has not yet been determined (Table 4). Two of the translocations warrant special comment. The karyotype termed A/G translocation is shown in Figure 2. At first glance, the karyotype appeared to be that of an infant with a G/G translocation, and one might have anticipated the clinical appearance of Down's Syndrome. The child, however, was phenotypically normal. In group A there appeared to be three no. 2 chromosomes. It seems most likely, therefore, that a reciprocal translocation has occurred between A-1 and the G chromosomes. The parent's karyotypes were normal.

TABLE 4. Clinical and cytogenetic findings in 22 infants with major chromosomal variants

Type of variant	Days in special care unit	Birth weight (grams)	Maternal age	Anomalies and course	Parent's karyotype
TRANSLOCATIONS					
D/D ♀	0	3250	24	None	Mother: normal Father: normal
D/D ♀	0	3090	32	None	Mother: unknown Father: unknown
A/G ♂	0	3570	24	None	Mother: normal Father: normal
D/G ♂	0	3600	26	None	Mother: D/G Trans. Father: normal
B/? ♀	0	3930	26	None	Mother: normal Father: same Trans.
A/B ♀	0	3460	24	None	Mother: normal Father: normal
AUTOSOMAL TRISOMY					
D ♀	16	2010	28	Characteristic Anomalies (Expired)	Mother: normal Father: normal
E ♂	90	2070	23	Characteristic anomalies still living age 10 mths.	Mother: normal Father: normal
G ♂	15	2990	32	Characteristic appearance. Patent ductus	Mother: unknown Father: unknown
G ♂	0	2600	23	Characteristic appearance.	Mother: normal Father: normal
G ♀	0	3640	40	Characteristic appearance. Disloc. Hip	Mother: normal Father: normal
47,XYY	0	3470	26	None	Mother: normal Father: normal
	35	1170	36	?Megalocephaly, Inguina Hernia	Mother: normal Father: unknown
	13	2370	25	Strabismus, Pectus Carinatum	Mother: normal Father: normal
47,XXY	0	2410	23	None	Mother: normal Father: normal
	0	2610	27	None	Mother: normal Father: normal
	0	2750	38	None	Mother: normal Father: normal
	0	3360	36	None	Mother: normal
47,XXX	0	3300	38	None	Mother: normal Father: normal
	18	2160	20	Ears folded, Heart defect	Mother: normal Father: normal
	5	3560	21	None	Mother: normal Father: normal
45,X	0	2680	19	None	Mother: normal

QUANTITATIVE KARYOTYPY

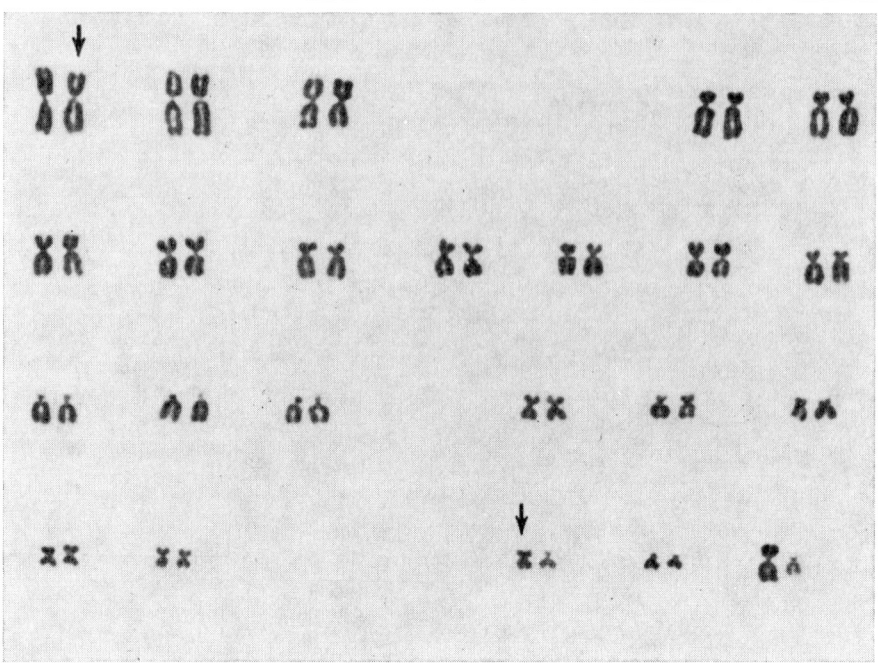

FIGURE 2
A/G reciprocal translocation. The phenotype of this infant was normal

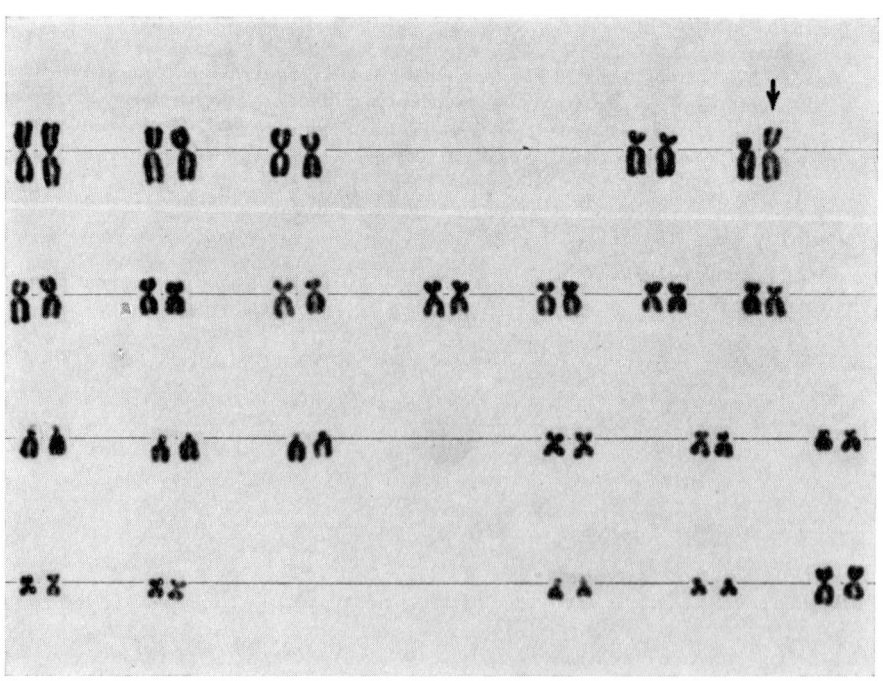

The B/? translocation, which was one of the two inherited translocations, is shown in Figure 3. The karyotype of both the father and the newborn were identical in respect to the apparent presence of only three B group chromosomes and an extra large metacentric chromosome. The normal phenotypic appearance of both father and child was unexpected. The extra chromatin material may be derived from an X chromosome, but this remains to be proven. These two examples of major chromosome rearrangement are especially instructive. They demonstrate, together with the remaining four translocations, that an appreciable number of major autosomal variants exist which are not ascertained phenotypically. Such rearrangements pose increased risks for the carriers and their offspring. The phenotypes of the infants with autosomal trisomies were all consistent with previous descriptions, and will not be commented upon further.

The XYY infants deserve special comment (Table 4). Two of the three XYY infants had neonatal complications. The first XYY infant was one of dizygous male premature twins and had a very low birth weight of 1,170 gm. The only anomaly present was an inguinal hernia. The second XYY infant also weighed less than 2,500 gm, and was diagnosed as having placental insufficiency and dysmaturity. His course was complicated by apnoea, and he was found to have both pectus carinatum and strabismus. Each of these three infants, however, is now more than one year of age and doing well. A developmental examination is being carried out on the XYY babies at ages 1, $2\frac{1}{2}$, and 5 by members of the Yale University Child Study Centre (Drs S. Provence, G. Landy, and J. Schowalder) who have no knowledge of their karyotypes. Most of these infants have now been evaluated after one year and their developmental examinations were within normal limits. It is likely, therefore, that severe mental retardation is not present. One of the XYY infants was regarded by the parents as a particularly difficult child and as completely different from their other children. On the other hand, another of the three XYY infants appears to be a very happy and normal one-year-old.

At Yale–New Haven Hospital each infant with neonatal complications was admitted to a Newborn Special Care Unit. The number of days spent in this unit is shown in Table 4. Some estimate of the cost of detection and of the cost of care of these infants can be made. At $50 a day, the 222 days amounted to more than $11,000-worth of special care during the first year for this group of infants. Their subsequent burden to society can only be surmised. The cost of preparing two karyotypes for each newborn was

---

FIGURE 3 (lower, facing)
B/? translocation. In view of the large size of this translocation, it is unlikely that there is an undetected reciprocal translocation although the phenotype of this infant was also normal. The father's karyotype was identical

estimated at $10 per newborn, or a case finding cost of about $2,000 per newborn having a major chromosome variant.

An effort was made to identify a high risk group of parents. To date, this has not proven successful, but certain factors were present more frequently in the infants with major sex chromosome variants or in their parents than in other parents or infants in the study (Table 5). These should not be regarded as more than possible clues. Four of eleven affected newborns (36%) were less than 2,500 gm. Only four were more than 3,000 gm. The mother is more likely to be 35 or older and/or to weigh more than 150 pounds. Two of the eleven mothers of infants with sex chromosome variants as well as one mother of an infant with Down's Syndrome had significant thyroid disease in contrast to only 1 per cent of control mothers (two hyperthyroid and one hypothyroid). The control figures were obtained from the first 1,000 families in the study. Lastly, there was an increased frequency in affected families of three or more abortions, one or more mentally retarded relatives, and three or more relatives with congenital anomalies.

TABLE 5. Clinical and family data

	% Control	% Sex chromos. 11 variants
Birth wt. < 2500 Gms.	10	36
Minor anomaly	9	18
Major anomaly	2	9
Maternal age > 35	7	36
Maternal Thyroid disease *	1	9
Maternal wt. > 150 lbs.	11	27
Family history ≥ 3 abortions	13	27
Family history ≥ 1 mental retardation	4	18
Family history ≥ 3 congenital anomalies	1	9

* Mothers treated for hyperthyroidism or hypothyroidism

Certain conclusions can be reached from this survey with respect to the major variants. The minimum frequency of major variants observed in these 4,400 newborns was 1/220, which is comparable to other studies of consecutive newborns. The majority of such infants can only be recognised by cytogenetic screening, since a combination of phenotypic diagnosis and Barr body determination would have detected only about half of the infants. Determination of the karyotypes of large numbers of consecutively born infants is both technically and financially feasible and is the method of choice if ascertainment of these infants is to be effected routinely in the future. All of the infants with translocations and the majority of the infants

TABLE 6. Frequency of chromosome variants in 2444 newborns

GROUP A		CRITERIA	#	%
1	Long arm : long	>> Long arm of No. 2	15	0·62
2	Long arm : long	>> Homologue	4	0·17
3	Short	≤ No. 5	7	0·29
GROUP B				
5	Long arm : short	< Long arm of No. 3	1	0·04
GROUP C				
6	Unusual 2° constriction		1	0·04
11	Unusually metacentric	Short arm = long arm	3	0·12
GROUP D				
13–15	Short arm : long	= Short arm of No. 18	333	13·3
13–15	Short arm : very long	> Short arm of No. 18	7	0·29
13–15	Giant satellites	> Short arm	77	3·1
13–15	Tandem satellites		1	0·04
13–15	Streaked satellites		1	0·04
13–15	Short arm : deleted		2	0·08
GROUP E				
16	Short	≤ No. 18	20	0·82
16	Long	> No. 13 including satellite	3	0·12
16	Long arm : long	> Short arm of No. 6	80	3·3
GROUP F				
19	Unusual 2° constriction		1	0·04
19	Long arm : long		1	0·04
GROUP G				
21–22	Short	< Short arm of No. 16	2	0·08
21–22	Long	≥ No. 20	2	0·08
21–22	Short arm : long	= Short arm No. 18	84	3·5
21–22	Short arm : very long	> Short arm No. 18	1	0·04
21–22	Giant satellites	> Short arm	61	2·5
21–22	Streaked satellites		1	0·04
Y	Long	≥ No. 19	155	6·3
	Very long	> No. 18	8	0·33
	Short	< No. 22	6	0·25
	Metacentric		1	0·04

> Greater than   ≥ Equal to or greater than
>> Much greater than   ≤ Equal to or less than

with X or Y chromosomal variants had a normal phenotype at birth. The autosomal trisomies together with a portion of the sex chromosome variants, however, constituted a significant medical problem and the cost for their hospitalisation in the first year was more than $11,000. The precise burden imposed by the translocations and remaining sex chromosome variants remains to be determined.

Minor variants. A preliminary analysis of minor variations in arm length and satellite size has been completed for 2,444 infants. Twenty-seven categories of variants were defined, and the criteria upon which ascertainments were based are given in Table 6. Intracellular standards for minor variant detection were employed. Such standards, on the basis of our previous experience, should be relatively independent of compaction errors. An example will illustrate the method. A Y chromosome was described as long if greater than or equal to F–19 in total length and very long if greater than E–18. A two-cell sample from each infant was employed, and the variant was judged as positive if present in both cells. Scoring was done on idiograms by means of manual measurement with calipers.

The frequencies of minor variants are given in Table 6 and Figure 4.

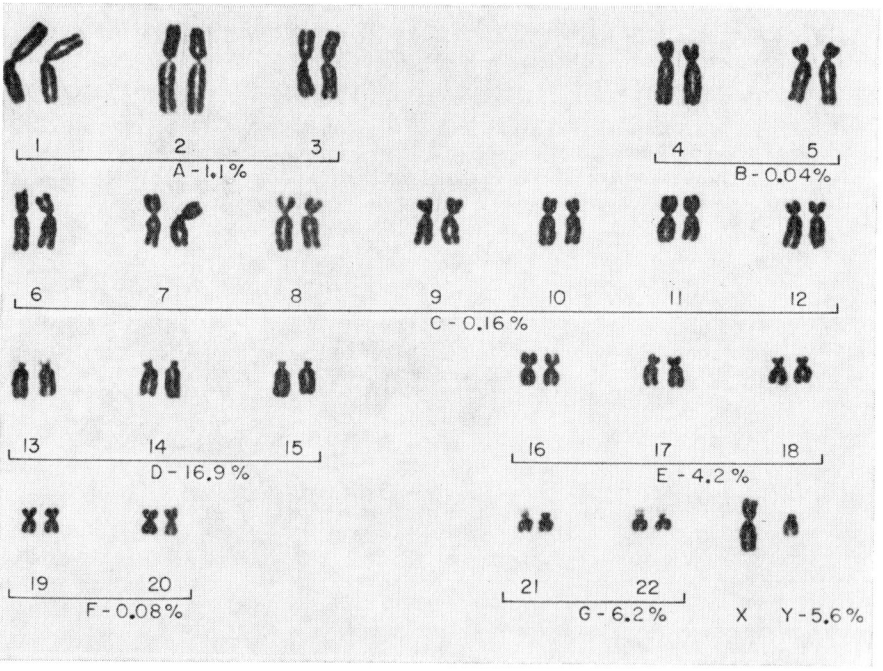

FIGURE 4

The percentage of newborns with a minor variant in each group is shown below the group. Minor variants were detected most frequently in the short arms in groups D and G. 5·6% of males had a Y chromosome equal to or greater than F-19

Variation in length occurred most frequently in the short arms of the two acrocentric groups with 17 per cent of individuals in group D having a short arm variant, and 6 per cent having a similar variation in group G. Examples and criteria are shown in Figure 5. The short arms of a D chromosome were found to be very long in 0·3 per cent of the babies (Fig. 6), and of a G chromosome were very long in only 0·04 per cent.

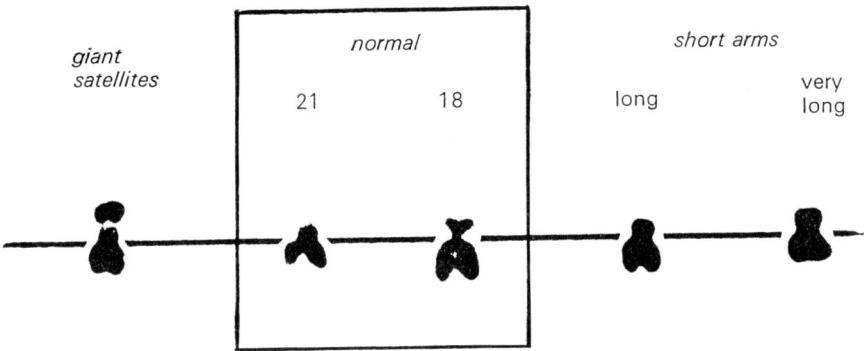

FIGURE 5. Group D and G variants
The criteria for minor variation in the length of the short arm of the acrocentric chromosomes are shown. Satellites larger than the short arm of the same chromosomes were described as giant. If the short arms of an acrocentric were equal to the short arms of E-18, they were described as long and as very long if they were clearly greater than the short arms of E-18

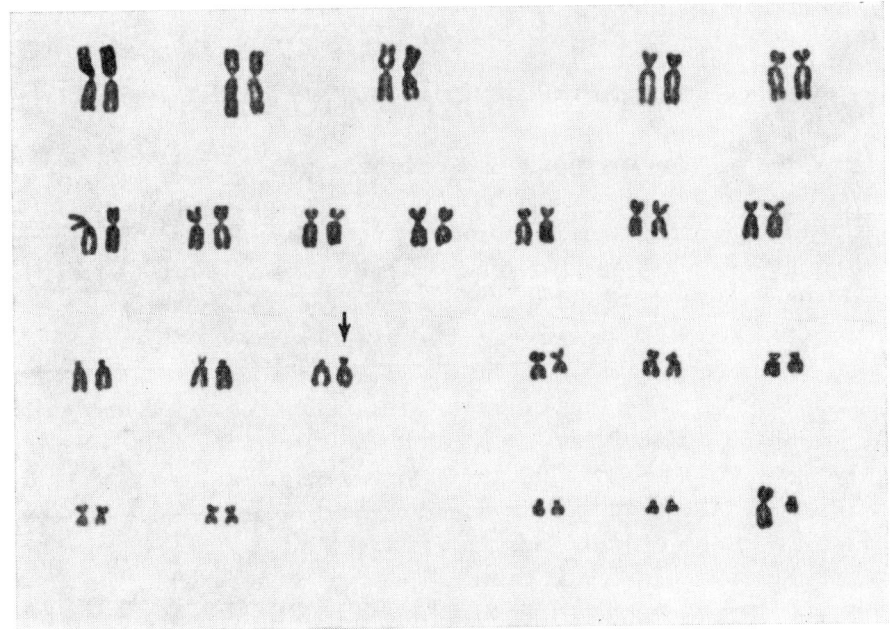

FIGURE 6
Very long short arms in group D (longer than the short arms of E-18)

Three per cent of infants had a giant satellite in D group chromosomes, and 2·5 per cent had a giant satellite in group G. A long E–16 chromosome was present in 3 per cent of infants, and an example is shown in Figure 7. The Y chromosome was judged as long in 6 per cent of males.

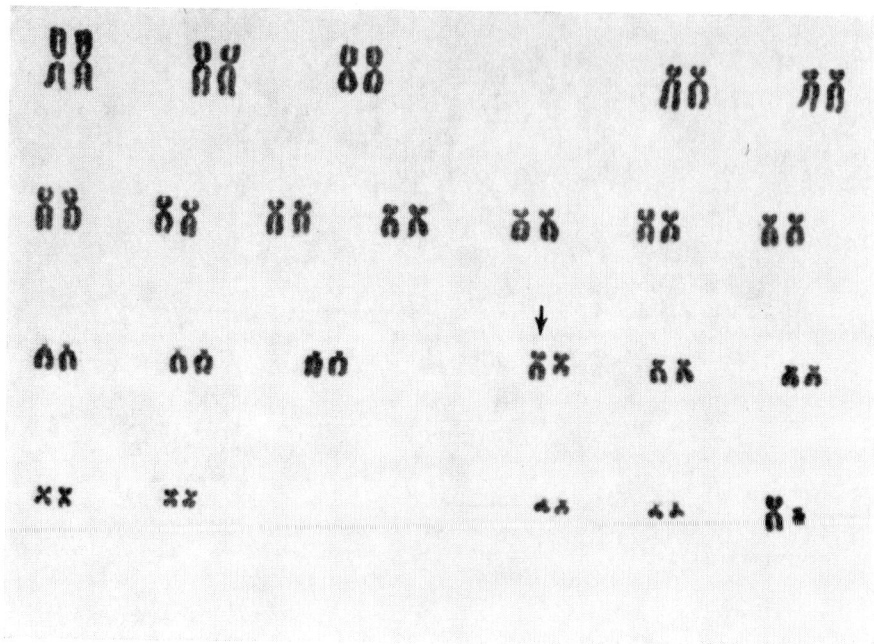

FIGURE 7
Increased length of the long arms of E-16 (equal to or greater than the short arms of C-6)

Correlations by means of two way contingency tables were carried out between each of the 27 types of variants and the clinical and other parameters that were recorded, including birth weight, major anomalies and the frequency of abortions and abnormalities in the newborn's family. In several cases, significant clinical correlations were found. Infants with giant satellites in Group G, for example, had a 4-fold increase in major congenital anomalies (Table 7). The highest correlations, however, were found between certain of the variants and race. Of the first 2,444 newborns, 448 (or 18%) were black and 1,929 (or 77%) were white. The remaining 5 per cent were of other racial groups or mixed racial groups, and are not included in the present data because of their relatively small numbers. The most common variants with significant differences in racial distribution were in groups D, E, and G and are summarised in Table 8. Each of these variants was roughly twice as frequent in black newborns as white newborns. The differences were all significant and highly significant in two instances. Giant satellites in Group G were not significantly different in

TABLE 7. Correlation between giant satellites G21–22 and major anomalies

	Major Anomaly	No major Anomaly	
Giant satellites 21	4	58	(6%)
No giant satellites 21	31	2344	(1·3%)

$$\chi^2 = 11·3, P < ·001$$

TABLE 8. Common minor variants and race. 2444 newborns

Chromosomal variant		White (1929)	Black (448)	χ^2 (1 D.F.)	P
D13–15	Short arm : long	12·3%	20·0%	17·4	·00002
D13–15	Giant satellite	2·8%	4·7%	6·5	·01
G21–22	Short arm : long	2·8%	6·2%	13·2	·0003
E16	Long	2·4%	4·7%	5·4	·02

TABLE 9. Rare variants and race. 4000 newborns

		# White	# Black	χ^2 (1 D.F.)	P
Metacentric C		2	9	28·9	·00000006
D13–15	Short arm : very long	6	8	14·2	·0001
G21–22	Short arm : very long	1	1	–	–

their racial distribution although the percentage was slightly higher in blacks. Variation in chromosomal length was found in more than 40 per cent of black males, but only 25 per cent of white males. It should be emphasised that these preliminary correlations were based on only two cells per subject. Better estimates will be derived from projected quantitative studies in progress, and by the examination of more cells.

Correlations with respect to the less common variants have been completed in 4000 infants. Certain of these less common variants, such as an unusually metacentric C chromosome, were even more markedly different in their racial distribution (Table 9). Usually, there is no chromosome in group C that is completely metacentric. In eleven newborns, a metacentric C with a characteristic secondary constriction adjacent to the centromere was observed in each of the two cells sampled per person (Fig. 8). Nine of eleven metacentric C chromosomes were present in black infants, a highly significant difference. Short arms in group D longer than the short arm of 18 occurred in fourteen of 4,000 infants (Fig. 6). The majority of the fourteen unusually long D 13–15 short arms were present in black infants. This difference was also highly significant. One of the two babies with very

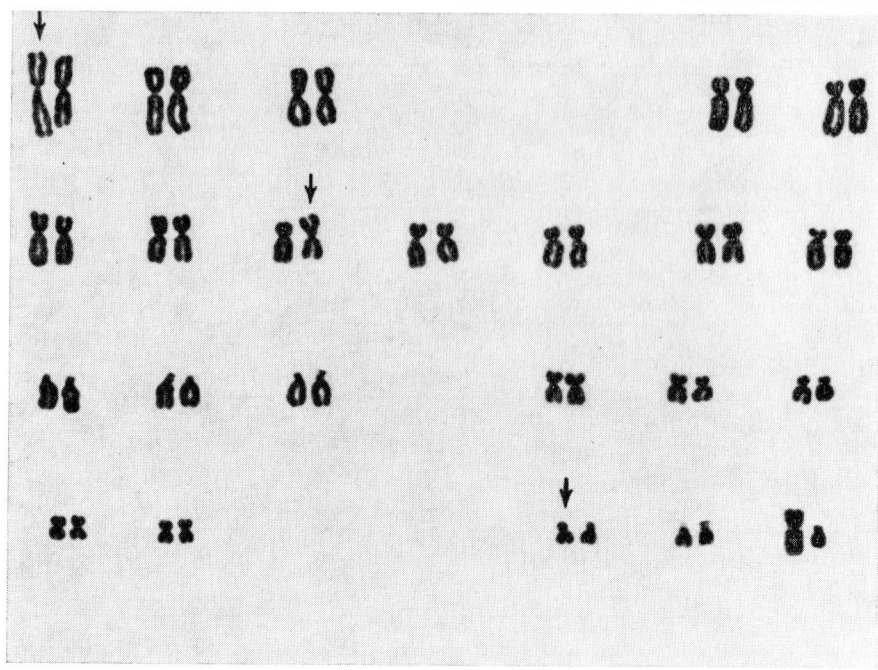

FIGURE 8
Three minor variants were present in this infant's karyotype. The long arm of A-1 was significantly greater than A-2. One chromosome in group C (shown here paired with number 8) was exactly metacentric and showed the same characteristic secondary constriction adjacent to the centromere that was seen in each of the newborns with this variant. There is also an increase in length of the short arms of one G chromosome

long D short arms was also black. The remaining rare variants did not show significant differences in their racial distributions.

Two previous studies have suggested that chromosomal racial polymorphisms occur. Cohen and Shaw, using hand measurement data, found a greater length of the Y chromosome in twenty Japanese than in other racial groups [9]. Starkman & Shaw [10] also studied the same short arm acrocentric variants discussed here and found a higher frequency of longer short arms and satellites in blacks (9 of 40) than in whites (2 of 40). Although a final conclusion must wait upon completion of the quantitative studies, these results suggest strongly that racial polymorphism in the human karyotype occurs, and that it accounts for a significant proportion of length variation in human chromosomes. These observations may be useful both in population genetics and clinical medicine, including the definition of normal. Race-specific norms are not customary in medicine and in 1969 are not popular, since many hospital-admitting offices in the u.s. have stopped recording race. This is an irrational practice and may do the patient a disservice.

By dividing the clinical material into subgroups by race, inheritance, and other parameters, different probabilities for the variant being associated with a clinical abnormality can be determined for each of the subgroups. Last year, three infants with metacentric C chromosomes were detected in the diagnostic laboratory at Yale-New Haven Hospital. Each was referred because of multiple anomalies, and all three were white. In contrast, none of the infants ascertained as having metacentric C chromosomes in the newborn study were clinically abnormal, and nine of eleven were black. It is, therefore, possible, but not proven that the same minor variant has a different probability of being associated with clinical abnormality in each race. Alternatively, a seemingly identical chromosome variant in whites and blacks may have a different origin and genetic consequence. More data will have to be acquired before uncertainties such as these can be resolved.

Automated data Extraction and analysis. It is becoming increasingly apparent to us as we pursue our population studies on chromosome variation that the older methods of chromosome mensuration and analysis are inadequate. If we are to proceed along epidemiological lines then it is necessary to extract data from large populations of karyograms, and to devise methods for the statistical analysis of such information. Data processing problems of such a magnitude can only be solved by the application of automated technics.

We are now employing two automated systems for chromosome analysis. The first system is semi-automated, and consists of an X,Y digitiser, Thirty-five-mm negatives of metaphase cells are projected onto the digitiser table, and an operator trained in cytogenetics types the chromosomes according to group and then records the X,Y coordinates of ends of arms, centromere, satellites, secondary constrictions, etc. for each chromosome.

The data is recorded onto magnetic tape and is recalculated by a high speed digital computer. In final form the chromosomes are typed according to groups, and listed with their arm length and arm ratio data. The data for individual cells is stored on magnetic tape and punch cards and can be retrieved easily for various statistical operations. The time required for a trained operator to analyse a single cell is approximately twenty minutes, which is between 5–10 times faster than conventional manual data extraction. The error rate which is monitored by a number of computer checks is also very low – probably considerably lower than by hand methods. The cost per cell is estimated at between 3 and 5 dollars.

The second method of data extraction is completely automated. This is the Fidac-Fidacsys system, already adequately described elsewhere [3]. It consists of a high-speed flying-spot scanner (Fidac) coupled to a high speed digital computer. Fidac reads and transfers the optical density information of a karyogram in the form of a 35-mm negative into the core memory of the computer in 0·1-0·2 seconds. The pictorial information is extracted by means of a complex pattern recognition program termed Fidacsys. This system analyses a 46-chromosome karyogram in approximately two minutes. The level of accuracy is good and improving as the Fidacsys program is optimised. In order to utilise Fidac-Fidacsys at its present level of capability, we have instituted a procedure which we term verification. In addition to producing a numerical representation of the karyogram, the computer also draws a pictorial (Cal-Comp) representation of the karyogram. The picture shows the profile of each chromosome as it appeared in the original metaphase spread. Superimposed on each chromosome are the calculated end of arm and centromere positions as well as an identification number. A trained operator examines this picture and compares it to a photograph of the original metaphase spread. Incorrectly analysed chromosomes are identified, and eliminated from the data. Since the chromosome data will be compiled with respect to variously defined categories and then statistically evaluated, it is not required that every cell be chromosomally intact. Verification on a routine basis also serves an important role in the optimisation of the Fidacsys program.

Replication errors of both technics have been compared to manual measurements and these results are shown in Table 10. Three cells were measured on five separate occasions by each of the three technics. Average coefficients of variation for the three methods are shown here for the short arm, long arm, and total length. The coefficients of variation were identical for each of the three technics in respect of total length measurements (under 2%). With all procedures there is greater variation in placement of the centromere than for the ends of arms. The coefficients of variation are, therefore, greater for arm measurements than for total length. It is greatest for the smallest measured objects.

Values for an average chromosome in each group obtained from the X,Y plotter and from Fidac analyses are shown in Table 11. These values are based on 50 cells analysed by Fidac and verified using the original metaphase, and 500 cells analysed by the X,Y digitiser. The measurements were normalised by expressing each chromosome as a proportion of the ten longest chromosomes, as discussed previously. The significance of these values is twofold: (1) Comparable values can be obtained from each of these technics, and both in turn were comparable to those obtained by manual measurement; (2) both sets of data were obtained from what might be termed the first production runs of quantitative karyotypy.

Two major problems in data analysis are: (1) development of the optimal mode of normalisation, and (2) development of an automated karyotyping system. The most frequently used normalisation base is either the total

TABLE 10. Variation between replicate measurements

Type of measurement	Coef. of variation *		
	Short arm	Long arm	Total
Manual : length	3·3	1·8	1·4
Plotter : length	3·9	2·3	1·4
Fidac-Fidacsys : length	7·7	4·0	1·3
Fidac-Fidacsys : area	8·6	4·6	1·1

* C.V. $=\dfrac{S.D.}{Mean} \times 100$

TABLE 11. Comparison of quantitative technics. Average chromosome in each group expressed as proportion of ten largest chromosomes

Chromos. Group	Total length		Total area
	X,Y Digitiser	Fidac	Fidac
1	·1240	·1226	·1205
2	·1050	·1133	·1161
3	·0960	·0975	·0971
4–5	·0850	·0897	·0909
6–12	·0710	·0726	·0704
13–15	·0550	·0528	·0497
16	·0520	·0502	·0425
17–18	·0470	·0465	·0396
19–20	·0420	·0401	·0309
21–22	·0350	·0312	·0230
Y	·0360	·0364	·0278

length or total autosomal length. Both are complicated by sex differences and by possible differential contraction between groups. In the preliminary phase of the present study, we have used the ten largest chromosomes as the normalisation base. This permits comparison of all parameters in both sexes, and is based upon the chromosomes with the smallest error in measurement. It produced undue variation in the values for the smaller chromosomes, however, and is probably not the optimal form. To date, no one has published a reliable program for automatically converting measurements of a cell into a form comparable to an idiogram. A number of chromosomes in each cell are usually misclassified because of overlap between groups. Both of these problems must be solved before an automated data acquisition system can be utilised in routine work.

Quantitative karyotypy. In the absence of technics for automated data extraction and analysis, cytogenetics has proceeded in a Procrustean fashion in its first decade. Procrustes was a legendary Greek robber who subjected his guests to an unusual form of hospitality. They were required to sleep in one

QUANTITATIVE KARYOTYPY

of two beds. One was very long, the other was very short. If they chose the short one, their legs were cut off during the night so that the bed would fit. If they chose the long bed, they were stretched to the point of fitting. Medical cytogenetics has customarily operated in a comparable fashion by insisting that each of the 46 chromosomes fit into one of 46 slots of predetermined size. Individual variation was generally not studied and even more rarely measured. It is particularly appropriate that this conference should be held in Edinburgh since work here has contributed importantly to the recognition that considerable chromosomal variation exists in the population.

The major effort in the development of technics for automated chromosome study has generally been directed toward data acquisition rather than analysis of the quantitative data. The information below is presented as an example of one approach to quantitative karyotypic analysis. When two karyograms from each infant were scored manually for variation in arm length, 18 of 4,400 newborns showed an increase in length of the long arm of chromosome no. 1 (Figures 8 & 9). This is almost one infant in 200, or approximately the same frequency with which major chromosomal variation occurred. Moreover, the added length is about half that of a G group chromosome. It is, therefore, important to know the significance of a variation of this frequency and magnitude. Three of these infants were compared to babies who were not found to have this variation. Thirty cells

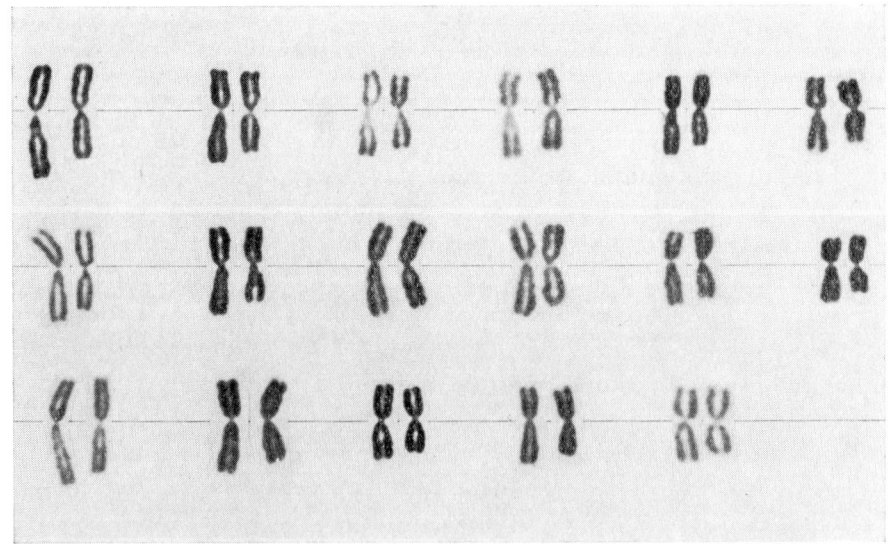

FIGURE 9
An A-1 pair from each of the 17 infants found to have an increased length of one A-1 long arm is shown. Although the secondary constriction is prominent in several of the A-1's with increased length of the long arm, in every case the appearance is that of increased chromosome material compared to its homologue

from each infant were measured using the X,Y digitiser. The longer chromosome 1 was selected in each cell. The means of these 30 cells were determined and the distribution of these means is shown in Figure 10. The means of the three individuals preselected because of increased length of the long arm of one chromosome no. 1 are indicated by the cross-hatched bars. The three individuals ascertained as having long no. 1 chromosomes showed an 11–19 per cent increase in the length of the long arm of this chromosome, and their long arm length means were between 10 and 15 standard errors away from the overall sample mean. The P value for the infant with the highest mean was 10^{-50}, a level of significance with which we are generally not accustomed. It is interesting to note that means of two of the seventeen control infants also departed significantly from the sample

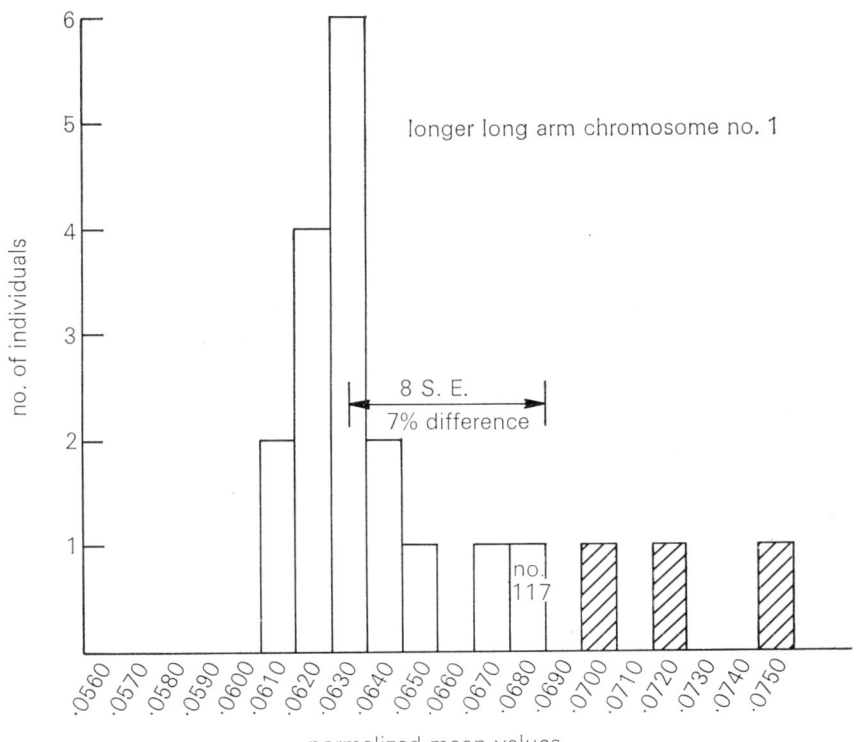

FIGURE 10
Distribution of mean normalized long arm measurements from each of 17 randomly selected infants and 3 infants found to have an increased size of A-1 long arm. Thirty cells were measured in each infant. The normalized intervals are shown on the abscissa and the number of individuals whose means fell within these intervals are indicated on the ordinate. In this study only the longer of the two homologous A-1 long arms were compared, which produces an apparent skew to the right. This is artefactual and can be eliminated by including both homologues. Each of the 3 A-1 variants were more than 8 standard errors from the sample mean and the length of their A-1 long arms was 11 to 19% greater than the sample mean

mean. Quantitative karyotypy, therefore, has the capacity to detect significantly smaller variations in length than can be detected by eye. We believe quantitative karyotypy can be extremely useful in solving a variety of cytogenetic problems related to variation in chromosome length. Similar arguments have been presented elsewhere by Ruddle [11].

What do these variations in length mean? Either they represent differences in coiling or they represent significant differences in the amount of chromatin material which is present. The type of variation described here for chromosome no. 1 has been attributed to differences in coiling [12]. In Figure 9, one pair of A-1's is shown from each of seventeen individuals with an unusual length of the long arm of A-1 observed in the preliminary survey. The longer homologue is shown to the left. In a few instances, length variation can be attributed to uncoiling at the area of the secondary constriction adjacent to the centromere, but in the great majority of these seventeen cases, the appearance is that of an increased amount of chromatin material. Fidac-Fidacsys studies of area and integrated density are in progress and these data may permit us to discriminate between chromosome coiling or differences in chromatin.

The karyotype from the eighteenth infant with an increased length of the long arm of A-1 is shown in Figure 11. The long chromosome A-1 appeared identical to those shown in Figure 9. This newborn, however, was one of six with translocations found in the newborn study. The cri du chat deletion in group B was also present. Since the infant was normal in appearance and development, it is likely that a balanced translocation between the short arm of a B and long arm of A-1 had occurred, How many of the other infants with long A-1 chromosomes had a balanced translocation that we were unable to detect? It might also be asked whether this infant really has a balanced translocation. No meiotic studies have been done. Quantitative studies, however, may prove an equally reliable, and certainly a more feasible way of documenting a reciprocal translocation in view of the statistical power of quantitative karyotypy on the one hand, and the problems of interpreting meiotic preparations on the other.

It is likely that variations in length have multiple explanations. Some are undoubtedly due to differences in coiling. Others may be due to translocations or duplications. The cytogeneticist, however, is still faced daily with the necessity of making decisions about the significance of these variations in length. Is a long no. 1 or a metacentric C related to the multiple abnormalities in a child referred for study? Is a metacentric Y related to hypospadias or hypogonadism? It is now possible to approach these questions statistically and quantitatively. Over the next year we plan to measure chromosomes from each of the 4,400 newborns, and to correlate the statistical outlyers with the clinical information which we have pre-

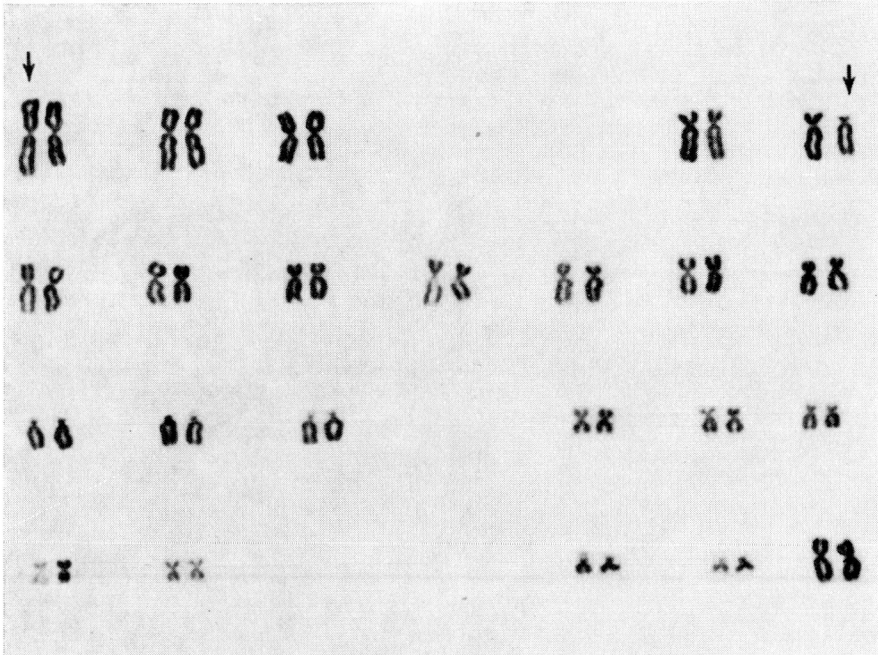

FIGURE 11

A karyotype from the 18th infant with an increased A-1 long arm is shown. In addition, a deletion of a B short arm is present. In view of the normal phenotype of this infant it is most likely that a balanced translocation between A-1 and a B chromosome has occurred. The long arm of A-1 in this infant was not detectably different from the 17 infants shown in Figure 10. Quantitative measurements of the latter 17 infants may disclose presently undetected deletions in other chromosomes and a portion of those infants may, therefore, also have reciprocal translocations or other structural rearrangements

viously recorded. In this fashion, it should be possible to begin to determine the significance of the minor variants and perhaps induce cytogenetics to grow beyond its Procrustean adolescence.

REFERENCES

[1] Hungerford, D.A. *Stain Techn.,* **40,** 333, 1965.
[2] Ruddle, F.H. and Lubs, H.A., unpublished data.
[3] Ruddle, F.H., Smith, S., Ledley, R.S. and Belson, M., *Ann. N.Y. Acad. Sci.,* **157,** 400, 1969.
[4] Sergovich, F. *et al., N.E.J.M.* **280,** 851, 1969.
[5] Court Brown, W.M. and Smith, P.G., *Br. med. Bull.*, **25,** 74, 1969.
[6] Penrose, L., *Down's Anomaly,* Chapter 10. Boston : Little, Brown & Co., 1966.
[7] Robinson, A. and Puck, T.T., *Science,* **2,** 83, 1965.
[8] Robinson, A., Goad, W.B., Puck, T.T. and Harris, J.S., *Am. J. hum. Genet.* (in press, Sept. 1969).
[9] Cohen, M.M., Shaw, M.W., and MacCluer, J.W., *Cytogenetics,* **5,** 34, 1966.
[10] Starkman, M.N. and Shaw, M.W., *Am. J. hum. Genet.,* **17,** 162, 1967
[11] Ruddle, F.H. and Ledley, R.S., *Symp. Int. Soc. Cell Biol.,* **3,** 273, 1964.
[12] Donahue, R.P., Bias, W.B., Renwick, J.H. and McKusick, V.A., *P.N.A.S.,* **61,** 947, 1968.

Chromosome Studies of Normal Newborn Infants

PARK S. GERALD *and* STANLEY WALZER

Harvard Medical School
Boston, Massachusetts

¶A CHROMOSOME survey of phenotypically normal newborn infants was begun in 1965 for the purpose of identifying infants with sex chromosome abnormalities [1]. The abnormal infants identified in this survey are being assessed for their behavioural development–these results will be described at a later time. Chromosome analysis, rather than buccal smear, was chosen as the ascertainment technique to permit the additional benefit of determining the extent of chromosome variation in 'normal' individuals.

Chromosome analyses in the number desired were economically feasible only by severely limiting the procedure used. It was found practicable to employ a microculture technique with capillary blood collected in the newborn nursery [2]. The cultures were harvested in the usual fashion but the chromosome examination was restricted to the microscopic examination of two high quality metaphase figures. If a deviation from normal was detected during the examination of these two cells, further cells were examined. The number of additional cells examined varied, depending upon the deviation initially encountered. If the initial observation included a hyper-diploid count, then a total of ten or more cells were scrutinised. Unless at least two of the total cells examined exhibited a similar deviation, the result was classified as normal. In general, we do not consider numerical variations as abnormal unless at least three identically abnormal cells per hundred are detected. Morphological abnormalities are more difficult to treat objectively since the degree of abnormality, as well as the per cent of abnormal cells, influences the decision.

An abnormality was not considered as definitely proven until a thorough study was completed, including photographic karyotypes, examination of a second blood specimen, and, where possible, a family survey. The limited procedure used made it possible for a single technician to collect and process blood specimens from 25–30 infants a week. Of necessity, this limited examination technique will be less adequate for detecting aberrations than a more exhaustive approach. The errors to be expected may be of two kinds – failure to detect mosaic abnormalities and diminished sensitivity for structural rearrangements.

It is a mathematical necessity that only a fraction of mosaic individuals will be detected when the examination is limited to two cells. Our experience with more thorough examinations in other populations has strengthened our belief that mosaic individuals, with an abnormal population representing 20 per cent or more of the cells, are relatively uncommon except among the sex chromosome abnormality group. This belief is consistent with the identification of only one mosaic abnormality among the aberrations observed.

The second possible error, that of decreased sensitivity for structural rearrangements, is difficult to assess. When only two metaphases are being examined, they are carefully chosen for their quality and are scrutinised closely. The limitation to only two cells may not of itself be very desensitising. The use of a microscopic study unsupported by photographic karyotypes is possibly more significant. Again our experience with photographic karyotyping of other populations is the basis for our belief that this is a limited, though definite, source of error.

The neonatal population studied was not a random sample, since they were delivered on a non-private obstetrical service. Infants with a major physical defect or definite morbidity were excluded. Finally, no formula for randomisation of the selection was employed – the first infants born on a given day were chosen. Ninety-six per cent of the cultures attempted were successful. Most of the failures occurred in clusters and could be attributed to a definable technical defect.

At the present time, 3,543 infants (1,931 males and 1,612 females) have been successfully examined; a report on the first 2,400 infants in this series has been published elsewhere [1]. Eighteen abnormalities considered 'major' have been detected in this population (see Table 1). Family studies have been completed for all but three of these individuals. The observed frequency of the X X Y karyotype and of the D/D centric fusion are in accord with the findings of other chromosomal surveys [3–5]. The remaining aberrations are sufficiently distinctive to warrant commenting upon individually.

TABLE 1. 'Major' chromosomal abnormalities observed in 3543 phenotypically normal infants (1931 males and 1612 females)

Chromosome complement	No. observed		No. familial
	Male	Female	
47, XXY	5		0
46, X inv(Yp + q−)	1		1
46, X small metacentric Y	1		–
45, D−,D−,t(DqDq)+	2	0	2
45, D−,G−,t(DqGq)+	1	0	1
47, extra small metacentric chromosome	2	1	2
46, break in long arm of a single C	1	0	0
46, t(2q−;Cq+)	1	0	1
46, t(16 ?−;F ?+) ?*	0	1	–
46, inv(2p−q+)	1	0	1
46, 'median C'*	0	1	–
Total	15	3	8

*Examination incomplete, diagnosis tentative

One male lacked a typical Y chromosome and possessed instead a metacentric, G-sized chromosome. This apparent Y chromosome was assumed to result from a pericentric inversion. This belief was supported by the presence of a similar Y chromosome in the father, who otherwise was an apparently normal male. The normal phenotype of the propositus was confirmed by a repeat physical examination in the neonatal period. The father has sired a total of three pregnancies – a miscarriage, a phenotypically normal male (chromosomes not examined) and the propositus. This is in keeping with the normal reproductive history of the other males with a normal-sized metacentric Y chromosome who have been described [6].

A second male infant with a metacentric Y was identified in the survey, but in this instance the Y was approximately half the size of a normal G chromosome. In the first description of this patient [1], this chromosome was said to be within the normal range for size of the Y. We now believe the Y chromosome in this particular infant is probably below the lower limit (a mensuration study is in progress). The stimulus to reassess the normal range in size of the Y was precipitated by our recent examination of an adult male referred to us with the diagnosis of a Klinefelter-like syndrome. The karyotype in this abnormal male [7] is essentially identical to that of the foregoing infant. It seems reasonable to conclude that the abnormally small Y in this man is responsible for his developing the Klinefelter-like syndrome. Our prognosis for this infant is therefore quite guarded.

A single centric fusion translocation carrier of the D/G type has been encountered so far in this survey. The same abnormality was found in the child's father (Fig. 1). This variety of centric fusion translocation has previously been ascertained almost solely through patients with Down's syndrome. The frequency of D/G translocation Down's syndrome provides a means of estimating the minimum incidence of the D/G translocation carrier. If it is accepted that Down's syndrome occurs about once in 700 births [8], that D/G translocations account for approximately 3 per cent of all individuals with Down's syndrome [9], that 38 per cent of these are familial [9], that the propositi inherit the translocation largely (84 per cent of the time) from the mother [10–12], and that about one ninth of the liveborn progeny of such mothers have Down's syndrome [10], then the minimum incidence of the carrier state in newborns should be: (1/700) (·03) (·38) (1/·84) (9) (2) or about one in 2,900 births. The observed occurrence of one D/G translocation in this survey is in keeping with this estimate. There does not seem to be any significant incidence of D/G translocations of a type incapable of causing Down's syndrome.

Three individuals with a metacentric chromosome smaller than a G, in

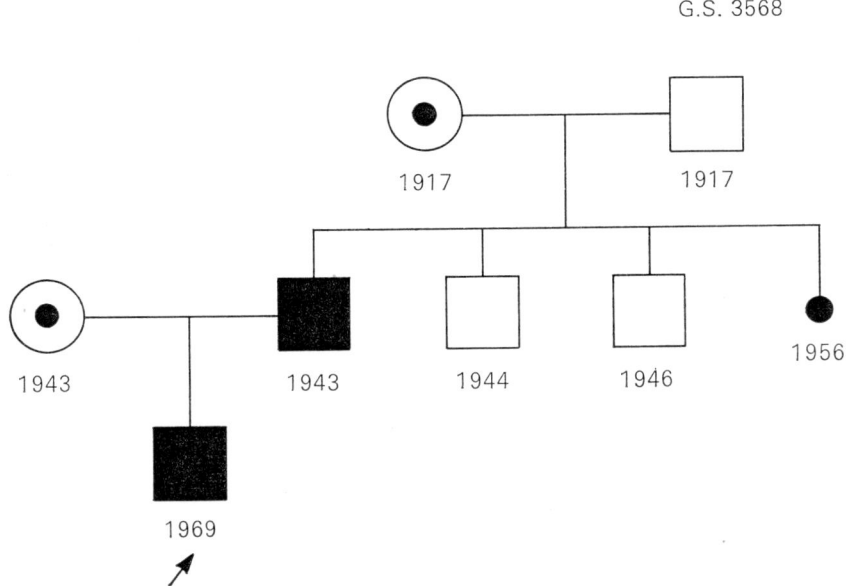

FIGURE 1
Kindred of D/G translocation carrier propositus. Propositus and his father are carriers (■), propositus' mother and paternal grandmother (⊙) have normal karyotypes

addition to an otherwise normal karyotype, were observed. In all three of these infants, at least one end of the chromosome was occasionally found in proximity to the satellites of a D or a G chromosome, suggestive of 'satellite association'. A similar extra chromosome was observed in two antecedent generations of one of these infants, in mosaic form in the mother of the second child and was absent from both parents of the third. Since the individual arm lengths in this abnormal chromosome approximate that of the short arm of a D or G, and since these chromosomes probably possess satellites on at least one end, it is tempting to consider that they arise during the formation of a centric fusion translocation and represent either a t(DpDp), t(GpGp) or t(DpGp). No independent evidence has yet been obtained to support this possibility, however.

Individuals possessing a morphologically similar small, extra, metacentric chromosome – often with satellites – have been relatively frequently reported [13–23]. In one instance, this abnormality was ascertained in a phenotypically normal individual [23]. In all other reports, the propositus has been examined because of some overt abnormality. In several instances [21, 22], however, one or more relatives of the propositus possessed the extra chromosome but lacked any phenotypic abnormality. In most instances, extra chromosomes of this type do not seem to cause disease. Conceivably, such chromosomes lack functional genes.

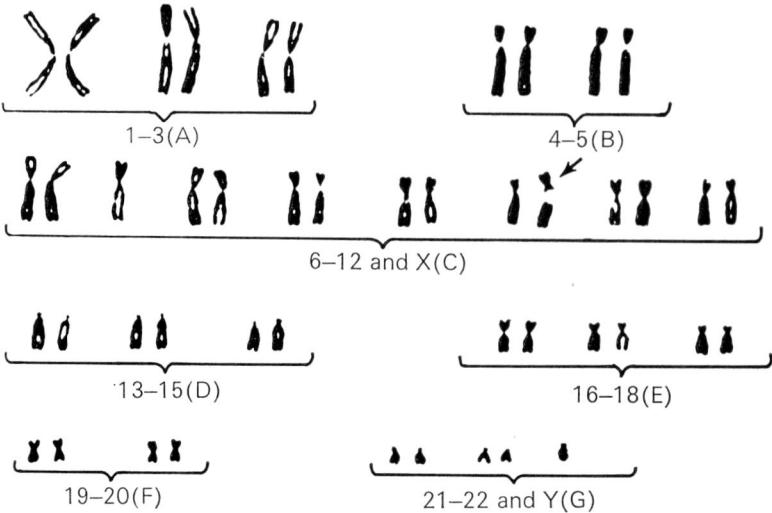

FIGURE 2
Karyotype of propositus (GS 2742) showing a chromosome break in the long arm of a C (10 ?) chromosome

A chromosomal break in the long arm of a single C chromosome (possibly a no. 10) was observed in 65 per cent of the cells of one infant (Fig. 2). The break occurred in the same chromosomal region in each abnormal metaphase. The distal fragment was separated a variable extent, ranging from simple angulation of the fragment to complete dislocation. In view of the unusual nature of this defect, the infant was re-examined physically, but no gross abnormality was noted. Repeat chromosomal examination four weeks later (at five weeks of age) revealed the same chromosomal abnormality, but at that time it was present in only 10 per cent of the cells. Neither parent of this infant had any chromosomal abnormality. The classification of this aberration as a major abnormality is obviously quite arbitrary, since in the present series it was not associated with any obvious defect in phenotype and was probably not persistent.

The presence of a consistent chromosomal break in a single and constant C chromosome has been observed in this laboratory in one other patient. This woman was examined because of primary amenorrhoea and congenital heart disease. Twenty-six per cent of her cells possessed a chromatid or chromosome break in the long arm of a C (possibly a no. 10) and a further 14 per cent of her cells possessed a marked secondary constriction in the same region of apparently the same chromosome. None of the twenty cells examined on a repeat culture five weeks later had a similar abnormality.

This variety of abnormality may be reasonably selective for a member of the C group. Gooch and Fischer observed a similar and transient abnormality of a C chromosome (no. 10?) in a normal individual, a volunteer blood donor [24]. Emerit et al. [25] observed a comparable abnormality in a presumptive no. 8 chromosome in a portion of the cells of two unrelated individuals. The first of these was a malformed child and the second was the phenotypically normal father of two malformed children. In each case, one or more relatives exhibited an increased incidence of chromosome breakage, though not with the degree of specificity noted in the propositus. Dekaban et al. [26] observed a persistent break confined to a no. '9' chromosome in an older woman who had received therapeutic irradiation.

Although reciprocal translocations are not necessarily different from the so-called centric fusion type, they are discussed separately here. Two reciprocal translocations have been observed so far. The first of these, $t(2q-; Cq+)$, was also present in two antecedent generations. There is no history of abnormal progeny or increased frequency of miscarriages in this family. The second apparent reciprocal translocation, $t(16?-; F?+)$, has been only recently discovered and studies are not yet complete.

Of the five translocations observed, including both the reciprocal and centric fusion types, all four of those for whom family studies have been completed were inherited. The combined experience from other surveys [3–5] confirms that a minority of balanced translocations are new mutations.

The last category of abnormalities observed is that of autosomal pericentric inversion. One inversion in a no. 2, $inv(2p-q+)$, has been encountered. This aberration is present in at least the two immediately antecedent generations. No procreative abnormalities resulted from this aberration. This is not surprising since, in man at least, autosomal pericentric inversions in general very rarely produce genetic imbalance. Chromosome no. 2 is frequently involved in a pericentric inversion process – we have encountered three others, of the $inv(2p+q-)$ type, among a total of eight pericentric inversions ascertained through phenotypically abnormal individuals.

One of the most recent abnormalities identified, a metacentric C chromosome in an otherwise normal karyotype, is probably also a pericentric inversion. Since family studies have not yet been completed, this is necessarily questionable. It is of probable significance that this infant is Negro, since Lubs has noted that a metacentric C chromosome is relatively frequent in this racial group [4].

Before concluding this discussion, it is worth noting that certain expected abnormalities were not observed. In the combined series [3–5, 27] of random, unselected infants (which therefore excludes the present data), 12 males with an XYY karyotype were observed in a total of nearly 6,700 males, or 1·8 per 1,000. Since at least three of the 12 had an abnormal phenotype [3, 4], the frequency of XYY among phenotypically normal males is 1·3 per 1,000, or less. The absence of XYY males in the present series therefore does not differ significantly from expectation.

Finally, it should be noted that no example of testicular feminisation has been reported in any of the newborn surveys, including those using a sex chromatin technique [28], although over 90,000 infants have now been examined. This is consistent with the estimated prevalence of one in 62,400 genetic males, derived from studies of inguinal hernia in girls [29].

SUMMARY

A survey of over 3,500 phenotypically normal newborn infants has identified eighteen with a 'major' chromosomal abnormality. These included five infants with XXY, two with an abnormal Y, three with a balanced centric fusion translocation, two with a balanced reciprocal translocation, two with a probable pericentric inversion of an autosome, three with an extra small metacentric chromosome and one with a consistent break in a single C chromosome. Family studies have been completed for most of these and no phenotypically abnormal relatives were discovered, although most of the translocations were inherited.

ACKNOWLEDGEMENTS

We gratefully acknowledge the expert technical assistance of Julie Gardinier, Kathleen Dale, Toni Acker, John Tripodi and especially Germaine Breau.

This study was supported in part by grants from the Children's Bureau (H-89), the John A. Hartford Foundation, the National Foundation (CRCS-55), and General Research Support Grant FR 5481 issued by the National Institutes of Health to the Boston Hospital for Women. One of us (S.W.) was the holder of a Public Health Service Fellowship (2 F3 HD-25, 070-01) from the National Institute of Child Health and Human Development during part of this investigation and is currently the holder of a Research Scientist Development Award (1 KO1 MH 25070-01A1) from the U.S. Public Health Service, National Institute of Mental Health.

REFERENCES

[1] Walzer, S., Breau, G. and Gerald, P.S., *J. Pediat.,* **74,** 438, 1969.
[2] Arakaki, D.T. and Sparkes, R.S., *Cytogenetics,* **2,** 57, 1963.
[3] Sergovich, F., Valentine, G.H., Chen, A.T.L., Kinch, R.A.H. and Smout, M.S., *New Engl. J. Med.,* **280,** 851, 1969.
[4] Lubs, H.A., Jr. This volume, p. 119.
[5] Jacobs, P.A., personal communication.
[6] Jacobs, P.A., *Br. med. Bull.,* **25,** 94, 1969.
[7] Murdock-Pilon, R., Sherwood, L. and Gerald, P.S., unpublished studies.
[8] Penrose, L.S. and Smith, G.F., *Down's Anomaly.* London: Churchill, 1966.
[9] Wright, S.W., Day, R.W., Muller, H. and Weinhouse, R., *J. Pediat.* **70,** 420, 1967.
[10] Hamerton, J.L., *Cytogenetics,* **7,** 260, 1968.
[11] Hamerton, J.L. In *Chromosomes Today,* p. 237. Edinburgh: Oliver and Boyd, 1966.
[12] Gerald, P.S., unpublished studies.
[13] Frøland, A., Holst, G. and Terslev, E., *Cytogenetics,* **2,** 99, 1963.
[14] Gustavson, K.H., Atkins, L. and Patricks, I., *Acta Paediat. Stockh.,* **53,** 371, 1964.
[15] Hart, Z.H., Cohen, M.M., Dietze, M.R. and Reisman, L.E., *J. Pediat.,* **66,** 120, 1965.
[16] Taft, P.D., Dodge, P.R. and Atkins, L., *Am. J. Dis. Child.,* **109,** 554, 1965.
[17] Hultén, M., Lindsten, J., Fraccaro, M., Mannini, A. and Tiepolo, L., *Lancet,* **2,** 22, 1966.
[18] Tamburro, R.F. and Johnson, C.E., *J. med. Genet.,* **3,** 295, 1966.
[19] Ishmael, J. and Laurence, K.M., *J. med. Genet.,* **5,** 335, 1968.
[20] Mukherjee, A.B., Partington, M.W., Simpson, N.E. and Walmsley, K.A., *J. med. Genet.,* **5,** 329, 1968.
[21] Smith, K.D., Steinberger, E., Steinberger, A. and Perloff, W.H., *Cytogenetics,* **4,** 219, 1965.
[22] Armendares, S., Buentello, L., Cuevas-Sosa, A. and Cantu-Garza, J.M., *Cytogenetics,* **8,** 177, 1969.
[23] Book, J.A., Kjessler, B. and Santesson, B., *J. med. Genet.,* **5,** 224, 1968.
[24] Gooch, P.C. and Fischer, C.L., *Cytogenetics,* **8,** 1, 1969.
[25] Emerit, I., Grouchy, J. de and Vernant, P., *Annls Génét.,* **11,** 22, 1968.
[26] Dekaban, A., *J. nucl. Med.,* **6,** 740, 1965.
[27] Wald, N., personal communication.
[28] Mikamo, K., *Obstet. Gynec.,* **32,** 688, 1968.
[29] Jagiello, G. and Atwell, J.D., *Lancet,* **1,** 329, 1962.

Chromosome Patterns in a General Neonatal Population

J. H. TURNER *and* NIEL WALD

Department of Biostatistics and
Department of Radiation Health
University of Pittsburgh

¶ THE PRINCIPAL objective of this investigation was to study a large sample of cells derived from randomly selected newborn infants of the general population. Each cell was classified according to the number and morphological characteristics of its chromosomes. The derived composite of the karyotypes of the sample was then used to define the range of variation to be expected among normal individuals. This information provides a standard for proper evaluation of genetic and environmental influences which may act to increase the frequencies of aberrant chromosomal complements.

One of the important features of the project was that the individuals involved in the preparation of slides and karyotypes had no knowledge of any of the characteristics of the newborn from whom the blood had been obtained. This feature was made possible by the fact that the collection, storage, and analysis of all information pertaining to the newborns, their parents and their families, were carried out by individuals not involved in the laboratory operation.

The study population was a random sample of 1,000 infants (517 males, 483 females) born alive, by normal or surgical delivery, on the private and ward services of Magee-Women's Hospital, Pittsburgh, Pennsylvania. Neonates born in the delivery suite were selected for study by use of a random sampling frame in which one of the 6 four-hour time periods of each of the first four days of the week was chosen by random number. The first four deliveries of the selected periods were studied. Surgical deliveries were similarly selected by a random sampling procedure. In order to facilitate follow-up, as well as to permit a sharper definition of the population to which the findings were to be generalised, the sample was limited to the residents of Allegheny County, Pennsylvania, in which the hospital is located.

At the time of birth, peripheral blood was taken from the umbilical cord under supervision of the Department of Obstetrics and Gynaecology of the University of Pittsburgh School of Medicine. This department also obtained information concerning the parents' age, race (approximately 15% were Negro), the father's occupation, the mother's parity and a history of previous pregnancies, sibling status, a history of the present pregnancy (including x-ray exposure), and complications of delivery. With the assistance and co-operation of the attending obstetrician and gynaecologist, the name of the family physician and other pertinent information necessary to establish a basis for follow-up were obtained from the mother.

Information concerning the physical status of the newborn was obtained from the attendant at delivery, and was supplemented by a pediatric examination under supervision of the Department of Pediatrics of the

University of Pittsburgh School of Medicine before the infant left the hospital. The examination focused on the baby's height, weight, and head circumference. In addition, a complete physical examination was performed routinely on these infants in the same manner as it was carried out on all babies in the newborn nursery, without the examiner being aware of any special interest in these particular patients. Funduscopic and rectal examinations and blood pressure determinations were carried out only if there were clinical indications for them. Ancillary laboratory and x-ray facilities were also employed in cases of a physical anomaly that required further examination.

The manifestation of a striking malformation or other unusual pathological condition introduced a special type of problem. These conditions often result in early postnatal deaths and the individuals afflicted require more careful study. Because of the double-blind aspect of the project, however, personnel in the routine cytogenetics laboratory could not be acquainted with diagnostic information pertaining to any individual in the study. The solution to the problem was to establish a research laboratory and assign it two major responsibilities (i) the development of new and better technical methods which would not disturb the current routine; and (ii) the complete study of unusual cases, including family follow-up as required. Final assessment of these cases also included the independent karyotype analysis prepared by the routine cytogenetic laboratory.

When the heparinised blood obtained from a liveborn infant was brought to the cytogenetics laboratory, the leukocytes were inoculated into three aliquots of culture medium, which were then processed according to well established short-term culture procedures. Slides from all aliquots of a particular specimen were pooled and examined in a manner that provided equal representation of cells from each aliquot. The first 50 cells found to be suitable for photomicrography were photographed with a polaroid camera, and the number of chromosomes in these cells were counted. Thirty of these cells, chosen at random, were photographed on 35-mm film. These photographs were then enlarged and used to prepare karyotypes and to study chromosome morphology. Whenever abnormalities in number or morphology of chromosomes were observed, further study was made (i.e. autoradiography, etc.).

The quantitative and qualitative information derived from study of the first 300 diagnostically normal neonates [1] was used to establish a 'standard karyotype' which was then used as a baseline and as an indication of the variation to be expected.

The numerical characteristics of each cell member of the entire *population of cells* were classified according to the following scheme:

I. NUMERICAL
   N^2 = diploid cell
   N^{2q} = quasidiploid cell
   N^{2+} = hyperploid cell
   N^{2-} = hypoploid cell
   N^3 = triploid cell
   N^4 = tetraploid cell

II. STRUCTURAL characteristics were classified according to the following factors:
(1) Measurement deviations (i.e. 2 or more standard deviations from the expected).
(2) Minor structural features including: $B_{(a)}$, $B_{(b)}$, $B_{(c)}$ cells; unusual satellites, elongated secondary constrictions, etc.
(3) Major stable structural variants.
(4) Major unstable structural variants (all rings and centric fragments were included in this category).

The karyotypic analysis of an individual was classified as abnormal in Table 1 if the analysed cells deviated *consistently* from the range of normal variability as ascertained in the 300 control group individuals. A non-mosaic abnormality required a consistent observation in all cells analysed. Minor structural characteristics were not utilised in the classification of an individual as abnormal. In order to be considered as evidence of mosaicism, the presence of the minor cell line in 5 of the 30 cells analysed was required. Tissues other than peripheral blood, although studied, were not utilised in developing the table. Mixed numerical and structural variants are also not included in the table. Individuals classified as diagnostically abnormal are those whose condition, regardless of severity, was clearly attributable to faulty pre-natal development. The general population characteristics are summarised in Table 1.

TABLE 1. Association between the chromosomal and clinical characteristics of 1000 newborns

Medical diagnosis of newborn	Karyotype analysis	Number of individuals	Number of cells analysed *	Percent of sample
Normal	Normal	934	48,568	93.4
Normal	Abnormal	12	636	1.2
Abnormal	Normal	33	1,650	3.3
. Abnormal	Abnormal	12	720	1.2
Normal	Mosaic	7	420	.7
Abnormal	Mosaic	2	120	.2
		1000	52,114	100.0

*30 Karyotypes of each newborn – plus a minimum of 20 additional cells counted by polaroid method

Note that according to our criteria, 4·7 per cent of our newborn population were congenitally abnormal and of those 3·3 per cent could not be associated with deviant karyotypes. Thus, among the phenotypically abnormal newborns 1·4 per cent were associated with consistent chromosomal aberrations.

In our search for consistent karyotypic abnormalities, we observed a frequency of 3·3 per cent among the total study population. 1·9 per cent of these occurred in newborns who were considered normal on medical examination.

A listing of the major autosomal and sex chromosome anomalies observed is of interest. For the group of infants having no clinical abnormality but an abnormal karyotype, the autosomal aberrations found are listed in Table 2 and the sex chromosomal aberrations in Table 3. For the babies with abnormalities of both phenotype and karyotype, the autosomal aberrations are given in Table 4 and those involving the sex chromosomes in

TABLE 2. Autosomal aberrations in normal appearing neonates

Number	Finding
1	Excess N4 cells (9·4% vs. expected ·26%)
2	Non-pairing D's (?)
2	45,XX,D−,G−,t(DqGq)+
1	45,XY,G−,G−,t(GqGq)+
6	

TABLE 3. Sex chromosome aberrations in normal appearing neonates

Number	Finding
4	46,XX/45,X
2	47,XYY
2	47,XXY
2	46,XY/45,X
3	47,XXX
13	

TABLE 4. Autosomal aberrations in abnormal neonates

Number	Finding
1	47,XY,G+ (Down's Syndrome)
1	46,XX,D−,t(DqGq)+ (Down's Syndrome)
1	47,XY,E+ (E Trisomy Syndrome)
2	47,XY,D+ (D Trisomy Syndrome)
1	46,XY,Bp− (Cat Cry Syndrome)
1	46,XY,t(Bq+ ;Bq−) (?? Cat Cry Syndrome)
7	

Table 5. The phenotypic abnormalities are included parenthetically in the two tables. A more detailed analysis of these data is in preparation.

TABLE 5. Sex chromosome aberrations in abnormal neonates

Number	Finding
1	45,X/46,XX (intersex)
2	45,X/46,XY (intersex)
1	45,X/46,XY (multiple anomalies including intersex)
1	48,XXYY (intersex)
1	45,X (Turner's Syndrome)
1	45,Xqi (Turner's Syndrome)
—	
7	

ACKNOWLEDGEMENTS

This investigation was supported in part by Public Health Service Grant Rg-9622 by Public Health Service General Research Support Grant FR05451; by Pittsburgh Health Research and Services Foundation Grant H-48; and by Public Health Service Training Grant ES00117 from the National Institute of Environmental Health Sciences.

REFERENCES

[1] Turner, J.H., Li, C.C., Wald, N. and Borges, W., Preliminary Reports on a Continuing Study of Chromosome Patterns in a General Neonatal Population. *Research Methodology and Needs in Perinatal Studies: Proceedings of the Conference on Research Methodology and needs in Perinatal Studies,* held in Chapel Hill, N.C., Chapter 7, p. 176. Ed. S.S. Chipman *et al.* Springfield: Charles C. Thomas, 1966.

Incidence Studies of constitutional chromosome abnormalities in the Post-Natal Population

P. G. SMITH *and* PATRICIA A. JACOBS

MRC, Clinical and Population Cytogenetics
Research Unit, Western General Hospital
Edinburgh

¶ KNOWLEDGE of the incidence of persons carrying chromosome abnormalities is still very sparse, with the exception of those abnormalities of the sex chromosomes which are detectable by nuclear sexing. Until comparatively recently the literature has been dominated by the reporting of single patients or small groups of patients, usually with no detailed reference to the size and structure of the population from which they have been drawn. Such reporting has enabled a fairly good clinical picture to be constructed of the effects of certain specific chromosome aberrations and this has been particularly true for those aberrations which are associated with congenital abnormalities and are detectable at birth or soon after birth. However, as in medicine generally, it is the clinically unusual patient who is most likely to be noticed and reported. Such patients may represent a very biased sample of those carrying the particular chromosome abnormality and, this in turn, may lead to an overestimate being made of the morbidity or mortality risks amongst carriers. In order to obtain unbiased estimates of such risks it is necessary to undertake soundly based epidemiological studies aimed, in the first instance, at identifying a representative sample of all persons with a specific abnormal karyotype. The cytogenetic study of populations of newborn infants represents an important step in this direction.

Ideally, in order to ascertain the effects of possession of an abnormal chromosome complement, persons with an abnormal karyotype would be identified at birth and then, together with suitably chosen controls, would be prospectively followed throughout the course of their lives. In this way it would be possible to assess accurately those risks associated with the abnormality which might lead to an increased susceptibility to morbidity, mental subnormality, antisocial behaviour or mortality. The difficulties associated with such studies are clear and unless the effects are gross, it will take a long time, perhaps a generation or more, to obtain accurate estimates of the risks. Also, cytogenetically abnormal infants are a relatively rare occurrence and it would be necessary to examine many thousands of newborn children to construct samples of sufficient size. Using the blood culture technique such undertakings are unlikely to be feasible until some form of automated processing and analysis of blood samples is functioning on a large scale. It is much easier, however, to survey large populations using the buccal smear technique, and Maclean [1] has been able to build up a small group of about 40 children, consisting mainly of 47,XXY males and 47,XXX females, who have been identified through nuclear sexing surveys of the newborn in Edinburgh. These children are currently aged up to eight years and are the subject of a prospective study.

Unfortunately it will be some time before individuals with chromosome aberrations, which are detectable only through complete karyotypic analysis, are available in sufficient numbers to make this type of study possible. It has, therefore, been necessary to investigate other means of assessing effects. To date most effort has been concentrated on incidence studies. A particular population is defined, for example, a mental subnormality hospital, and is screened for persons with an abnormal karyotype. If the incidence of persons with an abnormal chromosome constitution is higher in this population than in some suitable control population it is presumed that the individuals with a chromosome abnormality are at special risk with regard to mental subnormality. It is the choice of the control population that often creates considerable difficulty.

The population on which there exists most information on the incidence of individuals with an abnormal karyotype is the newborn. It is a very easy population to define and the practical problems associated with obtaining a blood sample are minimal. The information from the newborn is important as this enables an estimate to be made of the rate at which persons with an abnormal karyotype are being introduced into the population as a whole. However, the validity of using the incidence in the newborn as a control for comparison with studies on adult groups is sometimes questionable, as there are a number of assumptions inherent in so doing. For example, a direct comparison of adult and newborn populations might be misleading if there had been a temporal change in the incidence of the chromosome abnormality. Figure 1 shows the annual proportions of births in Scotland in which the mother was aged greater than 35 years or aged less than 20 years. The data are plotted in yearly intervals from 1939, the first year in which the maternal age distribution was given by the Registrar General for Scotland. Since 1945 the proportion of mothers aged greater than 35 years has almost halved, dropping from 20 per cent in 1945 to 10·5 per cent in 1967. Matsunaga [2] has drawn attention to an even more marked change in the Japanese population, where, between 1947 and 1966, the corresponding proportion fell from 20 per cent to 6 per cent. Such changes will have a considerable influence on the incidence at birth of those kinds of chromosome aberrations which are most common in the children of older mothers – this effect being marked for autosomal trisomies and, to a lesser extent, for 47,XXY males and 47,XXX females.

Thus comparison of the incidence of abnormalities in the newborn with that in, say, a mental subnormality hospital may lead to an overestimate of the risk associated with possession of an abnormal karyotype.

Conversely, the risk may be underestimated if the childhood mortality of affected individuals is higher than that of unaffected children. It seems very likely that this is true for females with 45,X complement and there is a suggestion of such an increase for chromatin positive males in the first year of life [3].

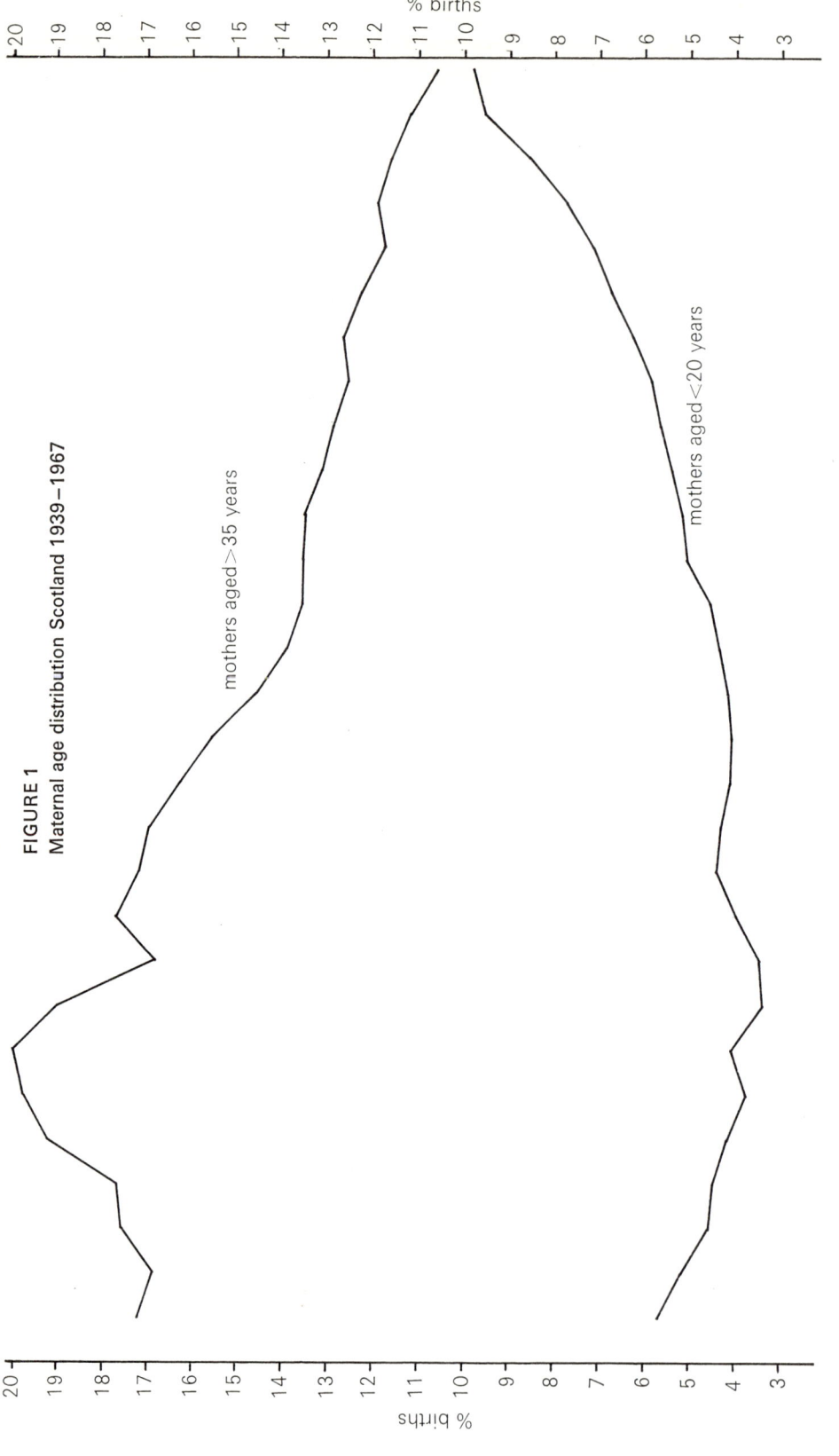

FIGURE 1
Maternal age distribution Scotland 1939–1967

The incidence of chromosome abnormalities in some special group might best be compared with the incidence expected in a group of similar age and sex drawn from the 'general population'. Definition of the general population proves difficult and in practice it is usually necessary to use populations which are far from ideal. In this paper we present data relating to studies on a number of different populations some of which may be taken to be representative of the general population and others, which in some respects clearly may not. We will be considering mainly the incidence in various populations of those constitutional chromosome abnormalities which are detectable only by a full karyotypic analysis. Court Brown [4] has recently reviewed the large number of nuclear sexing surveys of various populations.

NEWBORN SURVEYS

The Medical Research Council Unit in Edinburgh has undertaken the chromosomal examination of all newborn males in two Edinburgh hospitals and the findings are summarised in Table 1. Since April 1967 in the Western General Hospital and October 1967 in the Eastern General Hospital, samples of blood have been taken from all liveborn males. Up to 1st April 1969, 2,644 male babies had been examined for chromosome abnormalities. The decision to concentrate resources solely on male babies was influenced by the desire to obtain some estimate of the frequency at birth of males with an extra Y chromosome. In parallel with the chromosome study, all infants at each hospital, of both sexes, are routinely buccal

TABLE 1. Edinburgh newborn surveys

		Liveborn males (Hospitals E and W)		Randomly selected liveborn males (Hospital I)		Randomly selected liveborn females (Hospital I)
Number of cells examined/person		2		10		10
Number of individuals		2644		266		236
Aneuploid	Sex chromosome	47,XXY 47,XYY	3 3	0		0
	Autosome	47,XY,G +	4	0		0
Structural abnormalities	Sex chromosome		0	0		0
	Autosome	45,XY,D/D 46,XY,t(Cq +;Cp −)	1 1	45,XY,D −,E −,C + / 46,XY,E −,mar +	1	0
	Other	46,XX	1	0		0

POST-NATAL POPULATION

TABLE 2. Surveys of liveborn males

			Edinburgh		Ontario [5]	
	Number of individuals		2,910		1,066	
Aneuploid	Sex chromosome		47,XXY 47,XYY	3 3	47,XXY 47,XYY	1 4
	Autosome		47,XY,G+	4		0
Structural abnormalities	Sex chromosome			0		0
	Autosome		45,XY,D/D 46,XY,t(Cq+ ;Cp−) 45,XY,D− E− C+/ 46,XY, E− mar+	1 1 1	46,XY,D+,D/D	1
	Others		46,XX	1		0

TABLE 3. Surveys of liveborn females

		Edinburgh	Ontario		Boston		New Haven	
	Number of individuals	236	1015		1068		2178	
Aneuploid	Sex chromosome	0	0		0		45,X 47,XXX	1 3
	Autosome	0	47,XX,G+	2	0		47,XX,D+ 47,XX,G+	1 1
Structural abnormalities	Sex chromosome	0	0		0			0
	Autosome	0	45,XX,D/D 46,XX,Bp−−	1 1	0		45,XX,D/D 46,XX,B ? 46,XX,t(A?−;Gp+)1	2 1
	Others	0	0		47,XX,mar+	1		0

smeared. Thus most abnormalities of the sex chromosomes should be identified, although the autosomal data will be missing for the females.

Two cells are counted and analysed for each male baby and if either or both cells are found to be abnormal in any way a further 8 cells are counted and analysed. If the karyotype is abnormal, or there is still doubt, a further 20 cells are examined. Thus we can be sure of detecting those abnormalities which are presumed to be present in every cell, but will tend to underestimate the frequency of mosaicism where one cell line has a normal male constitution.

Also shown in Table 1 is an earlier survey conducted at the Elsie Inglis

Boston [6]		New Haven [7]	
1,332		2,222	
47,XXY	4	47,XXY	4
		47,XYY	3
	0	47,XY,E+	1
		47,XY,G+	2
46,X,inv(Yp+q−)	2	46,X,inv(Yp+q−)	2
45,XY,D/D	2	46,XY,t(A;G)	1
46,XY,t(2q−;Cq+)	1	45,XY,D/G	1
46,XY,inv(2p−q+)	1		
47,XY,mar+	2		0

Maternity Hospital Edinburgh (Hospital 1), from January 1964 to January 1965. Blood samples were taken each week from the first ten liveborn infants, irrespective of sex. The survey was conducted in order to obtain some base line data for the frequency of chromosome abnormalities in an unbiased population.

Relatively few chromosome studies have been carried out on the newborn female population in Edinburgh and in the small study at Hospital 1 no abnormal females were discovered.

In the survey of all consecutive liveborn infants we have found three 47,XXY males, three 47,XYY males, one 46,XX male, four mongols, one D/D translocation and a reciprocal translocation between two C group autosomes. (In all the tables we have used the Chicago nomenclature with the exception of that for the Robertsonian translocations.)

In the smaller survey in Hospital 1 the only abnormal found was a mosaic infant with two abnormal cell lines, both of which appear to be unbalanced.

Tables 2 and 3 compare the Edinburgh surveys with those from the other three centres which are undertaking large scale surveys of the newborn. The results shown do not exactly correspond with the most up-to-date findings from these centres which have been discussed earlier. With numbers yet small it is difficult to reliably assess the homogeneity of the findings in the four centres. There may indeed be marked ethnic and geographic variation in the incidence of specific kinds of abnormal chromosome complement but, taking the data as it stands, the differences observed are compatible with chance fluctuations. For comparison with other populations we have pooled the findings in these four newborn surveys in order to obtain better estimates of the incidence of abnormalities at birth. Most of the populations we will be discussing are Scottish and it may be that it will subsequently prove to have been invalid to have combined the Scottish and the North American newborn data to act as a base line for these groups. However, at present data are sparse and this seemed to us a not unreasonable procedure as there is nothing which yet indicates a substantial geographic variation in the incidence of chromosome abnormalities in the newborn. Amongst the 7,530 males there are ten 47,XYY's and twelve 47,XXY's, frequencies of 1·33/1,000 and 1·59/1,000 respectively, and there are eight babies with presumptively balanced translocations involving autosomes. In estimating the frequency of structural heterozygosity we have combined the male and female populations. If we add the four presumptively balanced autosomal translocations in the females to the eight observed in the males, we obtain, in the 12,027 babies examined, a frequency of 1·00 per 1,000 births, seven of these translocations being of the Robertsonian type, six D/D and one D/G.

GENERAL POPULATION STUDIES

The study of the general population is important for two reasons. First, to determine if selection is operating after birth on those individuals with abnormal chromosome complements, and secondly, to obtain control data for comparison with studies in special subgroups of the population, such as the mentally retarded. To achieve the first aim we might compare the incidence at birth with the incidence in age and sex specific groups drawn from the general population. Any changes might indicate a differential mortality between normal and affected individuals. However, it is at present rather wishful thinking to talk of age and sex specific incidences in the general population. One might speculate as to how a truly representative sample could be obtained from which mortality risks could be assessed but it would seem that all we might hope to do in the immediate future is to define groups which may be representative of certain sections of the community, for example the non-institutionalised resident population of an area. This is unsatisfactory as far as the study of mortality is concerned as the possibility of selective emigration is difficult to account for, but it does enable groups to be defined which may justifiably be used as control populations for incidence studies amongst, for example, special groups of institutionalised subjects, where the incidence of persons with abnormal karyotypes might be raised.

Table 4 lists a number of groups which have been studied in Edinburgh which might be taken to be reasonably representative of the 'normal' adult population. Again, as with the newborn, there has been a tendency to concentrate on the study of male populations.

TABLE 4. 'Normal' adult populations

	No. of cells examined	Males			Females		
		No.	Mean age	S.D.	No.	Mean age	S.D.
General practice survey	30	207	57·3	19·4	231	59·4	19·4
Workers at risk of radiation exposure	100	77	42·5	9·8	66	43·9	7·8
Workers handling toxic substances	100	112	48·9	11·3	–	–	–
Non-blood relatives of persons with a familial chromosome abnormality	2	87	39·7	19·6	88	37·9	17·2
Survey of tall ($\geq$ 183 cms.) workers in Atomic Energy establishments	2	629	40·6	10·7	–	–	–

1. *General practice survey.* This survey has been described by Court Brown et al. [8]. The subjects, all aged 15 years or over, were taken at random from the lists of four general practices in the Edinburgh district, three urban and one rural. There is a bias towards older subjects as one of the major purposes of the study was to examine a possible rise with age in the proportion of aneuploid cells in an individual's peripheral blood. The sample derived is probably not an unreasonable sample of the Edinburgh population except that the mean age will tend to be higher in the sample and the social class distribution may not be entirely representative.

2. *Workers at risk of radiation exposure.* This group consists of a mixed population of individuals from a number of industrial populations. Some of the subjects are liable to have suffered an increased amount of radiation exposure. They were primarily examined to determine if the exposure was associated with an increased number of chromosome breaks and there would seem to be no a priori reason to expect an increased frequency of constitutional aberrations.

3. *Workers handling toxic substances.* This is a similar group to the above employed in occupations which it was thought might lead to an increased level of chromosome breaks in the peripheral blood. Sixty-eight men were concerned with the industrial use of benzene and 44 were involved in the application of mercurial seed dressings.

4. *Non-blood relatives of familial chromosome abnormalities.* These individuals have been studied during the course of investigations of families in which an abnormal chromosome was known to be segregating. The group consists almost entirely of spouses of affected individuals and the spouses of their normal relatives. Blood samples were obtained for genetic linkage studies and chromosomes were investigated routinely on the same sample.

TABLE 5. Chromosomal findings in 'normal' adult populations

		General practice survey	Industrial workers
	Number of individuals	207 Males / 231 Females	189 Males / 66 Females
Aneuploid	Sex chromosome	0	0
Aneuploid	Autosome	0	0
Structural abnormalities	Sex chromosome	46,X,inv(Yp+q−) 1	0
Structural abnormalities	Autosome	46,XX/46,XX,t(Dq+;Dq−) 1	45,XY,D/D 1
	Others	0	0

5. *Tall workers in atomic energy establishments.* This group includes all men of height 6 feet (183 cm) or more who were taking part in a survey on ischaemic heart disease. The 209 men were all aged between 40 and 55 years at the time the survey began in 1967. Also included are 434 men of 6 feet in height or more who were given a routine medical examination in the course of one calendar year. Only seven men refused to provide a blood sample and all of the men in the heart disease survey provided blood. The reason for examining a group of tall men is fairly obvious as it was desired to determine the incidence of males with an extra Y chromosome in what might be taken as a random sample of the tall adult male population. It was thought that this population might be reasonably representative, although clearly it is not a completely unselected population. It was chosen as the heights of all the employed men were already known.

The findings in the normal adult population (Table 5). In so far as can be judged with such relatively small numbers of abnormalities most of the findings in the groups are comparable. A possible exception to this is the finding of four XXY males in the group of 629 tall workers, an incidence of 6·36/1,000. To obtain an estimate of the incidence at birth we have combined the newborn surveys shown in Table 2 with the results from buccal smear surveys. In Court Brown's summary of these [4] he estimates, from the studies in which chromosome findings are available, that 24 XXY's appear in 18,953 male babies examined in the two major buccal smear surveys. This incidence (1·27/1,000) is not significantly different from the incidence derived from Table 2 of 1·59/1,000 male babies studied. Combin-

-blood tives		Tall workers in A.E. establishment		Total males	Total females
Males Females		629 Males —		1,111	384
XY	1	47,XXY	4	5	0
	0		0	0	0
	0		0	1	0
XY,D/D X,t(Bq−;Eq+)	1 1	45,XY,D/D	1	3	2
	0		0	0	0

ing the two gives an estimated incidence at birth of 1·36/1,000 newborn and thus in the tall workers we might expect only 0·9 XXY's. The finding of four XXY's may represent a real increase (Poisson test, $P < \cdot 03$) and might be explained by the height selection in the population. The study of Stewart et al. [9] suggests that XXY males tend to be of above average height.

The absence of any XYY males in these groups is not a significant finding compared simply with the newborn incidence of 1·33/1,000. However, the fourth group in the table was selected on the basis of height as it has been suggested that XYY's may tend to be unusually tall. In the original study by Jacobs et al. [10] of a maximum security hospital, five of the nine XYY males identified were over 6 feet in height. In order to obtain an approximate estimate of the incidence expected in a 'normal' adult population, assuming no selection is operating, we might assume that 50 per cent of XYY males will be greater than 6 feet. The estimate of the incidence at birth is 1·33/1,000 and approximately 10 per cent of the adult English population is greater than 6 feet tall [11]. Hence, assuming that XYY's are not selectively segregated out of the normal population we would expect an incidence of about 6·7/1,000 in men over 6 feet tall and would, therefore, expect to find approximately 4·2 XYY males amongst the 629 tall atomic energy workers. We have evidence that some selection does occur which takes XYY's out of the non-institutionalised population and into maximum security hospitals but as yet it is difficult to estimate the magnitude of this selection.

It is also possible that the tall atomic energy workers may not be a very representative sample of the normal male population in terms of the social class distribution. Also, if, for example, XYY males show a greater tendency towards behaving in an antisocial or unusual fashion, in a way which might prejudice their chances of employment with the Atomic Energy Authority but at the same time which is not sufficiently extreme to cause them trouble with the Police, then our sample will be biased and will not give a good estimate of the frequency of XYY's in the non-institutionalised population. This is certainly a possibility as the Atomic Energy Authority is likely to be somewhat more selective in its choice of employees than perhaps would be some other employers.

Four constitutional presumptively balanced autosome translocations and a similar mosaic have been found in male and female groups combined. The predominant cell line in the mosaic was that of a normal female and, as such a person might have been missed in a survey in which only two cells were examined, she will be omitted for incidence comparisons. The frequency of balanced autosome translocations (4/1,495) is 2·68/1,000 which is somewhat higher than that in the newborn (1·00/1,000) but not significantly so. The only other abnormal individual was a male with a pericentric inversion of the Y chromosome.

HOSPITAL PATIENTS

Hospital patients may be representative of the 'normal' population but are somewhat less satisfactory in this respect than the above groups. A number of studies have been undertaken on groups of patients with particular diseases and these groups are listed in Table 6.

1. *Ankylosing spondylitics.* Patients with ankylosing spondylitis have been examined over a number of years in a variety of studies relating to the damaging effects of X-irradiation on chromosomes. There was no reason to suspect that they might show an increased incidence of constitutional

TABLE 6. Surveys of hospital patients

	No. of cells examined	Males			Females		
		No.	Mean age	S.D.	No.	Mean age	S.D.
Non Malignant							
Ankylosing spondylitics	50–100	206	39·2	11·7	—	—	—
Patients receiving thorotrast	100	29	53·0	11·6	25	54·1	11·4
Malignant							
Survey of cancer patients	2	661	62·4	13·8	659	60·5	14·3
Miscellaneous cancer patients studied for radiation damage	50–100	33	54·6	14·2	19	57·2	10·4

chromosome aberrations, aside from the ever present possibility that particular aberrations may predispose towards specific diseases.

2. *Patients receiving Thorotrast.* These were a group of patients who had been given Thorotrast in radiographic diagnostic procedures between 1926 and 1955. This group were also studied in an investigation of the damaging effects of radiation.

3. *Survey of cancer patients.* This survey was initiated by Professor Harnden and his colleagues and we are grateful to him for allowing us to quote his as yet unpublished findings. Blood samples were obtained from all patients with a malignant disease admitted as in-patients to a radiotherapy department which served the whole of the South East Region of Scotland. The study started on 30th January 1968 and was set up specifically to examine whether individuals with a constitutional chromosome abnormality were at greater risk of developing a malignant disease than persons with a normal karyotype.

4. *Miscellaneous cancer patients.* This is a miscellaneous group consisting of patients with cancer who were studied because they were to be treated with X-irradiation. Twenty-one had cancer of the lung, twelve had cancer of the testes, fourteen had cancer of the breast and five had cancer of the cervix.

The findings in surveys of Hospital patients (Table 7). Court Brown [4] estimates the frequency at birth of XXX females to be 1·0/1,000 and this figure is in good accord with the reported chromosome surveys of the newborn (three XXX in 4,497 babies examined). The finding of two 47,XXX females and a 46,XX/47,XXX mosaic in the 659 female cancer patients is more than might be expected but is not a statistically significant increase. The other sex chromosome findings do not deviate from those which might be expected on the basis of the newborn surveys. The XYY cell line predominated in the 45,X/47,XYY mosaic found amongst the thorotrast patients and, for purposes of comparison only, he will be regarded as an XYY. Six presumptively balanced translocations in 1,632 patients (3·68/1,000) is in good accord with the incidence of these abnormalities found in the 'normal' adult population (2·68/1,000) but is higher than that observed in the newborn (1·00/1,000), and the difference is statistically significant ($\chi^2 = 5\cdot93$; $P < \cdot02$). The findings in these groups as a whole do not appear to be markedly different from that which might be expected on the basis of studies on presumed normal groups.

TABLE 7. Chromosomal findings in surveys of hospital patients

		Ankylosing spondylitics	Thorotrast patients	Surv
	Number of individuals	206 males	29 males 25 females	661 males
Aneuploid	Sex chromosome	0	45,XY,D/D 1	
Aneuploid	Autosome	0	0	0
Structural abnormalities	Sex chromosome	0	0	46,X,inv(Yp+q−
Structural abnormalities	Autosome	45,X/47,XYY 1	0	46,XY,inv(Cp−q−
	Others	0	0	47,XY,mar+

SOCIALLY DEVIANT INDIVIDUALS

Following the discovery of an increased incidence of XYY males in a maximum security hospital, the possibility was immediately raised that possession of this chromosome complement might in some way be associated with antisocial behaviour. This led the Edinburgh Unit on a search through the Scottish Penal system and certain other groups where it was thought there might be an increased incidence of XYY males. These groups are listed in Table 8.

1. *Approved schools.* Approved schools are part of the Scottish Educational system rather than of the Penal system. They are specifically for children over the age of 7 years and children are sent to such schools either when they are judged by a Court to be beyond parental control or if a Court decides the child is in need of care and protection. No child is admitted after the age of 16 years but some stay on, once admitted, until they are aged up to 18 years.

All the major approved schools in Scotland were approached for permission to undertake chromosome examinations and 16, out of 19, schools agreed to co-operate, the schools studied including about 92 per cent of all approved school boys in Scotland. It was hoped that by studying this population young persons with an XYY complement might be identified.

2. *Borstals.* Borstals are corrective training institutions for young persons aged between the ages of 16 and 21 years. Admission is through the Courts for those persons who it is felt would benefit more from a Borstal training

er patients	Miscellaneous cancer patients		Total males	Total females
559 females	33 males 19 females		929	703
xX 2 x/47,XXX 1	46,XY/47,XXY 1 45,X/46,XX 1		2	4
0		0	0	0
0		0	1	0
x,D/D 2 x,t(2q−;Cq+)1	46,XX,t(Dq+;Cq−) 1		2	4
x/47,XX,mar+ 1		0	1	1

TABLE 8. Surveys of 'socially deviant' individuals

	No. of cells examined	Number of Males	Mean age (years)	S.D.
Approved schools	2	1119	14·2	1·8
Borstals	2	607	15–21	
Prison allocation centres	2	302	30·7	8·8
Tall prisoners (≥ 178 cms.)	2	510	29·6	9·4

than from committal to a prison or detention centre. Sentences are not given a duration other than that a youth must be detained for no more than two years. Between March 1966 and February 1967, 617 males were admitted as new entrants to Scottish Borstals and blood samples were obtained from 607 individuals, 10 being not prepared to co-operate in the study.

3. *Prison allocation centres.* All persons who are sentenced by the Courts to a period of imprisonment in excess of one year are supposed to first enter an allocation centre where it is decided to which prison they would be most suited. In practice, because of an overcrowded prison population, some prisoners do not pass through such a centre. Between March 1967 and January 1968, 325 passed through the allocation centre studied and 302 agreed to provide a blood sample, 20 refused and three were unavailable for study.

TABLE 9. Chromosomal findings in surveys of 'socially deviant' populations

		Approved schools		Borstals	
	Number of individuals	1,119		607	
Aneuploid	Sex chromosome	47,XXY 47,XYY 46,XY/47,XXY	2 4 1	47,XXY 47,XYY	2 1
Aneuploid	Autosome	0		0	
Structural abnormalities	Sex chromosome	46,X,inv(Yp+q−)	2		0
Structural abnormalities	Autosome	46,XY,t(Bp+ ;Eq−) 46,XY,t(F ?+ ;Gq−)	1 1	46,XY,t(1 ?+ ;Cq−) 46,XY,inv(Cp−q+)	1 1
	Others	0		46,XX	1

4. *Prisons.* As other workers had reported finding males with an XYY complement in English prisons it was decided to examine a sample of men, chosen on a height basis, from every prison and young offenders institution in Scotland. 544 men of 178 cm or more were found in the prisons visited and of these 24 refused to co-operate and 10 were not available on the days of the visits.

The findings in surveys of socially deviant individuals (Table 9). These populations were examined primarily because it was expected that they might contain a raised frequency of XYY males. The striking finding is the relative absence of such individuals in these groups. The mean age of those in the approved schools was 14·2 years and the finding of four males with an XYY complement, two with an XXY complement and one XY/XXY mosaic (3 abnormal cells being found in the 30 examined) in the 1,119 boys examined is greater than the newborn frequencies but the differences are not significant. In the Borstals (age range 15–21 years) and the Prison Allocation Centre (mean age 30·7 years) the finding of two males with an XXY complement and one with an XYY complement in the 908 individuals examined is again in good accord with the newborn frequency. The 510 prisoners selected because they were over 5 feet 10 inches tall yielded only two XYY males. 149 of these prisoners were over 6 feet tall as were both the XYY males but, using the previously described calculation,

	Tall prisoners		Total
	510		2,538
0	47,XXY 47,XYY	2 2	14
0		0	0
0		0	2
XY,D/D 1 XY,D/G 1	46,XY,t(1 ?− ;16q+) 1		7
0	47,XY,mar+	1	2

we would expect about 1·0 men of the 149 men over 6 feet to be XYY (6·7/1,000) on the basis of the newborn frequency alone and hence the finding of two such men cannot be regarded as significant.

In these groups seven men were found with balanced autosomal translocations, an incidence of 2·75/1,000 a finding similar to that in the presumed normal population, but again somewhat higher than that observed in the newborn.

Considered as a whole the findings are perhaps suggestive of an increased frequency of XYY males in approved schools and penal institutions when compared with the frequency observed in the normal populations. However it is difficult to assess any differences as the sample of tall atomic energy workers included in the normal group may not be truly representative. Unfortunately if we exclude these men the sample becomes very small, consisting of less than 500 men, which is just not large enough when an abnormality with an incidence at birth of about 1·5/1,000 is being studied.

The height distribution by age of the boys in approved schools and the heights of those boys with sex chromosome abnormalities are shown in Figure 2. The mean height is shown at each age, together with 95 per cent confidence limits. It can be seen that in this population the XYY males again tend to be taller than the average for their age but no effect is apparent in the other sex chromosome abnormalities.

[175

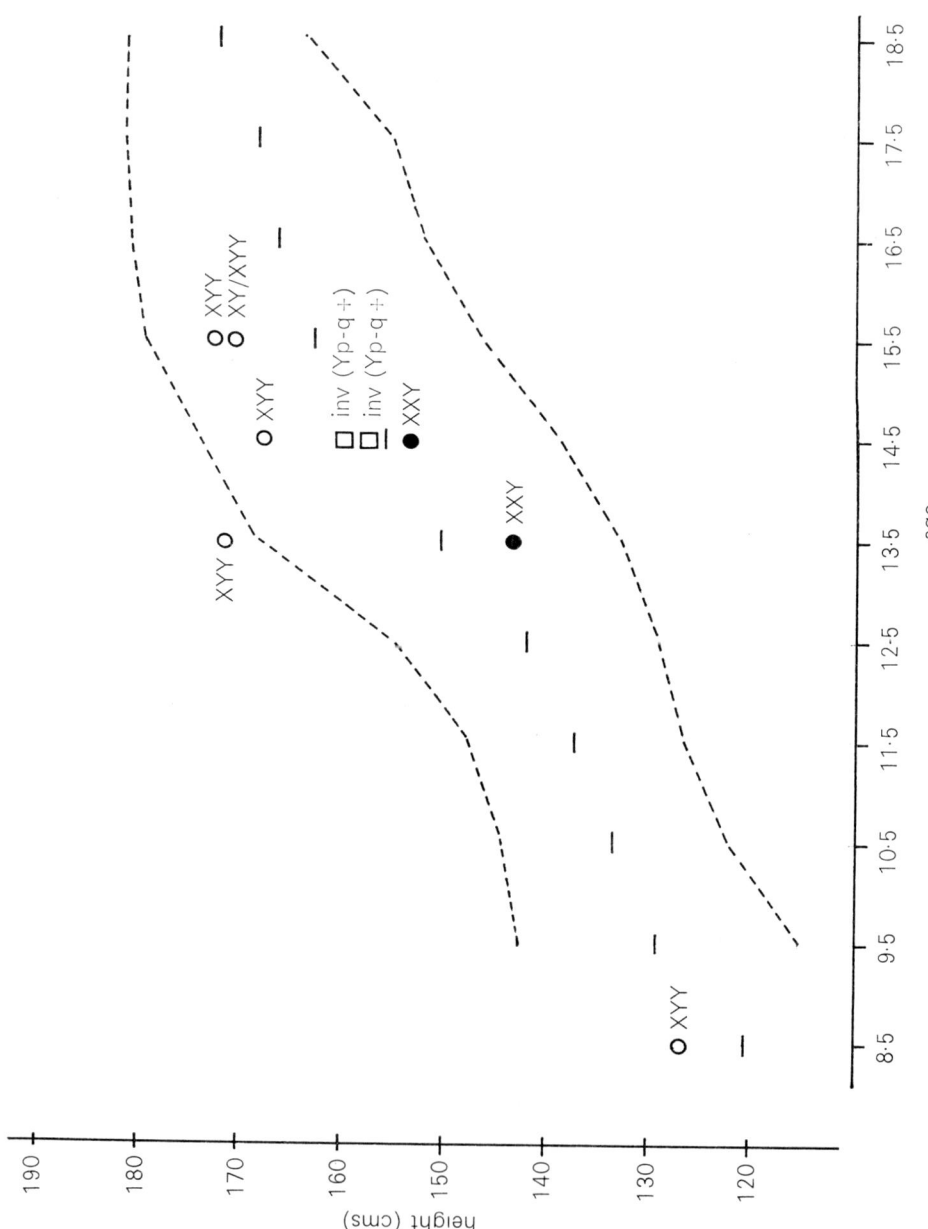

FIGURE 2
Height by age in approved schools and those with sex chromosome abnormalities (dotted lines represent 95% confidence limits)

PATIENTS IN MENTAL DISEASE AND MENTAL SUBNORMALITY HOSPITALS

Table 10 lists a number of special hospitals for the mentally subnormal or the mentally diseased in which chromosome surveys have been conducted.

1. *Maximum security hospital.* In 1965 Casey reported finding seven males with an XXYY complement following a nuclear sexing survey of 942 men in two English maximum security hospitals (published 1966) [12]. This finding was markedly different from previous studies on newborn males and in ordinary hospitals for the mentally subnormal. It thus seemed likely that possession of this chromosome complement might in some way influence behaviour such as to increase an individual's chance of being detained in a special hospital for dangerous, violent or criminal patients. If this was the case then it also seemed possible that the frequency of males with an XYY complement would also be increased. Jacobs and her colleagues therefore conducted a chromosome study on the inmates of the State Hospital, Carstairs in Scotland. The detailed findings in the initial survey have been published by Jacobs et al. [10]. Subsequently, blood samples have been obtained from a further 111 new male entrants and we are reporting the chromosome abnormalities found in the total of 426 men and 18 women examined to date.

2. *Court Order patients.* Close and his colleagues [13] suggested that in view of the discovery of an excess of male patients with an XYY complement in the maximum security hospital at Carstairs it was likely that an excess of such persons might also be found in those inmates of a mental subnormality hospital who were sent by the Courts under a Court Order. They therefore conducted a survey in the Darenth Park Hospital for the mentally subnormal which contains a large number of patients who had been admitted under a Court Order. 182 such men were chromosomally examined.

TABLE 10. Surveys of patients in mental disease and mental subnormality hospitals

	No. of cells examined	Males			Females		
		No.	Mean age	S.D.	No.	Mean age	S.D.
Maximum Security Hospital	10	426	32·3	10·6	18	33·2	7·9
Patients detained under Court Order in Mental Subnormality Hospital D	2	182	38·2	15·3	–	–	–
Mental Subnormality Hospital G	2	423	35·0	13·5	–	–	–
Mental Subnormality Hospital P	2	597	34·2	16·1	493	41·0	16·4
Hospital for Epileptics	2	76	31·0	16·0	–	–	–

TABLE 11. Chromosomal findings in survey of patients in mental disease and mental subnormality hospitals

			Maximum Security Hospital		Court Order patients		Hosp for epile
	Number of individuals		426 males 18 females		182 males		76 m
Aneuploid		Sex chromosome	47,XXY 47,XYY 48,XXYY 46,XY/47,XXY/48,XXXY	3 10 1 1	47,XXY 47,XYY	4 3	47,X 47,X
Aneuploid		Autosome		0		0	
Structural abnormalities		Sex chromosome		0		0	
Structural abnormalities		Autosome	45,XY,D/D 46,XY,t(1 ?– ;16 ?+) 46,XY,Gq–	2 1 1	45,XY,D/G 46,XY,t(Cq+ ;Cq–) 1	1	
	Others			0		0	

TABLE 12. Chromosomal findings in surveys of patients in mental subnormality hospitals

			Hospital G males		Hospital P males	
	Number of individuals		423		597	
Aneuploid		Sex chromosome	47,XYY 48,XXYY	2 1	47,XXY 47,XXY/48,XXXY 48,XXXY/49,XXXXY	
Aneuploid		Autosome	47,XY,G+	22	47,XY,G+ 46,XY/47,XY,G+	
Structural abnormalities		Sex chromosome		0		
Structural abnormalities		Autosome	46,XY,t(Cq+ ;Cq–) 1 46,XY,t(1 ?– ;Eq+) 1		46,XY,t(Dq+ ;1 ?–) Unbalanced	
	Others		47,XY,mar+	1		

*46,XY,D/D,G+   1
46,XY,F–,G–,mar1+,mar2+   1
46,XY,D/G,G+   1
46,XY,D–,C+   1
46,XY,Cq–   1

†46,XX,Bp–   1
46,XX,Fq+,mat   1
46,XX,D/G,G+   2
46,XX,Ep–   1

3. *Mental subnormality hospital G.* A study is currently in progress to examine the frequency of karyotypic aberrations amongst the male inmates of the Gogarburn Mental Subnormality Hospital which serves the Edinburgh area. 426 male patients have been examined. The Study was undertaken in order to attempt an assessment of the effects of chromosome aberrations on intelligence.

4. *Mental subnormality hospital P.* The Prudhoe Hall Mental Subnormality Hospital is a similar hospital to G, serving an area in northern England including Newcastle-upon-Tyne and parts of Cumberland, Durham and Northumberland. Both male and female adult patients are being studied and to date 597 men and 493 female patients have been examined.

5. *Hospital for epileptics.* Hambert [14] found 4 chromatin positive males amongst 512 inmates of Swedish residential institutions for epileptics. Following this finding a small study was undertaken of 76 males in a Scottish hospital for epileptics.

The findings in mental disease and mental subnormality hospitals (Tables 11 and 12). Jacobs *et al.* in 1965 reported an excess of males with an XYY complement in the State Hospital, Carstairs. In the 426 males examined to date, ten 47,XYY males, three 47,XXY males, one 48,XXYY and a mosaic 46,XY/47,XXY/48,XXYY, in whom the XXY line is predominant, have been ascertained together with four men with autosomal abnormalities, two D/D translocations and a reciprocal translocation between

POST-NATAL POPULATION

a number 1 and a number 16 chromosome and a presumed unbalanced autosomal abnormality. None of the eighteen detained females showed any chromosome abnormality.

The finding of four XXY's and three XYY's in the 182 Court Order patients is considerably more than expected on the basis of the newborn frequencies, both increases being statistically significant ($P < \cdot 02$). The number of epileptics is small but it is worth noting that an XYY and an XXY were found in this group. The incidence of balanced autosomal translocations is also high in the three groups considered together. Five men being found in the 702 persons examined, an incidence of 7·1/1,000. This is considerably higher than the frequency in the newborn and the finding is discussed later together with the incidence observed in other populations.

The height distribution of the patients in the maximum security hospital is shown in Figure 3 and it may be seen that the patients with sex chromosome abnormalities are considerably taller on average than the rest of the population, the effect being particularly marked amongst the XYY males.

There is no evidence that those with sex chromosome abnormalities in the maximum security hospital are of different intelligence to the rest of the hospital population (Table 13). It should be noted that a large number of the patients in this hospital are in the normal I.Q. range.

TABLE 13. Patients in Maximum Security Hospital

I.Q.	Total sample	XYY	XXY	Others
0–49	18	–	–	–
50–84	238	8	1	XY/XXY/XXXY XXYY
85–114	150	2	2	–
115 +	20	–	–	–

I.Q. distribution of total sample and of individuals with sex chromosome abnormalities

Figure 4 shows the height distribution for the patients admitted under Court Order to mental subnormality hospital D. Again the men with an XYY complement tend to be taller than the other patients as do the XXY's. Both XYY and XXY patients tend to have a higher I.Q. than the rest of the patients (Table 14). This finding is contrary to that at the State Hospital, Carstairs but in the mental subnormality hospital the overall distribution of I.Q. is much lower, and relatively few men have an I.Q. approaching the normal range found in the general population.

Table 12 shows the abnormalities found in the two mental subnormality

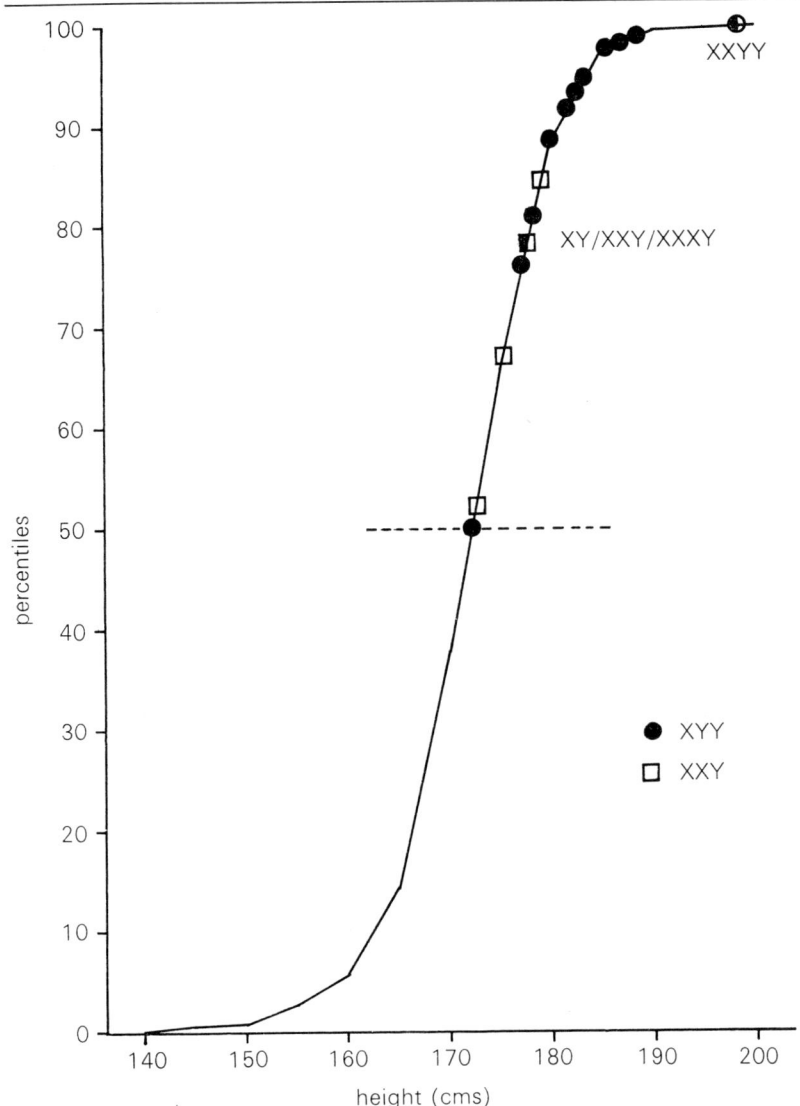

FIGURE 3
Height distribution of patients detained in the maximum security hospital and of those with sex chromosome abnormalities

hospitals; in Hospital G two 47,XYY males were found and one 48, XXYY male in 423 patients. It is of relevance that both the XYY patients were found in the hospital's maximum security ward containing 53 men, whereas Hospital P does not have such a ward. There is an apparent absence of males with extra X chromosomes in the Hospital G. The finding of 1/423 (2·4/1,000) contrast markedly with the 38/4,015 (9·5/1,000) found by

POST-NATAL POPULATION

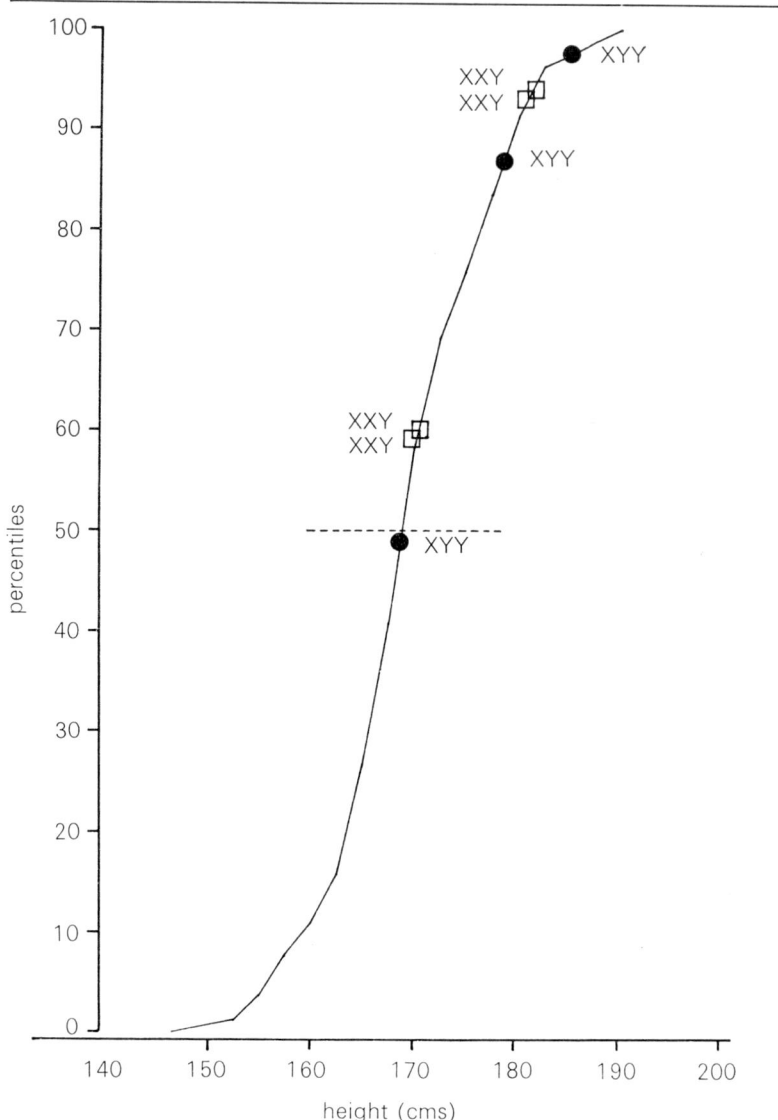

FIGURE 4
Height of Court Order patients in mental subnormality hospital D and of those individuals with sex chromosome abnormalities

Maclean [15] in Scottish mental subnormality hospitals, but the difference is not statistically significant.

There is a large difference in the incidence of mongols in the two hospitals and the reasons for this are not apparent; we can only speculate that it reflects a difference in the admission policies of the hospitals. There are four individuals with presumptively balanced autosomal translocations in

TABLE 14. Patients detained under Court Order in Mental subnormality Hospital D

I.Q.	Total sample	XYY	XXY
0–19	16	–	–
20–49	94	–	–
50–69	57	1	1
70–84	11	2	2
85 +	1	–	1
Not known	3	–	–
Total	182	3	4

I.Q. distribution of total sample and of patients with sex chromosome abnormalities

the 1,513 patients but to these we might add, for purposes of comparison with other groups, the male who had a D/D translocation and also an extra G chromosome. We clearly cannot do the same for the individuals with D/G translocations and an extra G group chromosome as these two conditions are known to be closely associated. The incidence of presumed balanced autosome translocations is then 3·30/1,000, an incidence of the same order as that in other adult groups.

In Hospital P, three males and three females had chromosome complements which are presumed to be unbalanced (other than those with trisomy 21) whereas no such individuals were found in Hospital G. This difference may represent the play of chance but the two hospitals do appear to differ in a number of respects regarding the chromosomal findings.

The I.Q. distribution in the Maximum Security Hospital amongst the Court Order patients and in Mental Subnormality Hospital P is shown in

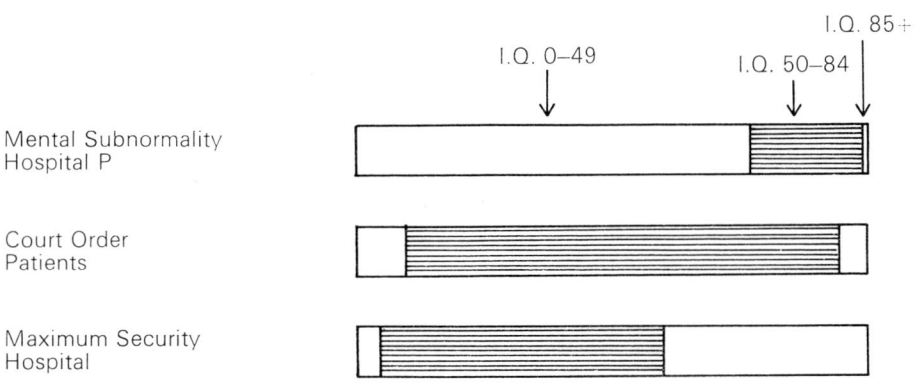

FIGURE 5
I.Q. distribution in mental subnormality and mental disease hospitals

Figure 5. It has been previously noted that males with an XYY chromosome complement do not appear to show a marked reduction in I.Q. and thus the finding of no XYY males in Hospital P may not be a very surprising finding. However, this did suggest that a difference in the I.Q. distributions between the two mental subnormality hospitals might help to explain the differences observed. Unfortunately information was not available on the I.Q.s of the men in Hospital G.

DISCUSSION

Sex chromosome abnormalities (Table 15). There is no doubt that the frequency of males with an XYY chromosome complement is considerably greater in the Maximum Security Hospital than in the newborn population and this indicates that at least some males with this abnormality are predisposed towards types of behaviour which necessitate that they be kept under conditions of strict security. However, there have been a number of reports of XYY males who are living an apparently normal life at large in the general population manifesting no obvious antisocial tendencies. It is, therefore, very important in assessing the genetic risk to determine the proportion of XYY males who have behavioural problems. A search through the Scottish prisons, borstals and approved schools has brought to light only seven XYY males in over 2,500 examined men, some of whom were selected on a height basis. This frequency is only a little higher than that which would be expected from the incidence observed amongst the newborn (the expected number in these groups would be about 4·2) and the difference is not statistically significant. It may be that men with an XYY complement are segregated out to mental subnormality and mental disease hospitals and this would explain the finding of three such patients

TABLE 15. Summary of sex chromosome and unbalanced autosome abnormalities (excluding trisomies)

	Baby survey (/1000)	Normal groups*	Hospital patients	Socially deviant groups
Males	7530	1111	929	2538
XXY	12(1·59)	5	(1m)	6(-)
XYY	10(1·33)	0	(1m)	7
XXYY	0	0	0	0
Y(p+q-)	4(0·53)	1	1	2
Other sex chromosomes	1	0	0	1
Males and females	9627	1495	1632	2538
Presumptively unbalanced autosomal (excluding trisomies)	4(0·42)	0	0	0

*Includes individuals selected on height basis   m Indicates mosaic

amongst 182 males admitted to a mental subnormality hospital under Court Order. The evidence to date suggests that the possession of an extra Y chromosome does not have a very marked effect on intelligence and thus it might be expected that only high grade patients in mental subnormality hospitals would yield an excessive number of XYY's. Supportive evidence has recently been obtained by Jacobs and her colleagues [16] who have studied 343 tall men in mental disease and mental subnormality hospitals in Scotland and have found eight XYY's, two XXYY's and one XY/XYY. However, there is still a lack of satisfactory control data from the non-institutionalised 'normal' population, which necessitates that the data from prisons, borstals and approved schools be interpreted cautiously. If the groups taken to be representative of the normal populations are combined with the surveys of hospital patients only a 45,X/47,XYY mosaic has been found in the 2,040 men examined. 629 of these men were selected on the basis of height and, taking account of this, it may be estimated from the newborn frequency that about 6·1 XYY men are to be expected in the total sample. The deficit may be explained by postulating that

(i) the sample of 2,040 men is biased against including men with an XYY complement who are living in the community

or (ii) persons with the abnormality are segregated from the normal population into special institutions.

or (iii) possession of the abnormality leads to increased mortality risks.

If (iii) is the case then the findings in approved schools, borstals and prisons may need to be reassessed, although at present there is little to suggest an increased mortality rate amongst XYY males. Reasons have been advanced in support of (i) and there is clearly still an urgent need to examine more non-institutionalised populations.

ximum curity spital	Court Order patients	Mental subnormality hospitals and epileptics
	182	1096
3 (+ 1m)	4	3
	3	3
	0	1
	0	0
	0	2
	182	1589
	0	6

N

TABLE 16. Summary of autosomal abnormalities (incidence/1000)

		Baby surveys	'Normal' groups	Hospital patients	Social deviant groups
Males and females		12,027	1,495	1,632	2,538
Presumptively Balanced Translocation	D/D	6	3	3	1
	D/G	1	0	0	1
	Others	5	1*	3	5
	Total	12 (1·00)	4* (2·67)	6 (3·68)	7 (2
mar+		3	0	1 (+1m)	1

* Does not include one mosaic 46,XX/46,XX,t(Dq+;Dq−)

In conclusion it might be said we still have much to learn about the selective mechanisms which are operating on individuals with an XYY complement and it is not possible to assess yet the probability that a carrier will suffer behavioural disorders.

In groups other than the newborn four males were found with pericentric inversions of the Y chromosome amongst 6,282 men examined. Combining this result with that from the newborn (4/7,530) gives an estimated frequency, in the general population, of this abnormality of 0·58/1,000.

Autosomal abnormalities. Presumptively unbalanced autosomal structural abnormalities would seem to predispose towards mental subnormality as all seven of the individuals found in the adult populations were in institutions catering for the mentally subnormal.

Table 16 summarises the findings relating to presumptively balanced autosome translocations and marker chromosomes.

The frequency of structural heterozygosity in the newborn is estimated to be 1·00/1,000. In all of the adult groups studied the frequency is higher than this and if the incidences in the different groups are considered together the variation between them is highly significant ($\chi^2_4 = 23·8$; $P < ·001$). There are a number of possible explanations for this finding:

(i) persons carrying such translocations may have a lower mortality rate than normal individuals

(ii) the birth incidence may have decreased with time and is lower now than it was when the individuals in the other populations were born, although this seems an unlikely explanation

(iii) possession of such abnormalities may predispose towards entry into one of the special groups studied

(iv) a biased estimate of the newborn frequency may have been obtained by pooling the data from the four newborn surveys.

We are only in a position to examine the last two of these possibilities. In the Edinburgh survey of livebirths, two balanced autosomal translocations

Mental disease and mental subnormality hospitals	Total
2,215	19,907
3	16 (0·80)
2	4 (0·20)
5	19 (0·95)
10 (4·51)	39 (1·96)
5	10 (+1m)

have been found in the 3,146 babies examined, an incidence less than that estimated from all of the surveys combined of 0·63/1,000. However, if all the groups studied in Edinburgh are tested for homogeneity, including the newborn, the differences observed are not statistically significant ($\chi^2_4 = 8 \cdot 42$; $P < \cdot 10$). Thus it is possible that the low incidence in the newborn *may* be a chance finding but it will be necessary to collect more data, particularly on the newborn population to ascertain whether there are geographic or ethnic differences in the birth incidence. It is not known, for example, how representative the Edinburgh newborn population will be of Scotland as a whole.

A second possibility is that the balanced structural autosomal abnormalities do predispose a carrier toward entering the special populations studied. However, the incidence in the presumed normal population is not very different from that in these groups and there would seem to be an urgent need to expand the size of the sample of the normal population to enable risks associated with carriers to be better assessed.

None of the preceding explanations seems particularly convincing, and on the face of it this result would seem to fall into the category which Professor Edwards has called 'statistically significant but biologically absurd'.

The finding of five individuals, each with an extra small chromosome amongst 2,215 patients in mental hospitals represents a significant increase over the three such individuals found amongst 9,627 newborn males and females (the findings from the Boston survey have been excluded as babies with congenital abnormalities were not studied in this survey). The five adult carriers in mental hospitals were all in mental subnormality hospitals and it seems likely that possession of an extra chromosome of this kind may affect intelligence. However, it is important to distinguish between those marker chromosomes which represent new mutations, as it may be that intelligence is only affected amongst the latter. Unfortunately the data presented in Table 16 is non-informative in this respect. Useful familial data has only been obtained for two of the five individuals in mental subnormality hospitals with marker chromosomes and in both these cases the abnormality was present in neither parent. It has not been possible to obtain any information on transmission for the two cancer patients or for the prisoner with an abnormal extra chromosome. Walzer et al. [6] report that in two of the cases found in the Boston newborn survey, the abnormal chromosome had been transmitted by a parent and for the other it had not.

SUMMARY

Data is presented from a number of cytogenetic studies on adult populations which have been conducted by members of the Medical Research Council's Clinical and Population Cytogenetics Research Unit. The populations studied include penal institutions, mental disease and mental subnormality hospitals and a number of groups of individuals who might be considered to be reasonably representative of the 'normal' population. In total, chromosome studies are reported on nearly 8,000 persons drawn from groups other than the newborn. A comparison is made of the frequencies of the different types of chromosome abnormality between populations in an attempt to establish the nature of any special risks associated with the possession of an abnormal chromosome constitution. An increased incidence of males with an XYY karyotype is observed amongst patients in a maximum security hospital and in patients admitted under Court Order to a mental subnormality hospital. The incidence of males with an XYY complement in approved schools and penal institutions is not significantly greater than that in the newborn population but interpretation of this finding is made difficult by an apparent deficit of such males in the non-institutionalised adult populations studied.

It is estimated that the incidence of carriers of balanced autosomal translocations in the general population, excluding the newborn, is about $3.4/1,000$. However, the incidence in the newborn is found to be considerably less than this ($1.0/1,000$). We can suggest no very satisfactory explanation of this difference and it may represent a chance finding.

The incidence of males with a pericentric inversion of the Y chromosome is estimated to be approximately $0.6/1,000$. The seven individuals found with presumptively unbalanced autosomal abnormalities in populations other than the newborn were all in hospitals catering for the subnormal, and mental subnormality would seem to be associated with possession of these aberrations. An increased incidence of individuals with an extra marker chromosome was also found amongst the mentally subnormal but it is not known what proportion of these patients had inherited the abnormal chromosome through their parents. It is important to obtain this information as it would seem likely that the effect of inherited and non-inherited marker chromosomes might be different.

ACKNOWLEDGEMENTS

Although Dr Jacobs and I are formally the authors of this paper, it really represents the work of the late Professor W. M. Court Brown and many members of the Edinburgh Research Unit. They are too numerous to name individually but we would particularly like to mention Dr W. H. Price who has been the clinician responsible for conducting a number of the surveys discussed and who has helped us greatly in the preparation of this paper. Some of the data presented have been previously published and some of the surveys to be discussed are not yet quite complete, but the detailed findings will be published elsewhere at a later date.

REFERENCES

[1] Maclean, N., personal communication, 1969.
[2] Matsunaga, E. In *Proceedings of the World Population Conference,* Belgrade, vol. 2, p. 481. New York : United Nations.
[3] Bochkov. N. P. In *Proceedings of the Symposium on the Mutagenic Process,* Prague, p. 121. Prague : Academia, 1965.
[4] Court Brown. W. M., *Int. Rev. exp. Path.,* **7,** 31, 1969.
[5] Sergovich, F., Valentine, G. H., Chen, A. T. L., Kinch, R. A. H. and Smout, M. S., *New Engl. J. Med.,* **280,** 851, 1969.
[6] Walzer, S., Breau, G. and Gerald, P. S., *J. Pediat.,* **74,** 438, 1969.
[7] Lubs, H. A., personal communication, 1969.
[8] Court Brown, W. M., Buckton, K. E., Jacobs, P. A., Tough, I. M., Kuenssberg, E. V. and Knox, J. D. E., *Eugenics Laboratory Memoirs No. 42.* London : Cambridge University Press, 1966.
[9] Stewart, J. S. S., Mack, W. S., Govan, A. D. T., Ferguson-Smith, M. A. and Lennox, B., *Quart. J. Med.,* **28,** 561, 1959.
[10] Jacobs, P. A., Price, W. H., Court Brown, W. M., Brittain, R. P. and Whatmore, P. B., *Ann. hum. Genet. Lond.,* **31,** 339, 1968.
[11] Tanner, J. M., Whitehouse, R. H. and Takaishi, M., *Arch. Dis. Child.,* **41,** 454 and 613, 1966.
[12] Casey, M. D., Segall, L. J., Street, D. R. K. and Blank, C. E., *Nature, Lond.,* **209,** 641 1966.
[13] Close, H. G., Goonetilleke, A. S. R., Jacobs, P. A. and Price, W. H., *Cytogenetics,* **7,** 277, 1968.
[14] Hambert, G., *Males with positive sex chromatin.* University of Göteborg I. Scandinavian University Books, 1956.
[15] Maclean, N., Court Brown, W. M., Jacobs, P. A., Mantle, D. J. and Strong, J. A., *J. med. Genet.,* **5,** 165, 1968.
[16] Jacobs, P. A., unpublished data, 1969.

Population Cytogenetics and Environmental Factors

H. J. EVANS

Department of Genetics
University of Aberdeen

¶IN CONSIDERING the influence of environmental factors in population cytogenetics, the geneticist may be concerned with one or both of two aspects. Firstly the influence of change in the environment in increasing, or decreasing, the incidence of chromosome structural change and the frequency of recombination. Secondly, with modifications of the environment that alter selection pressure and breeding habits which can sometimes be distinguished at the cytological level.

The first aspect of change, often referred to as primary change, is well exemplified in animal or plant populations exposed to mutagenic agents which induce chromosome structural changes : these primary changes can be studied in both mitotic and meiotic cells. The second, or secondary phenomena, have been extensively studied in populations of plants, invertebrates and indeed mammalian species : these changes are most often, but not invariably, studied at the meiotic level.

My concern here is with the first aspect of environmental modification and my brief is to consider the kinds of changes that can be induced in human chromosomes, the frequencies with which they occur and the nature of some of the environmental factors that we know or suspect to be causative agents. These environmental factors can be considered to fall into three classes : physical factors; chemical factors; and biological factors. However, before considering the effects of some of the agents that fall into these categories I would like to say a few words about the cell systems that we use to study the effects of mutagens on human chromosomes and consider the kinds of aberrations that we observe and their structure as influenced by the stage of cell development at the time of exposure.

The cell systems used

Studies on the actions of mutagens on mammalian chromosomes have been carried out on a variety of cells that are either normally involved in proliferation in the body, or can be made to proliferate *in vitro*. In the case of man, three sources of cells have been utilised, viz. skin, bone marrow and blood. For a variety of reasons skin and bone marrow cells are far from ideal materials for assessing the effects of environmental agents and they are of course quite impossible systems in relation to population studies necessitating the sampling of very large numbers of individuals. On the other hand peripheral blood leucocytes cultured by the method of Moorehead *et al.* [1], or some modification thereof, are ideal cells for study. Only small quantities of blood are required to yield large numbers of mitotic cells of excellent cytological quality [2] so that samples can be obtained easily, painlessly, and frequently from large numbers of individuals.

The leucocyte cells of interest are of course the small lymphocytes [3, 4, 5] and these cells offer two useful and indeed important advantages when we come to consider the action of mutagens. The first of these is that the small lymphocytes, at least in man, are long-lived cells [6–9] and significant yields of chromosome aberrations can be observed in these cells many years after their exposure to a mutagen. For example, there are numerous reports from Japan on aberrations in blood cells sampled from individuals up to 22 years after exposure to radiations from the Hiroshima and Nagasaki atom bombs [10]. The late Professor Court Brown and his colleagues at Edinburgh [11], and Norman and his group at Los Angeles [12], made use of the facts that many of the cells carrying chromosome aberrations in individuals exposed some years previously to radiation, appeared to be in their first mitosis post-irradiation when set up in *in vitro* culture and that these cells decreased in frequency with time after exposure. On the basis of these facts it was shown that the mean life span of these lymphocytes was somewhere in the region of 500 to 1,500 days.

The second advantage offered by the long lived lymphocytes is the fact that on the average they rarely undergo proliferation within the body. Exposure of unstimulated blood lymphocytes to H^3-thymidine reveals that normally less than one cell in a thousand undergoes DNA synthesis [3, 9, 13] and that virtually all these cells rest in an early interphase or G_1 stage [14, 15]. All the cells are therefore in the same stage of development, so that the complications of differential response to a mutagen due to differences in the sensitivity of different cell stages should not arise.

The types of chromosome anomalies induced and the influence of cell stage and time of observation after exposure

In very general terms, the chromosome anomalies that we observe are of two sorts; those that result in alterations in chromosome number giving aneuploidy or polyploidy, and those that result in structural changes within and between chromosomes. Alterations in chromosome number may result from errors in chromosome separation at anaphase and these errors are themselves quite often due to the presence of structural re-arrangements in the chromosomes involved. What I wish to consider now are the structural re-arrangements, since they can tell us something about the mode of action of environmental mutagens.

The types of structural re-arrangements induced in cells exposed to mutagens depend upon three factors: (i) the kind of mutagen that is used, (ii) the stage in the mitotic cycle of the treated cells at the time of their exposure, and (iii) the time of sampling of the cells after their exposure, i.e. whether the cells are observed at the first post-treatment mitosis or at subsequent divisions.

If we expose an asynchronously developing proliferating cell population to a mutagen such as ionising radiation, and then observe the cells as they come into their first post-treatment metaphase, the first cells that we see which contain aberrations are cells that were exposed in early prophase.

These cells contain structural changes in which the unit of chromosome that is broken and exchanged is a sub-unit of a chromatid; the aberrations are therefore referred to as sub-chromatid type aberrations and they are of little concern to us today. Following upon the cells with sub-chromatid aberrations are cells with chromatid-type changes. In these cells the unit of breakage and exchange is the whole chromatid, and chromatid aberrations are characteristic of cells irradiated either in the DNA synthesis (S) period or post-DNA synthesis (G_2) period or interphase (Fig. 1). Cells exposed to radiation in the early interphase of G_1 period have chromosome-type aberrations in which the unit of breakage and exchange is the whole chromosome, both chromatids being involved at identical loci in any breakage or exchange process.

The separation of the aberrations into these categories is well established [16] and dates back to work carried out on plant cells in the 1930s. Chromatid-type and chromosome-type aberrations are of course basically similar and although there is a slightly greater variety of chromatid type changes, the difference between these aberrations is simply a consequence of breakage and exchange either preceding or following after the process of chromosome replication (Fig. 2).

In the case of chromosome-type aberrations we can distinguish seven different kinds (Fig. 3), not all of which can be scored with equal efficiency in somatic cells. The asymmetrical, dicentric and centric ring aberrations can be easily scored in good preparations with an efficiency approaching 100 per cent, but the efficiency of scoring deletions will depend upon the size of the deleted fragments. In the case of symmetrical aberrations the efficiency of their detection ranges from zero to a maximum of 20 per cent. Inversions within a chromosome arm (paracentric) and inversions across the centromere (pericentric) that involve exchange points equi-distant from the centromere, cannot be detected in somatic cells. There is considerable evidence that symmetrical (reciprocal translocations) and asymmetrical (dicentrics) interchanges should occur with about equal frequency [17, 18], but examination of published data on human peripheral blood cells from irradiated subjects [11, 19–21] reveals that dicentric and ring aberrations are at least five times as frequent as 'abnormal monocentric' chromosomes (symmetrical inter- and intra-changes).

At the chromatid aberration level, the efficiency of detection of aberrations is very much greater and the efficiency of scoring asymmetrical and symmetrical interchanges approaches 100 per cent. This follows because the close pairing between sister chromatids holds together the association of chromosomes involved in an exchange.

Some of the aberrations, e.g. the dicentrics, may be lost or result in a

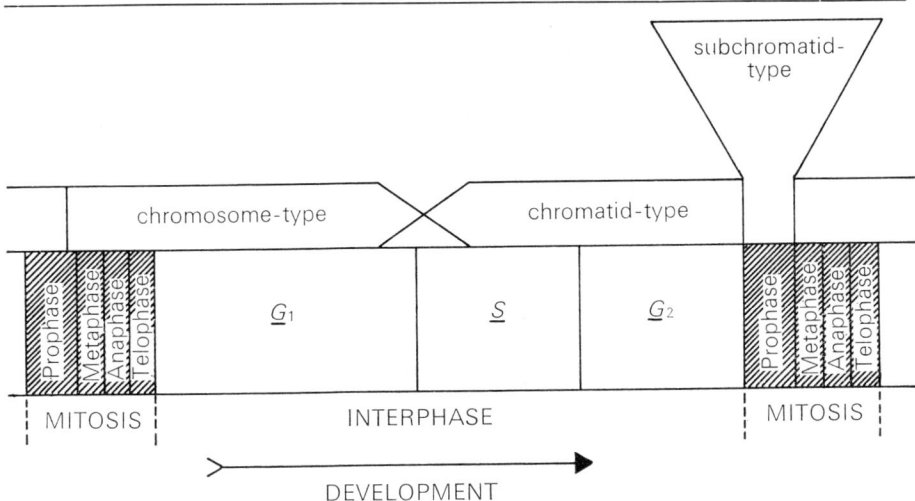

FIGURE 1
The relation between type of aberration induced by radiation and stage in the cell cycle at the time of exposure

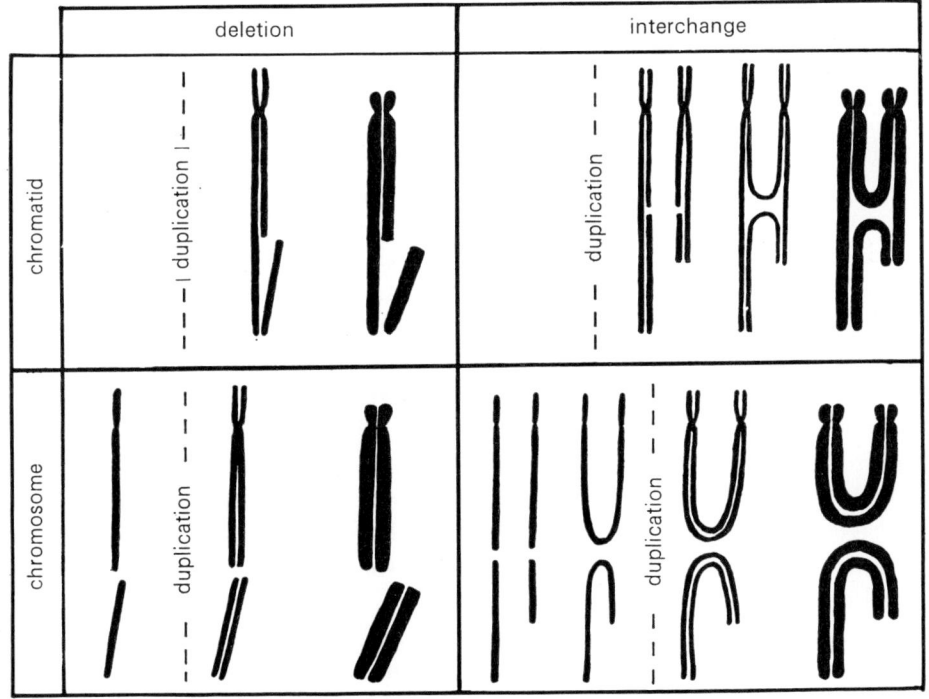

FIGURE 2
X-ray-induced chromosome and chromatid aberrations. Diagrammatic representation of the effect of time of duplication in influencing aberration type

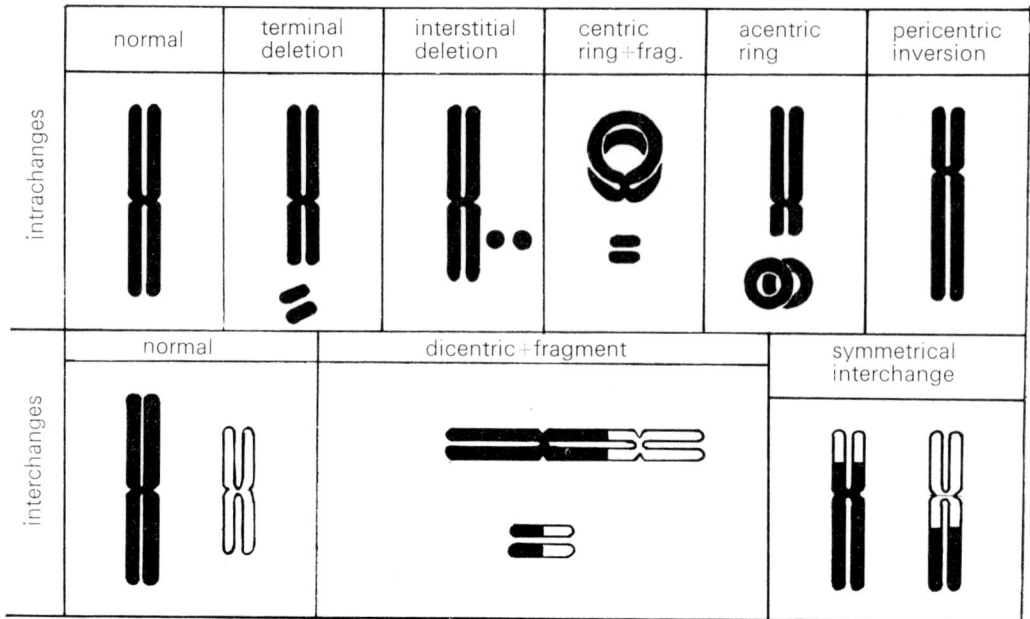

FIGURE 3
The seven kinds of chromosome-type aberrations that may be distinguished, with varying efficiency (see text), at metaphase of mitosis

mechanical failure of separation at anaphase and, if the cells are still viable, yield aneuploid or polyploid cells that can be distinguished as such at the next mitosis. However, aberrant chromatids that disjoin at anaphase, when observed at the second mitosis following their induction, will have undergone a further round of replication and appear as 'derived chromosome-type' changes. Thus a proportion of any of the chromatid-type aberrations produced by the action of a mutagen will, at the second mitosis following their induction be indistinguishable from 'primary' chromosome-type aberrations (Fig. 4). I wish to stress this rather obvious point, since it is of importance in considering the possible modes of action of certain environmental factors that appear to act as chemical mutagens. I should of course at the same time point out that the true nature of a proportion of the 'derived' chromosome-type changes will be clearly evident, e.g. by the absence of an accompanying fragment in the case of a dicentric chromosome.

Before going on to examine some of the data on the frequencies of these aberrations there is one further point of methodology that needs emphasis. A large number of laboratories that use peripheral blood leucocytes for chromosome studies culture the cells for 72 hours before fixation and slide preparation and this three days culture time has been used in many studies

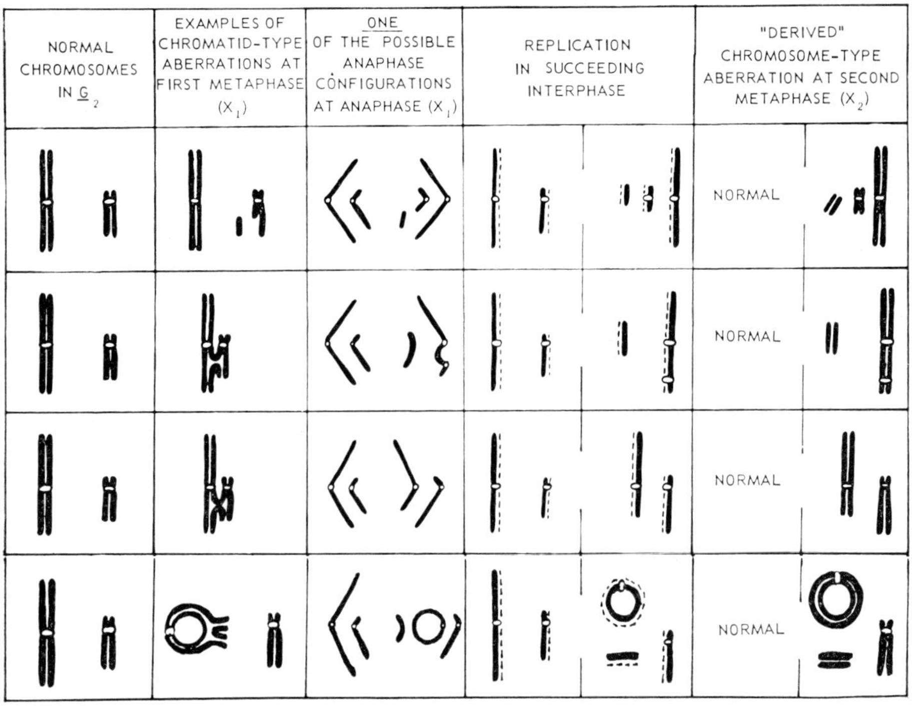

FIGURE 4

Diagrammatic illustration of examples of 'derived' chromosome-type aberrations, observed at the *second* mitosis (X_2) after exposure to a mutagen. The aberrations figured at the first division (X_1) are of the chromatid-type and only a limited number of the possible anaphase configurations are shown. It should be noted that in many instances acentric fragments may be lost and will not appear at the X_2 division with the chromosomes from which they were derived

on the induction of chromosome aberrations by mutagens. Unfortunately, it is known that in good cultures maintained at 37°C, only a proportion of the cells seen in mitosis at around 72 hours are in their first mitosis in culture; indeed at this time many of the cells are in their second and a small proportion may even be in their third division in culture.

Accurate information on the number of chromosome aberrations induced by a given exposure or treatment can of course only be obtained if the aberrations are scored at the first cell division following their induction. The customary practice used to obtain cells in their first mitosis in culture is to sample at around 48 hours and this problem of the influence of culture sampling time on aberration yield was considered in some detail at a symposium held here in Edinburgh two years ago [22]. However, in considering sampling time, we also need to bear in mind that a proportion or in some cases all, of the apparent chromosome-type changes observed in three or four days cultures may be 'derived' chromosome-type aberrations.

[197

This, as we shall see later, may be of particular importance in relation to some of the reported findings on the actions of certain chemical agents and viruses.

So much then for the types of chromosome aberrations that we have to consider. However, before turning to examine some of the data that we have on the environmental agents that can induce these changes, we need firstly to know something about the background levels of such changes in peripheral blood lymphocytes of so-called 'normal healthy' human beings.

The spontaneous frequencies of aberrations

There are many reports which quote aberration frequencies in peripheral blood lymphocytes of individuals who have not been unduly exposed to radiations or to other known environmental mutagens, and the reported frequencies (often, unfortunately, referred to as total breaks) vary widely between laboratories.

One of the problems which is associated with this variation is the difficulty of distinguishing between achromatic lesions, or gaps, in the chromatids, which are not discontinuities or breaks but are non-staining zones, and true chromatid or chromosome deletions. The real criteria for distinguishing between gaps and breaks can at present only be defined through observing cells at both metaphase and anaphase, and in this way determining the types of metaphase aberrations that release free fragments at anaphase. To my knowledge there are no reported attempts at defining these criteria in human cells and only a few laboratories have adhered to the criteria adopted by cytologists studying aberrations in plant chromosomes.

In addition to the difficulties of interpretation in relation to gaps and true breaks, there are two other problems of importance:

(i) Firstly, it is known from studies on plant chromosomes [16] that gaps are weak points on the chromosomes that are very prone to mechanical breakage during cytological processing. The same is also true for normal secondary constrictions in both plant and human chromosomes, so that some of the variations in the frequency of 'breaks' between laboratories may be a consequence of differences in cytological technique, including pre-fixation treatments, fixation itself and, in particular, the methods used in slide preparation.

(ii) Secondly, since the small lymphocyte is normally in a 'resting' G_1 stage in the body, then it must be recognised that any chromatid type aberrations that we see in cultured small lymphocytes must have been produced during the S or G_2 phases of development that the cells proceed through in culture. These aberrations result from errors in replication or repair, although the lesions that are the initial cause of these errors may well have been present in the unreplicated G_1 chromosomes – as has been demonstrated in the case of plant chromosomes exposed to alkylating agents [23, 24]. What is of great importance in this context is the fact that if we sample cells in their second, or later, mitosis in culture, then any chromatid-type aberrations that we see, although possibly reflecting in some degree chromosome damage that may have been present in the cells *in vivo*, may reflect (a) the action of agents present in the blood *in vivo* and carried over into the culture where they will exert their effect, (b) the actions of any mutagenic agents inadvertently introduced into the culture medium, and (c) the effect of any other adverse culture conditions.

Although it is well known that a few rare syndromes in man, such as Bloom's syndrome [25, 26] and Franconi's anaemia [27, 28] are often associated with a high frequency of chromosomal aberrations in blood lymphocytes in culture, I believe that most of the variation that we observe in populations, particularly in the case of chromatid-type aberrations, is a consequence of the factors that I've outlined. I should mention in this context, that there are some reported cases [29, 30] of apparently normal healthy individuals with extremely high 'spontaneous' aberration frequencies (up to 20 per cent aberrant cells) in their cultured blood lymphocytes, the majority of the aberrations being achromatic lesions or simple deletions of one particular chromosome in the C group. In addition to these specific cases, we are all of us aware that occasionally in our laboratories, at any one given time, a number of cultures of unrelated bloods will give high yields of chromatid-type aberrations and this is often believed to be a consequence of viral contamination of the culture.

If we exclude the abnormal situations that I've referred to, then the kind of variation that one observes in the frequencies of 'spontaneous' chromatid-type aberrations is shown in Table 1. It may be seen from this table that the frequency of gaps ranges from around 0·01 per cell to around 0·1 per cell, but that the apparent deletions span a range of almost 200 (0·0033 per cell to 0·594 per cell). Interestingly enough, in all cases the frequency of interchange, i.e. the triradials, quadriradials, etc., is less than 0·001 per cell, and, taking into account all these data, a reasonable estimate of the frequency of interchange is that it is present once in around 3,000 cells. Before leaving chromatid-type aberrations, I should mention that the frequency of 'spontaneous' chromatid-type changes in comparable cultures maintained at 37°C is often greater in cells sampled after 72 hours as opposed to after 48 hours in culture.

In the case of chromosome-type changes, it is clear that the spontaneous frequency of these aberrations is very much lower than that of their chromatid counterparts. Furthermore, since all derived chromosome aberrations appear as chromosome-type, one suspects that the reported observed frequencies of chromosome-type changes in three-day cultures are higher

ENVIRONMENTAL FACTORS

TABLE 1. Frequencies of gaps and chromatid aberrations (per cell) in leucocytes from healthy donors

Authors	No. of individuals	No. of cells	Culture time	Chromatid gaps	Isochr gaps
Lubs and Samuelson [34]	10	3,720	72 hrs.	0.04	
Mouriquand et al. [131]	90	1,000	72 hrs.	0.077	
Court Brown [35]	414	12,420	48 hrs. 72 hrs.	0.022	0.007
Adams and Evans [132]	13	1,836	50 hrs.	0.0082	0.001

TABLE 2. Frequency of dicentric aberrations in 'normal' subjects not exposed to radiation other than routine diagnostic exposure

Authors	Norman et al. [42]	Ishihara and Kumatori [52]	Evans and Speed [129]	Norman [130]	Bloom et al. [49]
Number of individuals in sample	23	3	1,327	?	94
Number of cells scored	5,784	2,875	21,232	2,295	88,47
Number of † dicentrics observed	0	0	7	0	0
Frequency of dicentrics per cell	<1 in 5.8 × 10³	<1 in 2.8 × 10³	1 in 3.03 × 10³	<1 in 2.3 × 10³	<1 in 8.8 × 1

*The 47 individuals in column c were patients with ankylosing spondylitis, and the samples were prior to any therapy but *shortly after* the individuals had received diagnostic radiation exposures. T individuals in column b were individuals from a general population and served as controls for popul.
†Total of all samples excluding column c gives 16 dicentrics in 64,023 cells, i.e. 1 in 4,000

Chromatid deletions	Chromatid interchanges
0·594	<0·001
0·057	0
0·0033	<0·001
0·0038	0

	Court Brown [35]*		Sasaki and Miyata [53]
	b	c	
438	38	47	11
420	1,060	2,269	9,510
7	0	3	2
×10³	<1 in 1×10³	1 in 0·76×10³	<1 in 4·7×10³

than the spontaneous frequency that might be expected in first division cells.

A number of authors have reported the complete absence of chromosome-type changes, but the presence of chromatid-type changes, in samples of around 1,000 cells or more [31, 32], whereas others report frequencies ranging from 1 in 1,000 cells [e.g. 33] to around 1 in 100 [34]. However, what stands clear from all the observations that have been reported, is the fact that in individuals that have not been unduly exposed to radiations or other mutagens, all observers have noted that the frequency of the dicentric and ring exchange aberrations is extremely low. It is for this reason that many authors interested in the effects of mutagens – and in particular ionizing radiations – tend to emphasise their results in terms of the frequencies of dicentric aberrations. The rarity of dicentric aberrations (around 1 in 4,000 cells), is clearly illustrated by the eight sets of data summarised in Table 2.

Finally, before leaving the question of spontaneous aberration frequencies, I should mention that work here in Edinburgh [35] indicates that neither sex nor age has much influence on the spontaneous frequency of chromatid structural changes (Table 3) although recent Russian work [36] has disputed the fact that age may be unimportant.

Let us now turn to consider briefly the kinds of mutagens that can induce chromosome structural changes in human peripheral blood lymphocytes.

TABLE 3. 'Spontaneous' frequency of chromatid lesions in peripheral blood leucocytes of males and females of various ages in a general population [35]

Age group	Total cells	Chromatid gaps*	Isochromatid gaps*	Chromatid breaks*	Chromatid interchanges*
15–24	1050	0·024	0·011	0·002	0·000
25–34	1380	0·030	0·005	0·004	0·002
35–44	1016	0·017	0·010	0·003	0·001
45–54	1050	0·027	0·004	0·000	0·000
55–64	1288	0·024	0·005	0·002	<0·001
65–74	4169	0·023	0·006	0·003	<0·001
75+	2467	0·020	0·008	0·005	<0·001
All ages	12420	0·022	0·007	0·003	<0·001

* Mean no. per cell

Physical agents

A variety of physical agents are either known (e.g. U.V. light) or suspected (e.g. ultra-sound), to be mutagenic, but in so far as man is concerned the most important known agents are of course the ionising radiations. In addition to background radiation from the natural environment, individuals may receive additional exposures to ionizing radiations from a variety of sources including: clinical exposures – both for diagnostic and therapeutic purposes, occupational exposures, accidental exposures and exposures consequent to nuclear testing or to nuclear warfare. Ionizing radiations are therefore an every-day part of our modern environment and, since they are powerful mutagens, there is a considerable amount of information from both *in vivo* and *in vitro* studies on the response of human peripheral blood lymphocyte chromosomes to these agents. Most of this information has been discussed at some length in a recent review carried out by the United Nations Scientific Committee on the effects of Atomic Radiation [10] and all that I wish to comment on today are the dose levels and types of exposure that result in significant increases in the frequencies of chromosome-type aberrations in peripheral blood cells.

(a) *Clinical exposures.* Tough et al. in 1960 [37] were the first to report on the presence of aberrations in blood cells of individuals exposed to x-rays, in this case from patients with ankylosing spondylitis who were treated with fairly high doses, totalling up to around 2,000 rads, but given to only a limited part of the body. These observations were quickly followed by a spate of papers reporting similar kinds of observations in blood cells from individuals exposed to ionising radiations for therapeutic purposes. Of greater interest in terms of population exposure, however, was the first suggestion by Stewart and Sanderson [38] of a possible effect of diagnostic x-ray exposures in inducing aberrations. These authors reported the presence of two dicentric aberrations in a total of 31 cells scored from a patient with Klinefelter's syndrome. In this patient blood cells were taken 8 hours after he had been exposed to a skeletal x-ray survey involving a skin dose of less than 2 rads from 60 kV x-rays.

Conen and his colleagues [39, 40] reported observations somewhat similar to those of Stewart and Sanderson and these were followed by a most important prospective study carried out by Bloom and Tjio [31]. Bloom and Tjio studied blood cells from six patients given diagnostic chest x-rays involving exposures of from 20 to 80 mr and found no dicentric aberrations in over 3,000 cells examined. However, these authors observed four dicentric aberrations in 300 cells taken from five patients after they had been subjected to gastro-intestinal examination by fluoroscopy. The exposures to these patients ranged from 12 to 35 r and the blood samples were taken 30 minutes after exposure.

Observations similar to those mentioned have now been reported by a large number of laboratories and it is clear that relatively low doses of x-rays, given to substantial regions of the body, can be easily detected through their effect in increasing the yield of chromosome aberrations. Moreover, the numerous *in vitro* studies in which blood cells have been exposed to x-rays *in vitro* show that one can detect significant increases in aberration yields with doses of as low as 5 rads [e.g. 33]. For instance, in our laboratory

TABLE 4. Aberration yields in peripheral blood leucocytes of individuals employed in the Atomic Energy industry [44]

Group	No. of men	Dose (rad) Range	Dose (rad) Mean	Age (yr) Range	Age (yr) Mean	Total cells counted	A cells No. modal	A cells No. non-modal
Randomly selected males	38	0	0	15–64	37·6	1060	966	45
A	13	1		25–45	35·6	1600	1456	37
B	19	1–10	3·8	23–62	38·3	1600	1440	72
C	14	23–34	27·5	31–52	40·3	1400	1242	37
D	7	15–37	24	36–58	52·8	1750	1567	36
E	9	75–98	84	36–54	44·9	1200	1069	34

10 rads *in vitro* gives around one dicentric in every hundred cells, whereas in control cultures the frequency is less than one in two thousand cells.

The one conclusion that we can certainly draw from these kinds of observations is that low doses of ionising radiations, that have no recognisable deleterious effect upon the individuals, clearly produce significant amounts of genetic damage in somatic cells.

(b) *Occupational exposure*. In considering clinical radiation exposures, I deliberately confined my remarks to the effects of low doses such as those that might be encountered by a small but significant proportion of a population in a modern 'medically orientated' society. In the same context we should consider the possible effects of the kinds of low dose exposures that are encountered by relatively small populations of people as a consequence of their work.

A number of laboratories have noted significant increases in the frequencies of chromosome-type aberrations in individuals receiving intermittent low dose exposures from external sources [41–47]. For example, Norman's group [42] studied 36 hospital radiation workers who had received cumulative doses of from 10 to 98 rads over a period of from 7 to 16 years at a median dose rate of 1·4 r/year. In this study no dicentric aberrations and no abnormal monocentric chromosomes were observed in 5,784 cells from control individuals but 14 dicentrics and 17 abnormal monocentric chromosomes were observed in 14,839 cells from the exposed group. These differences are highly significant and the authors comment that the exposure rates in their sample were on average, approximately ten times natural background.

Somewhat similar studies to those of Norman and his colleagues have been reported by the group working here in Edinburgh [44]. In the Edinburgh study 67 adult males working in Atomic Energy Establishments

s	Cu cells		Cs cells		Dicentrics and rings	
%	No.	%	No.	%	No.	per 1000 cells
3·2	6	0·6	9	0·9	0	0
5·0	10	0·6	16	1·0	1	0·6
4·5	18	1·1	15	0·9	6	3·7
5·9	26	1·8	12	0·8	7	5·0
5·3	36	2·1	19	1·1	14	8·0
5·2	21	1·7	13	1·1	6	5·0

were studied as well as 38 randomly selected males not employed in the Atomic Energy industry. The 67 men were divided into 5 groups: (A) A control group including men who had accumulated doses of less than 1 rad; (B) men with an average accumulated dose of 3·8 rads; (C) men with an average accumulated dose of 27 rads; (D) men with an average dose of 24 rads accumulated largely over the five years prior to study; and (E) a group with an average accumulated dose of 84 rads but whose rate of exposure was similar to the people in group D in the five years prior to sampling. A summary of the data obtained is given in Table 4. The data from the two control groups are similar and differ significantly from the exposed groups, particularly in the case of the dicentric and ring aberrations. It should be noted here that the individuals in group C had received average doses of around 2 rads per year over a twelve year period, i.e. a dose rate some 20 times that of natural background radiation and below the maximum permissible levels set for radiation workers.

In addition to the reports on aberrations in cells from individuals exposed to radiations from external sources, there are a number of reports of similar findings in cells from groups of workers exposed to internal emitters. For instance studies on radium clock-dial painters have shown significant aberration yields even in individuals having body burdens well below maximum permissible levels [48].

(c) *Nuclear explosion.* In terms of exposure of populations to radiations one of the most important and potentially disastrous sources is of course the radiation from nuclear explosion. A number of studies have been carried out on survivors at Hiroshima and Nagasaki who were exposed to radiation from nuclear explosion in 1945 [43, 49, 50, 51, 52, 53]. Most of these studies were undertaken around 20 years or so after the original explosions, but all report significant aberration yields in blood lymphocytes of exposed survivors relative to controls. Indeed, Bloom *et al.* [50] have recently reported a small but statistically significant increase in the frequency of lymphocytes with complex rearrangements, as compared with matched controls, in 38 *in utero* exposed survivors whose mothers received doses in excess of 100 rads at the time of the bombings.

In addition to these observations on individuals directly exposed to radiations from nuclear explosions, studies have also been made on individuals accidently exposed to radio-active fall-out from nuclear tests. The misfortune of the Japanese fishing vessel *Lucky Dragon* is well known and studies by Ishihara and Kumatori [54, 55] on 18 of the 22 fishermen on this vessel who were exposed to fall-out from the Bikini thermonuclear test have shown that all still have elevated aberration yields in their blood cells. Somewhat similar findings have also been reported by Lisco and Conard [56] on the Marshallese Islanders who were also the unlucky victims of nuclear fall-out from a test explosion.

I have no need to dwell on the genetic consequences of nuclear explosion, but wish merely to re-emphasise here that nuclear explosion and its consequent fall-out is a most potent source that can produce genetic changes in the cells of vast populations of individuals.

Finally, before leaving the topic of radiations I should mention that there are some populated regions on this planet in which the natural background radiation levels are extraordinarily high, e.g. in the monazite sand areas at Kerala in India and certain regions in Brazil. Since quite small doses of radiation appear to increase the aberration yields in blood cells it is anticipated the populations living in these regions might show elevated aberration yields and possible deleterious genetic effects. These populations are the subject of active study, but no published data are available as yet.

Chemical agents

In the last few years there has been a tremendous awakening of interest in the pollution of our environment by various chemical agents, some of which are undoubtedly mutagenic, at least to certain life forms. Herbicides, fungicides, insecticides, industrial effluents and the chemical by-products of modern society are now perhaps as much a topic for after dinner conversation as atomic radiations were ten to twenty years ago. Since the first reported discovery of chemical mutagenesis by Auerbach and Robson [57], a vast number of chemical agents have been tested for possible mutagenic activity using various test systems such as *Drosophila*, plant roots, bacteria and fungi. In the numerous cases where the induction of chromosome aberrations in plant root cells has been used as an index of mutagenic action, then it is almost true to say that virtually any chemical agent lying on a laboratory shelf is mutagenic provided it is applied at sufficiently high concentrations. Aberrations and mutations will almost certainly be produced in cells that are maintained in an environment containing an agent that is present at a concentration just below that which is lethal to the cells. This is simply a reflection of the fact that the maintenance and replication of the genetic materials of a cell are of course active processes, and they can be disturbed by a variety of environmental conditions. When I refer to chemical mutagens therefore I really intend my remarks to be limited to those agents that show mutagenic activity at low concentrations, i.e. at μg/ml levels.

Shortly after the development of the peripheral blood lymphocyte culture technique, Conen and Lansky [58] showed that the alkylating agent, nitrogen mustard, which was well known as a potent inducer of chromatid aberrations in a variety of cell systems, produced chromosomal aberrations in lymphocytes of individuals that were treated with this compound and that similar aberrations were also produced if the cells were exposed to this agent *in vitro*. Since that time, a variety of alkylating and other cytostatic drugs that are used therapeutically, particularly for certain forms of leukaemia, have also been shown to produce similar effects. These are listed in Table 5 and I should mention that my separation of compounds into the four groups in this table is simply for my convenience. Most of these alkylating agents had previously been shown to produce aberrations in plant or animal cells exposed for short periods (up to a couple of hours) at concentrations down to around 10^{-6} M. In some of these studies (e.g. 59) care was taken to determine the precise time in the cell cycle at which the aberrations were induced and the results confirmed earlier studies on plant cells [23, 24] in showing that the aberrations produced were of the chromatid type and that they were produced at the time of chromosomal replication.

The alkylating agents that have been studied are virtually all used only for clinical purposes, but we should note in passing that it has been demonstrated that many alkylating agents (e.g. many ethylenimine compounds [60]) are excellent insect sterilants.

A second group of compounds that are of particular interest are the antibiotics that have been shown to produce aberrations at μg/ml levels in *in vitro* cultures. The first four of these compounds listed under this heading in Table 5 are all substances isolated from *Streptomyces* species and all have been shown to inhibit DNA synthesis and produce single strand breaks in DNA. These agents appear to act during the DNA replication (S) or post-replication (G_2) phase and produce chromatid-type aberrations. The few chromosome-type aberrations that have been reported with Streptonigrin [61] and Daunomycin [62] may well be 'derived' chromosome-type changes. There are no reports on the production of chromosome aberrations by these compounds in *in vivo* studies.

The third column in Table 5 lists a variety of drugs, DNA inhibitors and

TABLE 5. Chemical agents that have been shown to induce chromosome aberrations in human peripheral blood leucocytes exposed *in vivo*[a] or *in vitro*[b]

Agent	Reference	Agent	Reference
Alkylating and other cytostatic agents		*Drugs, DNA-inhibitors, base-analogues and other agents*	
1. Ethylenimines (various incl. Trenimon, Chinon I, etc.)[b]	Hampel et al. [105] Chang + Elequin [59] Obe [106]	11. Deoxyadenosine[b] 12. Cytosine Arabinoside[b]	Kihlman et al. [119] Kihlman et al. [119] Brewen [63]
2. Triethylenemelamine[b]	Hampel + Gerhartz [107] Hampel et al. [105]	13. Thioguanine[b] 14. Hydroxylamine[b]	Engel et al. [120] Engel et al. [120]
3. Myleran[b]	Gebhart [108]	15. 5-Bromodeoxyuridine[b]	Engel et al. [120]
4. Imuran[a] *	Jenson + Soberg [109]	16. 5-Fluorodeoxyuridine[a]	Hampel et al. [105]
5. Nitrogen Mustard[a b]	Conen + Lansky [58] Nasjleti + Spencer [110]	17. 6-Mercaptopurine[a b]	Pederson [121] Nasjleti + Spencer [110]
6. Piperazine[a]	Nasjleti et al. [111]	18. 6-Azauridine[a]	Elves et al. [122]
7. Thiotepa[a b]	Hampel et al. [105]	19. Lysergic Acid Diethylamide[a b] (LSD-25)	Cohen et al. [67] Irwin + Egozcue [68] Cohen et al. [69]
8. Cyclophosphamide (cytozan)[a] only	Hampel et al. [105]	20. Benzene[a]	Tough + Court Brown [123]
9. Miracil-D[b]	Obe [112]	21. Plasma Factor[a b]	Goh + Sumner [124] Hollowell + Littlefield [125] Scott [126]
10. Pyrimethamine[b]	Bottura + Coutinho [113]		
Antibiotics		*Other plant products*	
22. Mitomycin C[b]	Nowell [114] Cohen + Shaw [115] Cohen et al. [61]	28. Caffeine[b]	Ostertag [70]
23. Streptonigrin[b]	Jacobs et al. [116]	29. 8-Ethoxy-Caffeine[b]	Jackson [127]
24. Phleomycin[b]	Grouchy + Nova [117]	30. Theophyllin[b]	Ostertag [70]
25. Daunomycin[b]	Vig et al. [62]	31. Heliotrine[b] †	Bick + Jackson [71]
26. Actinomycin D[b]	Ostertag + Kersten [118]	32. 8-Hydroxy-Quinoline[b]	Gebhart [128]
27. Proflavine[b]	Ostertag + Kersten [118]		

base analogues that are effective in producing aberrations. In those which have been studied in detail (e.g. Cytosine arabinoside, [63]), the aberrations that are produced are of the chromatid-type and are produced either in G_2 or S. All but the final three compounds in this column of the table are effective mutagens, but the activities of LSD-25, benzene and the 'plasma factor' are the subjects of some dispute and the reported aberration yields have generally been quite low. These three agents have been listed, however, since it has been claimed that they produce aberrations following *in vivo* exposure, although in the case of LSD-25 there are as many reported negative findings [64, 65, 66] as there are positive ones [67, 68, 69].

In the last column of Table 5, I have listed some plant products which appear to be effective in inducing aberrations, although in the case of Caffeine, effects are not readily demonstrable if solutions of concentrations below 1 per cent are used. As with most of the other agents, Caffeine produces chromatid-type aberrations that require replication for their formation [70]. The plant alkaloid heliotrine has been included in this column since it is a powerful mutagen in a variety of organisms and is a dangerous hepatotoxin to animal live stock. Indeed, Bick and Jackson [71] have shown that high yields of chromatid aberrations are produced in marsupial leucocytes exposed to this compound at levels of around 10^{-5} M.

In compiling the list illustrated in Table 5, I was struck by the absence of some notable compounds that have been shown to be highly mutagenic in non-human test systems and some of these compounds are in clinical use. Mercuric compounds are an obvious example; they are used in a variety of medicinal ways and they are forming an ever increasing component of our environment since they are liberally used as antifungal agents in agriculture and in the paper industry. Certain forms of mercury have long been known to produce chromosome aberrations in plants at very low concentrations, and Ramel [72, 73] has recently reported significant genetic effects in *Drosophila* at concentrations (around 1 ppm) well below toxic levels.

Hydrazines and related drugs are used as antidepressants (e.g. isocarboxazid and nialamide) and for the treatment of tuberculosis (isoniazid). Freese et al. [74] have shown that, in the presence of O_2, these compounds will react with DNA, produce single strand breaks and inactivate transforming principle and viruses. Moreover, they are known to produce chromosome aberrations in mouse ascites tumour cells [75] and *Vicia* roots [76]. The hydrazines are lipophilic and there seems no reason to doubt that they could find their way into cells and cell nuclei. I do not know if anyone has made an attempt to assess whether chromosome aberrations are induced in humans treated with these compounds, and we do not know if these compounds give rise to a genetic risk at the dose levels used.

The hydroxylamines behave in a similar way to the hydrazines and these are well known to produce aberrations in mammalian cells *in vitro* [e.g. 77]. Since any amines or amides that are converted into the hydroxy form will also be active, then, as Freese *et al.* [74] point out, the carbamates that are used as psychotropic drugs in man will be converted by N-hydroxylating enzymes into hydroxylamines and could be mutagenically active. In this connection we should note that ethylcarbamate, commonly known as urethane, was the first chemical mutagen shown to produce chromosome aberrations in plants [78] although Bateman [79] has been unable to show that it is mutagenic to germ cells of mice.

In summary, we can say that with the exception of the alkylating agents very few of the other chemical agents listed in table 5 have been clearly shown to produce chromosome aberrations *in vivo*. At the present time therefore our knowledge of the importance of various chemical agents in inducing mutations *in vivo* in man is minimal. Clearly we need to be much more aware of the possible mutagenic effects of chemicals in our environment and, although much more information is required, it is obvious that any attempts at extrapolation from *in vitro* studies to the *in vivo* state will be beset with a variety of difficulties.

Biological agents
The demonstrations by a number of workers that viral infections may produce chromatid deletions and chromatid shattering in human and in other mammalian cells, and the possible implications of virus infection in relation to carcinogenesis, has prompted a number of studies on this aspect of aberration production. A wide variety of both DNA and RNA viruses has been reported to be responsible for the production of chromatid aberrations in human peripheral blood lymphocytes and in human and other mammalian fibroblast cells maintained in continuous culture. The viruses that have been claimed to induce aberrations include: Sendai [80]; chicken pox [81]; measles [82, 83, 84]; yellow fever [85]; the Schmidt-Ruppin strain of the Rous sarcoma virus [83, 84, 86]; herpes simplex [87]; cytomegalovirus [88]; infectious hepatitis virus [81, 89, 90]; poliomyelitis [91]; and various human and simian adenoviruses [92].

Chromatid aberrations have also been reported in cultured human fibroblast cells exposed to Mycoplasma [93, 94] and similar aberrations have been found in *Drosophila* exposed to Rous sarcoma virus [95] and in other arthropods infected with a 'Rickettsia-like' organism [96].

Although there are conflicting reports on the presence or absence of aberrations in cultures of peripheral blood leucocytes from patients suffering from various virus infections [85, 97, 98] there is no doubt that viral and other infectious agents can induce chromatid-type aberrations in human cells under certain conditions. The general conclusion arrived at by many workers in this field is that the effects on chromosomes produced by these agents are very similar to the effects produced by chemical mutagens that interfere with DNA synthesis. This conclusion is supported by the observations of Nichols et al. [99] of a synergistic action of the Schmidt-Ruppin strain of the Rous sarcoma virus and cytidine triphosphate (a nucleoside triphosphate that induces chromatid-type deletions in human cells) in producing chromatid aberrations in human leucocytes treated *in vitro*. Moreover, the aberrations induced by nucleosides and by viral agents are both localised to particular chromosomes and chromosome regions [99, 100] and thus differ from radiation-induced aberrations which appear to be more randomly distributed.

Studies with viable and heat-inactivated virus [101] reveal that inactive virus does not produce aberrations and Stich and Yohn [92] have shown that, at least in the case of certain adenoviruses, aberrations are only produced by viruses that initiate but do not complete a full replication cycle. Indeed we should point out here that many of the observed effects of viruses on chromosome structure may either be the cause of, or the result of, inevitable cell death.

In studies on viral-induced chromosome aberrations in peripheral blood cells, aberrations observed in cultures of blood cells of persons showing clinical symptoms of viral infection are often considered to be demonstrations of 'chromosome breakage *in vivo*' [83]. These kinds of observations are contrasted with those demonstrations of so called '*in vitro* breakage' when virus is added to cultures of blood cells from uninfected individuals.

The observations of chromatid aberrations in cultures of peripheral blood cells taken from an infected individual cannot really be claimed to demonstrate that the aberrations are produced *in vivo*, unless it is shown that the affected cells have proceeded through an S or G_2 phase *in vivo*. This follows since Nichols and his colleagues [99] have shown that the aberrations are induced either at the time of DNA synthesis or in G_2. Thus, any aberrations observed in cultured small lymphocytes (i.e. cells that were in a G_1 state *in vivo*) will have been produced during the period of *in vitro* culture.

It is known that aberrations are only observed in cultured blood cells taken over a limited time period during the course of infection [102] and the evidence would indicate that perhaps no aberrations are actually induced *in vivo* in non-proliferating lymphocyte-type cells. I am unaware of any published information indicating that aberrations are to be observed in proliferating cells in the bone marrow during virus infection and feel that the possible importance of virus infection in inducing *in vivo* genetic

damage in the cells of man is still a problem requiring much more study.
The biological consequences of aberrations induced in somatic cells
The genetical consequences of aberrations in human germ cells are fairly well known as is evidenced by the fact that we are able to convene at a symposium such as this. However, our knowledge of possible genetical or biological consequences of the different chromosome structural changes that may be produced in a proportion of the somatic cells in the body is pitifully sparse. It is known that certain aberrations may result in cell death and because chromosome aberrations are sometimes associated with such disorders as neoplasia, auto-immune disease and 'non-specific ageing' some people incline to the view that one or all of these effects are quite often a consequence of chromosome structural change. There is, however, no good direct evidence in support of a relationship between a particular disease state and a chromosome structural change, except in one possible instance, that of chronic myeloid leukaemia and the Philadelphia or Ph^1 chromosome [103, 104]. At the present time, therefore, we cannot make any quantitative or qualitative statements regarding the clinical importance of aberrations in somatic cells.

CONCLUSION

Man has through the passage of time been able to a large extent to exert some control over, and to stabilise, his environment. However, at the present time man, in an attempt to increase his affluence, is continually changing his environment and contaminating it with agents which are either known or suspected to be harmful to good health. In the case of mutagenic agents, we are faced with many problems, not the least being our need to specify what agents are mutagenic to man and at what levels or concentrations. We are well aware of the deleterious effects of mutational changes in germ cells but have little knowledge of their effects in somatic cells, although any such effects will most certainly not be beneficial. From what knowledge we have, it is clearly evident that we simply cannot afford to pollute our environment with mutagenic agents, and that every effort should be made to specify what agents are mutagenic and to subject them to control whenever possible.

REFERENCES

[1] Moorehead, P.S., Nowell, P.C., Mellman, W.T., Battips, D.M. and Hungerford, D., *Expl Cell Res.,* **20,** 613, 1960.
[2] Hungerford, D., *Stain Tech.,* **40,** 333, 1965.
[3] MacKinney, A.A., Stohlman, F. and Brecher, G., *Blood,* **19,** 349, 1962.
[4] Carstairs, K., *Lancet,* **1,** 829, 1962.
[5] Tanaka, Y., Epstein, L.B., Brecher, G. and Stohlman, F., *Blood,* **22,** 614, 1963.
[6] Osgood, E.E., Seaman, A.J., Tiney, H. and Rigas, D.A., *Revue hémat.,* **9,** 543, 1954.
[7] Ottesen, J., *Acta physiol. scand.,* **32,** 75, 1958.
[8] Hamilton, L.D., *Brookhaven Symp. in Biol.,* **10,** 52, 1958.
[9] Little, J.R., Brecher, G., Bradley, T.R. and Rose, S., *Blood,* **19,** 236, 1962.
[10] United Nations, 'Radiation-induced chromosome aberrations in human cells'. Chapter in report of the U.N. Scientific Committee on the effects of Atomic Radiation 24th Session General Assembly Suppl. 13(A/7613), 1969.
[11] Buckton, K.E., Smith, P.G. and Court Brown, W.M. In *Human Radiation Cytogenetics,* p. 106. Ed. H.J. Evans, W.M. Court Brown and A.S. McLean. Amsterdam : North-Holland Publ. Co., 1967.
[12] Norman, A., Sasaki, M.S., Ottoman, R.E. and Fingerhut, A.G., *Blood,* **27,** 706, 1966.
[13] Bond, V.P., Cronkite, E.P., Fleidner, T.M. and Schork, P., *Science,* **128,** 202, 1958.
[14] Bender, M.A and Prescott, D.M., *Expl Cell Res.,* **27,** 221, 1962.
[15] Lima de Faria, A., Reitalu, A.J. and Bergman, S., *Hereditas,* **47,** 695, 1962.
[16] Evans, H.J., *Int. Rev. Cytol.,* **13,** 221, 1962.
[17] Evans, H.J., *Genetics,* **46,** 257, 1961.
[18] Heddle, J.A., *Genetics,* **52,** 1329, 1965.
[19] Buckton, K.E., Langlands, A.O., Smith, P.G. In *Human Radiation Cytogenetics,* p. 122. Eds. H.J. Evans, W.M. Court Brown and A.S. McLean. Amsterdam : North-Holland Publ. Co., 1967.
[20] Norman, A. In *Human Radiation Cytogenetics,* p. 53. Eds. H.J. Evans, W.M. Court Brown and A.S. McLean. Amsterdam : North-Holland Publ. Co., 1967.
[21] Suskov, I.I., *Genetika,* **7,** 112, 1967.
[22] Evans, H.J., Court Brown, W.M. and McLean, A.S. (Eds.) In *Human Radiation Cytogenetics.* Amsterdam : North-Holland Publ. Co., 1967.
[23] Evans, H.J. and Scott, D., *Genetics,* **49,** 17, 1964.
[24] Evans, H.J. and Scott, D., *Proc. R. Soc. B.* (in press).
[25] German, J., Archibald, R. and Bloom, D., *Science,* **148,** 506, 1965.
[26] Sawitsky, A., Bloom, D. and German, J., *Ann. intern. Med.,* **65,** 487, 1966.
[27] Schroeder, T.M., Anschutz, F. and Knopp, A., *Humangenetik,* **1,** 194, 1964.
[28] Swift, M.R. and Hirschhorn, K., *Ann. intern. Med.,* **65,** 496, 1966.
[29] Debaken, A., *J. nucl. Med.,* **6,** 740, 1965.
[30] Gooch, P.C. and Fischer, C.L., *Cytogenetics,* **8,** 1, 1969.
[31] Bloom, A.D. and Tjio, J.H., *New Engl. J. Med.,* **274,** 1341, 1964.
[32] Migeon, B.R. and Merz, T., *Nature, Lond.,* **203,** 1395, 1964.
[33] Schmickel, R., *Am. J. hum. Genet.,* **19,** 1, 1967.
[34] Lubs, H.A. and Samuelson, J., *Cytogenetics,* **6,** 402, 1967.
[35] Court Brown, W.M. In *Human Population Cytogenetics.* Amsterdam : North-Holland Publ. Co., 1967.

[36] Bochkov, N. P., Koslov, V. M., Pilosov, R. A. and Sevankaev, A. V., *Genetika*, **4**, No. 6, 93, 1968.
[37] Tough, I. M., Buckton, K. E. and Baikie, A. G., *Lancet*, **2**, 949, 1960.
[38] Stewart, J. S. S. and Sanderson, A. R., *Lancet*, **1**, 978, 1961.
[39] Conen, P. E., *Lancet*, **2**, 47, 1961.
[40] Conen, P. E., Bell, A. G. and Aspin, N., *Pediatrics*, **31**, 72, 1963.
[41] Sasaki, M. S., Ottoman, R. E. and Norman, A., *Radiology*, **81**, 652, 1963.
[42] Norman, A., Sasaki, M. S., Ottoman, R. E. and Veomett, R. C., *Radiat. Res.*, **23**, 282, 1964.
[43] Dioda, Y., Sugahara, T. and Horikawa, M., *Radiat. Res.*, **26**, 69, 1965.
[44] Buckton, K. E., Dolphin, G. W. and McLean, A. S., In *Human Radiation Cytogenetics*, p. 174. Eds. H. J. Evans, W. M. Court Brown and A. S. McLean. Amsterdam : North-Holland Publ. Co., 1967.
[45] El-Alfi, O. S., Ragab, A. S. and Eassa, E. H. M., *Br. J. Radiol.*, **40**, 760, 1967.
[46] Visfeldt, J. In *Human Radiation Cytogenetics*, p. 167. Eds. H. J. Evans, W. M. Court Brown and A. S. McLean. Amsterdam : North-Holland Publ. Co., 1967.
[47] Wald, N., Koizumi, A. and Pan, S. In *Human Radiation Cytogenetics*, p. 183. Eds. H. J. Evans, W. M. Court Brown and A. S. McLean. Amsterdam : North-Holland Publ. Co., 1967.
[48] Boyd, J. T., Court Brown, W. M. and Woodcock, G. E. In *Human Radiation Cytogenetics*, p. 208. Eds. H. J. Evans, W. M. Court Brown and A. S. McLean. Amsterdam : North-Holland Publ. Co., 1967.
[49] Bloom, A. D., Neriishi, S., Kamada, N. and Iseki, T. In *Human Radiation Cytogenetics*, p. 136. Eds. H. J. Evans, W. M. Court Brown and A. S. McLean. Amsterdam : North-Holland Publ. Co., 1967.
[50] Bloom, A. D., Neriishi, S. and Archer, P. G., *Lancet*, **2**, 10, 1968.
[51] Awa, A., Bloom, A. D., Yoshido, M. C. and Archer, P. G., *Nature, Lond.*, **218**, 367, 1968.
[52] Ishihara, T. and Kumatori, T., *Acta haemat. jap.*, **28**, 291, 1965.
[53] Sasaki, M. S. and Miyata, H., *Nature, Lond.*, **220**, 1189, 1968.
[54] Ishihara, T., and Kumatori, T. In *Human Radiation Cytogenetics*, p. 144. Eds. H. J. Evans, W. M. Court Brown and A. S. McLean. Amsterdam : North-Holland Publ. Co., 1967.
[55] Ishihara, T. and Kumatori, T., *Japan J. Genetics* (Suppl.) (in press).
[56] Lisco, H. and Conard, R. A., *Science*, **157**, 445, 1967.
[57] Auerbach, C. and Robson, J. M., *Proc. R. Soc. Edin. B.*, **62**, 271, 1948.
[58] Conen, P. E. and Lansky, G. S., *Br. med. J.*, **2**, 1055, 1961.
[59] Chang, T. H. and Elequin, F. T., *Mutation Res.*, **4**, 83, 1967.
[60] Smith, C. N., La Brecque, G. C. and Borkovec, A. B., *Ann. Rev. Entomol.*, **9**, 269, 1964.
[61] Cohen, M. M., Shaw, M. W. and Craig, P., *Proc. natn. Acad. Sci. U.S.A.*, **50**, 16, 1963.
[62] Vig, B. K., Kontras, S. B., Paddock, E. F. and Samuels, L. D., *Mutation Res.*, **5**, 279, 1968.
[63] Brewen, J. G., *Cytogenetics*, **4**, 28, 1965.
[64] Loughman, W. D., Sargent, T. W. and Israelstam, D. M., *Science*, **158**, 508, 1967.
[65] Sparkes, R. S., Melnyk, J. and Bozetti, L. P., *Science*, **160**, 1343, 1968.
[66] Bender, L. and Sankar, D. V. S., *Science*, **159**, 749, 1968.
[67] Cohen, M. M., Hirschhorn, K. and Frosch, W. A., *New Engl. J. Med.*, **277**, 1043, 1969.

[68] Irwin, S. and Egozcue, J., *Science*, **157**, 313, 1967.
[69] Cohen, M. M., Hirschhorn, K., Verbo, S., Frosch, W. A. and Groeschel, M. M., *Pediat. Res.*, **2**, 486, 1968.
[70] Ostertag, W., *Mutation Res.*, **3**, 249, 1966.
[71] Bick, W. A. E. and Jackson, W. D., *Aust. J. Biol. Sci.*, **21**, 469, 1968.
[72] Ramel, C., *Hereditas*, **61**, 208, 1969.
[73] Ramel, C. and Magnusson, J., *Hereditas*, **61**, 231, 1969.
[74] Freese, E., Skarlow, S. and Freese, E. B., *Mutation Res.*, **5**, 343, 1968.
[75] Rutishauser, A. and Bollag, W., *Experientia*, **19**, 131, 1963.
[76] Kihlman, B. A., *J. biophys. biochem. Cytol.*, **2**, 543, 1956.
[77] Somers, C. F. and Hsu, T. C., *Proc. natn. Acad. Sci. U.S.A.*, **48**, 937, 1962.
[78] Oehikers, F., *Z. Vererbungsl.*, **81**, 313, 1943.
[79] Bateman, A. J., *Mutation Res.*, **4**, 710, 1967.
[80] Aula, P. and Saskela, E., *Hereditas*, **55**, 362, 1966.
[81] Aula, P., *Hereditas*, **49**, 451, 1963.
[82] Nichols, W. W., Levan, A., Hall, B. and Ostegren, G., *Hereditas*, **48**, 368, 1962.
[83] Nichols, W. W., *Hereditas*, **55**, 1, 1966.
[84] Nichols, W. W., Levan, A., Aula, P. and Norrby, E., *Hereditas*, **54**, 101, 1965.
[85] Harnden, D. G., *Am. J. human Genet.*, **16**, 204, 1964.
[86] Kato, R., *Hereditas*, **68**, 221, 1967.
[87] Stich, H. F., Hau, T. C. and Rapp, F., *Virology*, **22**, 439, 1964.
[88] Selezneva, T. J. and Demidova, S. A. In *Symposium on the cytogenetics of humans and experimental animals*, Moscow, 1967.
[89] Kerkis, J. J., Sablina, O. W. and Radzhabli, S. J., *Genetica*, **5**, 183, 1967.
[90] Matsaniotis, N., Kiossoglou, F. and Maounis, F., *Lancet*, **2**, 1421, 1966.
[91] Bartsch, H. D., Habermehl, K. O. and Diefenthal, W., *Expl Cell Res.*, **48**, 671, 1967.
[92] Stich, H. F. and Yohn, D. S., *Nature, Lond.*, **215**, 1292, 1967.
[93] Paton, G. R., Jacobs, J. P. and Perkins, F. T., *Nature, Lond.*, **207**, 43, 1965.
[94] Fogh, J. and Fogh, H., *Proc. R. Soc. exp. biol. Med.*, **119**, 233, 1965.
[95] Burdette, W. J. and Yoon, J. S., *Science*, **154**, 340, 1967.
[96] Halkka, O. and Heinonen, L., *Hereditas*, **58**, 253, 1967.
[97] Gripenberg, U., *Hereditas*, **54**, 1, 1965.
[98] Tanzer, J., Stoitchkov, Y. and Harel, P., *Lancet*, **2**, 1070, 1963.
[99] Nichols, W., Levan, A., Heneen, W. and Peluse, M., *Hereditas*, **54**, 213, 1965.
[100] Nichols, W. W. and Heneen, W. K., *Hereditas*, **52**, 402, 1965.
[101] Bochkova, B. B., *Genetika*, **4**, 72, 1965.
[102] Nichols, W. W. and Levan, A., *Archiv gesamte Virusforschung*, **16**, 168, 1964.
[103] Nowell, P. C. and Hungerford, D. A., *Science*, **132**, 1497, 1960.
[104] Baikie, A. G., *Proceedings 11th Congress International Society Hematology*, Sydney, 198, 1966.
[105] Hampel, K. E., Kober, B., Rosch, H. G. and Meinig, K. H., *Blood*, **27**, 816, 1966.
[106] Obe, G., *Mutation Res.*, **6**, 467, 1968.
[107] Hampel, K. E. and Gerhartz, H., *Expl Cell Res.*, **37**, 251, 1965.
[108] Gebhart, E., *Mutation Res.*, **7**, 254, 1969.
[109] Jenson, M. J. and Soberg, M., *Acta Medica Scand.*, **179**, 249, 1966.
[110] Nasjleti, C. E. and Spencer, H. H., *Cancer Res.*, **26**, 2437, 1966.
[111] Nasjleti, C. E., Walden, J. M. and Spencer, H. H., *Cancer Res.*, **25**, 275, 1965.
[112] Obe, G., *Molec. Gen. Genetics*, **103**, 326, 1969.
[113] Bottura, C. and Coutinho, V., *Human Chromos. Newsl.*, **4**, 1965.

[114] Nowell, P. C., *Expl Cell Res.,* **33,** 445, 1964.
[115] Cohen, M. M. and Shaw, M. W., *J. Cell Biol.,* **23,** 386, 1964.
[116] Jacobs, N. F., Neu, R. L. and Gardner, L. I., *Mutation Res.,* **7,** 251, 1969.
[117] Grouchy, J. de and Nova, C. de, *Annls Génét.,* **11,** 39, 1967.
[118] Ostertag, W. and Kersten, W., *Expl Cell Res.,* **39,** 296, 1965.
[119] Kihlman, B. A., Nichols, W. W. and Levan, A., *Hereditas,* **50,** 139, 1963.
[120] Engel, W., Krone, W. and Wolff, U., *Mutation Res.,* **4,** 353, 1967.
[121] Pederson, B., *Acta Pathol. Microbiol. Scand.,* **61,** 261, 1964.
[122] Elves, M. W., Buttoo, A. S., Israels, M. D. and Wilkinson, J. F., *Br. med. J.,* **1,** 156, 1963.
[123] Tough, I. M. and Court Brown, W. M., *Lancet,* **1,** 684, 1965.
[124] Goh, K. O. and Sumner, H., *Radiat. Res.,* **35,** 171, 1968.
[125] Hollowell, J. G. and Littlefield, G. L., *Proc. Soc. exp. biol. med.,* **129,** 240, 1968.
[126] Scott, D., *Cell and Tissue Kinetics* (in press).
[127] Jackson, J. E., *J. Cell Biol.,* **22,** 291, 1964.
[128] Gebhart, E., *Mutation Res.,* **6,** 309, 1968.
[129] Evans, H. J. and Speed, R. M., unpublished data.
[130] Norman, A. In *Human Radiation Cytogenetics,* p. 53. Eds. H. J. Evans, W. M. Court Brown and A. S. McLean. Amsterdam : North-Holland Publ. Co., 1967.
[131] Mouriquand, C., Gilly, C., Patet, J. and Jalbert, P., *C.R. Acad. Sci., Paris,* **161,** 341, 1967.
[132] Adams, A. C. and Evans, H. J., 1969 (in preparation).

Human Chromosomes and Natural Selection
Introductory Remarks

L. S. PENROSE

Kennedy Galton Centre
Harperbury Hospital, St Albans

¶ I HAVE permission, even though I am chairman of this session, to make a few general remarks about natural selection and karyotype in man.

Population genetics (in so far as it concerns the effects of genes) is traditionally a field in which mathematics takes an important part. Human population cytogenetics, a subject originated by Court Brown, has not, as yet, required mathematics at all unless we include – as I think we may correctly – the exact study of such variables as parental age and chromosomal measurements.

The problems of evolution, in relation to chromosomes have not reached the stage where changes can be measured in terms of 'darwins', the units proposed by Haldane to measure the rate of change of a specific trait in a population per million years. What we see is the material for evolution in action at close quarters.

An important measurement in all evolution studies is rate of spontaneous mutation, using the term in the widest sense which includes changes of ploidy like those observed by de Vries. The rate of spontaneous mutation can be directly measured in man when the result is a lethal transmissible condition. A perfect example is Down's syndrome of standard trisomic type, and almost as good are the other trisomies (though they have not been proved to be transmissible). The rate, as compared with spontaneous gene changes, is very high. The incidence in human zygotes is at least 1/700 and, allowing for Carr's abortion observations, probably more frequent than 1/500. That is equivalent to 10^{-3} for G chromosome non-disjunction per gamete per generation. For sex chromosomes the rate is of the same order of magnitude. Spontaneous occurrence of translocations is a much rarer event. Direct measurement of their mutation frequencies could be obtained from the incidence at birth of instances with normal parents. The mutation rate of the D/G type can be estimated indirectly with comparative ease. The D/G translocation in the form of an unbalanced lethal occurs with a frequency of about one in 50 cases of Down's syndrome. Thus, in each generation, the rate of loss of D/G translocations in the population is

$$\frac{1}{500} \times \frac{1}{50} = \frac{1}{25,000}$$

If the population is in equilibrium for this aberration, a mutation rate of $1/50,000 = 10^{-4.7}$, per haploid set of chromosomes per generation, is implied. This value is not far from those found for mutation rates of some gene loci, calculated for dominant lethals, which usually lie between 10^{-5} and 10^{-6}. It raises the question as to whether diseases with slightly higher rates, like chondrodystrophy, are perhaps the results of small deletions caused by double breaks, as are translocations [1].

Estimates of mutation rates made indirectly may have to be modified if there is a compensatory advantage for carriers or, indeed, if they are at a disadvantage. Such modification does not seem to be necessary for the D/G or D/D translocation nor in other known examples because fertility of balanced carriers appears to be very close to the normal. It is interesting to note that if a translocation should have a slight advantage, it would have great difficulty in becoming established but, once it had a majority frequency, which might occur in a small inbred population, it would rapidly become universal in that group [2].

One important factor to be considered in this connection is the neglected possibility of gametic selection. This mainly concerns the question of advantage or disadvantage to sperm which carry abnormal genes or chromosome complements. Possible examples in man are few. One may be the relative failure of the male to transmit an unbalanced D/G complement. Differential selection with respect to X and Y sperm seems to occur in a manner which is associated with the father's age, for sex ratio is gradually reduced as the father becomes older [3]. The failure of transmission of Leber's disease, an autosomal dominant, through the male suggests that genes can also cause highly effective selection against sperm which carry them. Conversely, the ability of karyotypically abnormal sperm to effect fertilisation is shown by the frequent paternal origin of Turners, YY types of males and many types of translocation, so that no rules can yet be laid down.

REFERENCES

[1] Penrose, L. S., *Recent Advances in Human Genetics,* Chapter 1. London : Churchill, 1961.
[2] Penrose, L. S., *Acta Genet. med. Gemell.,* **11**, 303, 1962.
[3] Novitski, E. and Sandler, L., *Ann. hum. Genet., Lond.,* **21**, 123, 1956.

The Population Cytogenetics of Other Mammalian Species

C. E. FORD

MRC, Radiobiology Unit, Harwell

¶THE CHARACTERISATION of populations of animals in cytogenetic terms is far from new. The great work of Dobzhansky and his school in exploiting inversion polymorphism in natural and experimentally constructed populations of *Drosophila pseudo-obscura* [1, 2] is very widely known and so are the studies of White and his followers on other types of chromosomal polymorphism in natural populations of grasshoppers [3]. In comparison with these developments the cytogenetic study of mammalian populations has barely begun. For obvious reasons it can never hope to rival the *Drosophila* and grasshopper work in terms of numbers of individuals examined or in the possibilities of experimental manipulation. Nevertheless, the chromosomes of mammals are now particularly favourable for study – in soma and germ line, female and male meiosis and embryos as well as adults. This advantage, and the special place the class occupies, should eventually ensure that population problems receive attention.

The domain of mammalian population cytogenetics may be considered to embrace three problems, or groups of problems : the determination of the rate of origin of new chromosomal mutations (considered in the widest sense to include polyploidy and aneuploidy as well as structural changes); the investigation of systems of chromosome polymorphism; and the behaviour of somatic cell populations *in vivo*. The first of these is of considerable current interest in view of the apparently high frequencies of liveborn children with abnormal karyotypes [4, 5] and particularly of chromosomal abnormality in spontaneously aborted foetuses [6]. One may therefore ask whether the frequency of chromosome mutation in man really is uniquely high or whether it approaches an irreducible minimum common to eukaryotic species in general and set by the physical nature of the systems concerned. It appears, however, that any judgement regarding relative levels of chromosome mutation must be based on the accumulated experience of karyotypic normality of free-living, mostly adult individuals, since there is an all-but-complete absence of information of the chromosomes of embryos, even in classical species of the Acrididae and Diptera. The very limited data available on mammalian blastocyst and embryonic populations are given below and hint that frequencies of abnormal karyotypes may not only be lower in mouse, rat and pig than in human embryos but that different classes may be represented in different proportions. A satisfactory answer to the question must wait upon the gathering of further information.

Six different kinds of chromosome polymorphism have been described in invertebrate species [3, 7]. Four of these are now known in mammals. Examples of all four types are given below and their adaptive significance is discussed, though there is again a great shortage of suitable information.

Robertsonian changes at least have played a considerable part in mammalian karyotypic evolution and it would seem likely that they passed through a stage of balanced or transient polymorphism before fixation.

The analysis of the structure of populations of mammalian somatic cells *in vivo* in terms of karyotypically distinct classes has been developed to some extent, though many questions remain to be answered. In favourable situations it should be possible to learn something about both the rate of origin of new karyotypic variants and the operation of selective forces on them. Since the karyotypes of normal somatic cells in the normal tissues so far accessible for study are so stable, at least in young adults, the situations of special interest are: (i) in tissues of ageing animals, in which there may be increased aneuploidy, at least in lymphoid tissue; (ii) in mosaics and chimaeras, natural and artificial, where two and sometimes more karyotypically distinct cell lines may be considered to compete; (iii) in irradiated tissues, where extensive karyotypic changes are induced at a single point in time only, and the resultant populations of surviving cells are presumptively subject to selection for proliferative capacity; and (iv) in neoplastic tissues, where there appears to be a very high spontaneous rate of origin of variant genomes upon which selection can operate. These problems have been discussed elsewhere [8, 9] and will not be considered further here.

Natural incidence of chromosome mutation in liveborn mammals

The karyotypes of many different species of mammals have now been determined [10, 11, 12] but the information has mostly been derived from single specimens or small numbers of specimens. The most extensive observations are those of Gustavsson [13] on 1,134 cattle from a herd in Sweden, Matthey [15] on 213 pigmy mice of the *Mus (Leggada) minutoides* complex in Africa, and Hsu and Arrighi [16] on 190 individuals of 20 taxonomic species of deer mice (*Peromyscus*) in North America. Egozcu [17] lists 649 individual animals in a review of the karyotypes of primates. Only one abnormality is mentioned, a count of 49 instead of 48 chromosomes in cultured cells of a deer mouse by Hsu and Arrighi. But this abnormality could have arisen in culture so may not be referable to the animal from which the explant was taken. Among all the reports of normality and presumptive normality the one case of an apparent trisomic juvenile water vole (*Arvicola terrestris*) trapped by Fredga [18] stands out. This specimen was moribund when taken but no abnormality was recorded either externally or on dissection other than a relatively short tail and a minor dental anomaly. Counts could only be obtained from the two corneas, but all the cells analysed with confidence contained 37 chromosomes, including an additional small acrocentric chromosome apparently homologous with the shortest pair of autosomes. In the light of the only certain knowledge we have of trisomic mammals, namely the D, E and G trisomics of man and the tertiary trisomics reported in the progeny of two types of reciprocal translocation heterozygotes in the mouse by Lyon and Meredith [19] supplemented by the classical evidence of the deleterious effect of trisomy on the phenotype in other organisms, it is surprising that a true trisomic individual would survive in the wild. The possibility that the extra was an 'inert' supernumerary chromosome similar in nature to those identified in the marsupial glider, *Schoinobates volans* (see below) and many species of invertebrates and plants cannot be ruled out on present evidence. Nor can the possibility that the animal was a diploid/trisomic mosaic. A further possibility is that the pair concerned is largely heterochromatic, like three pairs of autosomes in the hedgehog, *Erinaceus europaeus* [20].

A planned survey of the karyotypes of large numbers of individuals of a single species just to determine the frequency of chromosome mutation would be tedious at best and is most unlikely to be undertaken. The only alternative is to make use of information obtained in the course of other investigations. Table 1 is a list of all the individual mammals used for cytogenetic investigations in my laboratory from 1954 to May 1969, principally in direct preparations of somatic tissues. The overwhelming majority of the mice and Chinese hamsters were animals with primary or transplanted neoplasms or irradiated animals bearing grafts of haemopoietic tissue. Individual animals varied in age from newborn to two-year-old mice and three- to four-year old Chinese hamsters, though most were young adults. Few failed to provide successful preparations, yet only three exceptional aneuploid individuals have been identified, all mice with sex chromosome abnormalities. Only a single structural mutant was identified. This was a Robertsonian translocation observed in bone marrow, spleen and thymus of a laboratory rat. Many structural mutations, however, such as paracentric inversions and translocations involving an exchange of equal-length segments, would not be detectable by examination of the somatic chromosomes alone.

TABLE 1. Numbers of individuals of different mammalian species used for cytogenetic investigation 1954–1969.

Species	Number
Mice (*Mus musculus*)	5460
Chinese hamsters (*Cricetulus griseus*)	289
Shrews (*Sorex araneus*)	234
Rats (*Rattus norvegigus*)	41
Cattle (*Bos taurus*)	27
Other species (19)	81
	6132

The three karyotypically abnormal mice were a 39,XO/41,XYY mosaic male reported by Evans et al. [21], a 41,XYY sterile male, and another sterile male with 42 chromosomes provided by Dr M. Lyon that was interpreted as having a presumptive XXXY sex chromosome constitution. The first was sired by an irradiated male, the second was irradiated *in utero* and the third was selected for examination because of its sterility combined with an exceptional variegated $Ta/+$, $Blo/+$ phenotype.

Numerical abnormalities of the sex chromosomes in mice

The 39,XO mouse is now well known as a fertile female with reasonably good post-natal viability [22] despite a high risk of pre-natal death [23]. The spontaneous incidence is reported to vary between 0·1 and 1·7 per cent in different stocks [24] and several individuals might have been expected among the mice listed in Table 1. Although there was a great excess of males there were still very many females and it may be that the CBA/H and CBA/H-*T6T6* inbred strains that were principally used have a naturally low incidence of 39,XO mutants.

A male mouse with a presumptive XXY sex chromosome constitution was first identified by McLaren [25] in an experiment specially set up for the detection of non-disjunctional exceptions. She found one such exceptional mouse in 1,662 male progeny classified but considered that this might be an underestimate of the true frequency. A similar exceptional mouse was reported by L. B. Russell and Chu [26] and in this case preparations of cultured cells derived from the tail tip showed 41 chromosomes. Although the X chromosomes of the mouse cannot normally be distinguished from similar autosomes the karyotype was consistent with the expected XXY sex chromosome constitution. Experiments similar in principle to McLaren's provided an estimated frequency of 41,XXY exceptions of 0·02 per cent, derived from the classification of 6,368 offspring [27].

Other abnormalities of the sex chromosomes appear to be very rare in liveborn mice. Only one example of a mouse with a 41,XYY karyotype has been reported and, as stated above, one more has been identified in this laboratory. Since both were sterile males but otherwise unremarkable it may be that more would be identified if the chromosomes of sterile males were examined regularly. At present, therefore, it is not possible to provide an estimate of the natural incidence of this type of abnormality.

There remain a small number of mice with mosaic karyotypes in respect of their sex chromosomes. These are listed in Table 2, which summarises present information on the status of sex chromosome anomalies in this species.

The quantitative genetic data on the natural incidence of the 39,XO and 41,XXY mice are sufficient to demonstrate a profound difference

TABLE 2. Numerical sex chromosome abnormalities in mice

Type	Incidence	Reference
39,X0	1/1000 to 1/60	[24]
41,XXY	1/5000	[27,25]
41,XYY	2 cases	[68,69]
42,XXXY	1 case	[70]
39,X0/40,XX	2 cases	[71,72]
39,X0/40,XY	1 case	[73]
39,X0/41,XYY	1 case	[21]

from the frequencies of the corresponding abnormal karyotypes among liveborn human infants [28, 4, 5]. The XO type is relatively much more frequent in mice, the XXY type much more uncommon. At present it is not possible to say to what extent these differences of incidence may reflect differences in the mechanism of origin or of selective loss during intrauterine life.

Genetic evidence that the single X chromosome of XO mice is nearly always derived from the mother [27] combined with cytogenetic evidence that non-disjunction at first anaphase of male meiosis is rare [29], and the fact that irradiation shortly after fertilisation increases the frequency of presumptive XO progeny led to the suggestion that XO mice may originate principally by loss of the paternal X chromosome subsequent to fertilisation [26]. Morris [23] has recently obtained evidence that there is a loss of XO embryos in XO mothers both before and after implantation, though the ratio of 539 XO daughters to 1,353 XX daughters at weaning when a 1 : 1 ratio would be expected shows that the loss cannot be appreciably greater than 60 per cent. In man, it is well known that a high proportion of spontaneously aborted foetuses have an XO karyotype and Carr [30] has estimated that only about 1 in 40 XO embryos survive to term. Since the incidence of liveborn XO infants is about 1 in 5,000 [28] the initial frequencies of XO zygotes in man and mouse may not be greatly different. Genetic evidence provided by the Xg antigen as an X chromosome marker however indicates that though there is a greater frequency of loss of the paternal X chromosome than of the maternal X chromosome, the difference is very much less than in the mouse. The best estimate for man on the basis of present information is 74 per cent X^MO and 26 per cent X^PO [31], whereas in a specially designed mouse experiment with marked maternal and paternal X chromosomes L. B. Russell [27] found 13 X^MO daughters but no X^PO daughters out of 1,338 offspring classified. This comparison is indicative of a real difference between the two species in respect of sex chromosome behaviour. The difference might be accounted for by a higher level of meiotic non-disjunction in man, which would also explain the higher frequency of XXY males, and a higher frequency of post-fertilisation X-chromosome loss in the mouse.

Autosomal trisomy in the mouse. Cattanach [32] reported a sterile but otherwise phenotypically normal male mouse with 41 chromosomes which he claimed to be an autosomal trisomic. The information presented, however, is insufficient to exclude the possibility that the extra chromosome is a Y chromosome or an X chromosome bearing a deletion. Griffen and Bunker [33, 34] have also claimed to have found autosomal trisomics in the mouse. The animals concerned were phenotypically normal progeny of irradiated males and were examined cytologically because of sterility or semi-sterility. The claim is based on observations of spermatocytes and spermatogonia in testis squash preparations. The supporting photographs are unconvincing and open to alternative interpretation. No counts of chromosomes in somatic cells were reported. Reasons for being sceptical of reports of autosomal trisomy in mammals unaccompanied by a more or less severe effect on phenotype or viability or both have been given above.

Undoubted autosomal trisomic mice were found by Lyon and Meredith [19] among the progeny of animals heterozygous for either of two reciprocal translocations, T158/H and T194/H. The somatic chromosomes of the latter included one exceptionally long marker chromosome and a very short one, perhaps one quarter of the length of the shortest normal autosome pair. There was a long marker chromosome in T158/H heterozygotes also and it may therefore be inferred that the segments exchanged in both translocations were of very unequal length. The second translocation chromosome in T158/H heterozygotes was not distinguishable in somatic mitosis. The points of great interest about the trisomic animals are that many examples were obtained from both stocks, that the T158/H trisomics presented an abnormal phenotype with reduced testis size in males and reduced litter size in females as well as a postural defect, and that though the T194/H trisomics exhibited no obvious disturbance of the phenotype, fertility was reduced in females and testis size in males. The extra chromosome in the T194/H trisomics was the exceptionally short translocation chromosome; the extra in some of the T158/H trisomics was also recognised as the shorter translocation chromosome by the occasional formation of trivalents in meiosis. Both types were therefore tertiary trisomics. These results show that the addition of even a very small autosome to the normal mouse genome can result in an adverse physiological effect.

The discovery of karyotypic mutant mice bearing Robertsonian translocations and the development of homozygous inbred lines from them (see below) raises the possibility of the regular production of specific primary autosomal trisomics. If non-disjunction of the trivalent configurations in F_1 heterozygotes should occur with frequencies comparable to those in human

Robertsonian translocation heterozygotes [35], two specific trisomic types and two complementary monosomic types could be present in the zygote populations derived from the cross of each type of heterozygote to normal. The possibilities have been enhanced by the recent discovery that wild mice from a single locality in a high-altitude Swiss valley have only 26 chromosomes as a result of Robertsonian changes, but are yet interfertile with normal laboratory stocks. It is of the greatest interest that the F_1 hybrids can be bred, though they are of reduced fertility, and produce some abnormal embryos [36, 37].

Spontaneous structural changes in the mouse. The somatic karyotype of laboratory mice as seen in preparations of bone marrow, spleen, lymph nodes, thymus, Peyer's patches, cornea, intestinal epithelium and ovary (granulosa cells) is remarkably stable [2, 38]. The evidence of stability has been extended in recent years by observations at meiosis in primary spermatocytes. Approximately 35,000 cells in stages from diakinesis to first metaphase have now been examined in control mice from irradiation experiments, in other normal mice and in mice known or suspected to be heterozygous for a reciprocal translocation [39–42]. Only nine of these were recorded as including a quadrivalent configuration that was not expected to be present. All these cells were in a preparation from one animal, which was considered to be mosaic for a reciprocal translocation.

TABLE 3. Spontaneous incidence of semi-sterility in the mouse

Source	Number of mice tested	Number of semi-steriles identified
Auerbach and Slizynski [74]	248	0
Charles et al. [75]	2640	2
Lyon et al. [76]	427	1
Phillips and Searle [77]	216	0
Total	3531	3

Notwithstanding this evidence of stability, spontaneous presumptive reciprocal translocations, identified genetically as semi-steriles, have been recorded in the mouse. The data given in Table 3 yield a mean frequency of one in 1,200 mice tested. The very much lower frequency of spontaneous reciprocal translocations detected in primary spermatocytes suggests that most of those identified as spontaneous semi-sterile mutants must arise either subsequent to meiosis in the male, or in the female germ line.

The most recent information on the spontaneous incidence of structural mutants in man is given by Court Brown and Smith [43] who reported 14 individuals with autosomal structural heterozygosity among 4,833 examined or approximately 0·3 per cent. But these data include Robertsonian translocations and presumptive pericentric inversions as well as presumptive reciprocal translocations. Only three of the latter are recorded in their tables, giving a frequency of approximately one in 1,600, which is not greatly different from the frequency in mice given above.

Pericentric inversions are likely to be exceptionally rare in the mouse, if they occur at all, but several Robertsonian translocations have arisen spontaneously, as already mentioned. Four stocks homozygous for such changes and therefore having 38 chromosomes in total instead of the normal 40 are known to exist [44–46] but no estimate of the frequency of occurrence is possible. Mutants of this kind have not been identified in the progeny of irradiated animals, but this is not surprising, since they would be expected to be phenotypically normal and fully fertile, or very nearly so, and would therefore be unlikely to be examined cytologically. However, no examples were found in an unselected sample of 150 sons of irradiated fathers, although five reciprocal translocation heterozygotes were identified [40].

The foregoing survey suggests that there is rather less karyotypic abnormality in adult mammals than in man. One then asks if the difference is attributable to a lower rate of origin of abnormality or a higher rate of selective death or both. This question can only be answered by examining the karyotypes of embryos.

Natural incidence of chromosome mutation in mammalian embryos

Fischberg and Beatty developed a method for counting the chromosomes in mouse embryos nearly 20 years ago. They discovered that embryos with triploid and other heteroploid karyotypes were common, that they constituted as many as 10 per cent of all embryos in the Silver stock, and that most would implant and survive to mid-term [48]. More recent studies on embryos of three species are given in Table 4. All the embryos examined were the products of natural mating between parents that had not been subjected to any experimental treatment.

Data presented in the report on the Standardisation of Procedures for Chromosome Studies in Abortion (World Health Organisation, 1966) show that 153 spontaneously aborted human foetuses found to be chromosomally abnormal included 32 (21%) with a presumptive XO karyotype, 63 (42%) trisomics, 34 (22%) triploids and tetraploids and 24 (16%) others making up a mixed group of mosaics, structural variants and undefined changes. In comparison with these figures the striking feature of Table 4 is the absence of trisomics and XO embryos. Another noteworthy feature is the relatively high total frequency (10%) of abnormality in the pig blastocysts, equal to the frequency found by Fischberg and Beatty in blastocysts of silver mice. The pig data provide the only estimated frequency of abnormality in embryos prior to implantation obtained by

TABLE 4. Frequency of prenatal chromosome abnormality in mammalian species

Species	Mouse [78]	Rat [79]		Pig [80]	
Material	14-day embryos	11-day embryos		10-day blastocysts	
Age of dams	8–45 weeks	3–6 months		Not recorded	
Types of abnormality		3n	3	3n	4
		42,XX/43,XX	1	4n	3
		42,XY/43,XY	1	2n/3n	1
		41,X0/42,XX	1	Deletion	1
Total abnormalities	0	6		11	
Total examined	419	410		88	

modern methods and there may well be some eliminations of karyotypically abnormal embryos during implantation. The observations on mouse blastocysts made by Chaganti and Madan suggest that there may be a selective elimination of karyotypically abnormal embryos before the fourteenth day of pregnancy in this species. It is already known that most of the unbalanced embryos produced in matings of translocation heterozygote to normal, die shortly after implantation [e.g. 49].

Obviously, many more observations must be gathered before any firm conclusions can be drawn regarding the incidence of chromosome mutations in mammalian embryos and the effects of selection as gestation progresses. When this is done it would be of value also to have counts of corpora lutea, of obvious dead embryos, and of the uterine scars that indicate implantation sites at which earlier death of the embryo had occurred, commonly referred to as 'moles' by mouse geneticists.

Intra-population chromosome polymorphism in mammals

Differences, sometimes very striking differences, have frequently been found between the karyotypes of animals of the same taxonomic species taken from different parts of its natural range. Recent examples include species of deer mice, *Peromyscus* [16, 50] and the cotton rat, *Sigmodon hispidus* [51] in North America, pigmy mice of the *Mus (Leggada) minutoides* complex in Africa [15] and the burrowing rodent, *Spalax ehrenbergi* in Israel [52]. Many of these differences can be explained in terms of homozygosity for Robertsonian tranlocations or presumptive pericentric inversions or both. The occurrence of chromosome polymorphism *within* populations should therefore be of considerable interest both to the evolutionist and to the population geneticist. Four types have so far been identified in mammalian populations and examples of each are listed in Table 5. An

TABLE 5. Intra-population systems of chromosome polymorphism in mammals

Species	Number of individuals studied	Reference
	Robertsonian translocations	
1. Common shrew (*Sorex araneus*)	42	[55]
2. – do –	51,7,8,5	[56]
3. – do –	40	[58]
4. Cattle (*Bos taurus*)	1134	[13]
5. Pigmy mouse (*Mus minutoides musculoides*)	11	[15]
6. Pig (*Sus scrofus*)	36	[53]
	Pericentric inversions	
7. Deer mouse (*Peromyscus spp*)	4,6	[50]
8. Pigmy mouse (*Mus minutoides*)	23	[15]
	Supernumerary chromosomes	
9. Greater glider (*Schoinobates volans*)	7	[81]
10. Silver fox (*Vulpes fulvus*) *	4	[14]
	Heteromorphic chromosomes	
11. Brown rat (*Rattus norvegicus*)	20,20	[65]
12. – do –	43	[66]
13. – do –	40	[63]
14. Black rat (*Rattus rattus*)	49,40	[67]
15. Guinea pig (*Cavia porcellus*)	15	[64]
16. Pigmy mouse (*Mus triton*)	39	[15]
17. Pigmy mouse (*Mus minutoides musculoides*)	12	[15]

*The interpretation is open to question in this species. See text

effort was made to ensure that this list was as comprehensive as possible, but it is unlikely to be complete and more examples are continually being discovered. In any case there is a greater need now to discover whether particular systems represent instances of balanced or of transient polymorphism [2]. No firm conclusion can yet be drawn from published data because of the small size of most samples studied, uncertainty about the nature of the population from which they were drawn or lack of repeated observations from the same population. Nevertheless, there is a strong supposition that balanced polymorphism is implicated in most of the examples listed.

Robertsonian translocation. The example reported by Gustavsson [13] demonstrates that a Robertsonian translocation can become established in a herd of domestic animals. An associated depression of fertility or viability would have been noteworthy, but there is no mention of either. The sample of 1,134 animals included 122 heterozygotes with 59 chromosomes instead of the normal 60 and four homozygous bulls with 58 chromosomes. Unfortunately, cattle herds are far from natural populations so that the data are unsuitable for comparison with expectation on the basis of the Hardy-Weinberg distribution, despite the large number of individuals studied.

The other examples of Robertsonian polymorphism listed were all observed in natural populations though some of the pigs studied by McFee et al. [53] were bred in captivity. The system in the common shrew was discovered by Sharman [54]. Polymorphism in respect of three different 'fusions' was then found within a relatively isolated population occupying a 4-acre thicket at Chilton in Berkshire [55]. Animals trapped subsequently in the Swiss and French Alps have provided evidence that three further elements may occur in the alternative metacentric and twin-acrocentric states [56, 57]. Since the common shrew has an XY_1Y_2/XX sex chromosome system [54] the diploid number has been found to vary from 20 to 32 in different individuals (the theoretical maximum number, 33 chromosomes in a male, has not yet been observed).

Some populations both in Britain and Switzerland appear to be monomorphic and in most others the polymorphism appears to involve only a single element, which however differs between one population and another. P. J. Ford and Graham [58], for example, found that a sample of 40 shrews from the Scottish island of Islay was dimorphic for autosome element number 8, whereas samples from the neighbouring islands of Jura (12), Gigha (7) and the adjoining mainland (11) were monomorphic. Animals trapped in south Sweden all have a diploid number of 20 (females) or 21 (males) all the autosomes being metacentric [59]; but three animals from Bialowieza, Poland were all heterozygotes, three different elements being involved [60]. Animals from lower altitudes in Switzerland, from France and from Jersey have a quite distinct karyotype and do not exhibit Robertsonian polymorphism [56, 61]. The karyotypic differences alone are sufficient to indicate that the two forms are cryptic biological species, but they are strongly supported by the discovery of co-existence without evident hybridisation at one locality in Switzerland [56]. The one is a northern, eastern and alpine form (Meylan's type B), the other a lowland western european form (Meylan's type A).

Further samples of shrews from the thicket at Chilton in Berkshire mentioned earlier were examined in the years 1957 and 1958 with the intention of testing agreement with the Hardy-Weinberg distribution. The 1956 data were analysed and tested in the same way. The comparisons were made separately for each of the variable elements and the results are given in Table 6 as χ_1^2 values obtained in goodness-of-fit tests. There was a small

TABLE 6. Common shrews trapped at Chilton, Berkshire. Hardy-Weinberg tests

Year	Number of animals	Autosome element number 6	7	8
1956	42	2·51 (E)	0·52 (D)	0·89 (D)
1957	57	1·12 (E)	0·90 (E)	0·49 (E)
1958	61	0·19 (D)	1·14 (E)	0·06 (E)

The main entries are χ_1^2 values calculated in goodness-of-fit tests of observed numbers of animals with alternative karyotypes to numbers expected from Hardy-Weinberg formula. Independent tests for autosome elements 6, 7 and 8
(E) Excess of heterozygotes: (D) Deficit of heterozygotes

excess of heterozygotes in six out of nine tests but none of the deviations in either direction was significant even at the 5 per cent level. These data strongly favour a balanced system and indicate that if there is a selective differential operating between karyotypic classes, it is small.

Pericentric inversions. The examples of polymorphism in respect of pericentric inversions are based on observations of somatic chromosomes in which the members of one or more autosome pairs exist in two forms, one with a nearly terminal centromere, the other with a median or sub-median centromere, the length of the two forms being apparently identical. This explanation of the morphological differences is not supported by observations of pachytene pairing or bivalent form in diakinesis or metaphase of meiosis in either of the examples quoted. The interpretation of the changes as pericentric inversions, though likely to be correct, must therefore for the present be regarded as presumptive.

Supernumerary chromosomes. The observations of variable numbers of supernumerary chromosomes in the marsupial greater glider and the silver fox are remarkably similar despite the great difference in taxonomic position between the two species. Chromosome counts in seven individuals of the greater glider trapped in a single location in New South Wales varied from 24 to 28 and included from 1 to 5 very small, apparently acrocentric chromosomes that were not certainly distinguishable from the Y chromosome in morphology. Four individual silver foxes from two farms gave modal counts of 35, 36, 36 and 37. In this species also the different counts were due to the presence of different numbers (1 to 4) of very small, apparently acrocentric chromosomes which were not distinguishable from the Y chromosome. The small chromosomes of both species were present in females as well as males and they were constant in number from cell to cell of the same individual. These observations appear to provide a close parallel to many observations of variable numbers of 'inert' supernumerary chromosomes in species of invertebrates and plants. However, in an earlier report on the chromosomes in testis preparations from two male silver foxes, Lande [62] gives the diploid number as 38 and states that the karyotype includes 'five dot-like chromosomes in striking contrast to the others in shape'. (One of these was presumed to be the Y). The observations of Gustavsson and Sundt [14] were made on preparations from tissue cultures established from lung, spleen, kidney or skin. Present information is therefore consistent with the possibility that two pairs of 'dot' chromosomes are normally present in cells of the germ line but that variation in number (principally or exclusively loss) occurs in somatic cells. Loss during culture is perhaps a less likely explanation in view of the few deviations from the mode in the counts reported. Until further information becomes available the status of the 'dot' chromosomes in the silver fox must remain uncertain.

It is now generally accepted that the normal Y chromosome of mammals is male-determining. The extra chromosomes of the great glider are therefore very unlikely to be normal Y chromosomes. However, their close morphological similarity raises the possibility that they may be derivatives of the normal Y chromosome from which the male determining factor or factors have been deleted. The minute size of the Y chromosome in some species of marsupials indicates that the essential genetic information necessary for initiating testicular differentiation could be located in a segment so short that its deletion from a Y chromosome like that of the great glider would not perceptibly change its length.

Heteromorphic chromosomes. The use of the adjective 'heteromorphic' is here arbitrarily restricted to instances where one form of a chromosome lacks or appears to lack a segment that is present in another. This implies a duplication (in the wide sense) or a deletion and therefore a difference in length between the two forms. Some of the examples listed involve the absence or apparent absence of the greater part of the short arm in a chromosome with a sub-terminal centromere, and since the change in total length is small, the possibility that they represent pericentric inversions cannot be excluded without supporting measurements. Some of the cases in Table 5 may therefore be wrongly classified. Bianchi and Molina [63], for example, studied a laboratory population of brown rats and found two forms of one of the shorter chromosome pairs, identified as pair 13. The one had a pronounced short arm, the other a barely detectable one. Deletion of the short arms seems the obvious explanation, but it might be pericentric inversion.

The observations of Cohen and Pinsky [64] on a laboratory population of the guinea pig *Cavia porcellus* dispel any doubt regarding interpretation. The normal longest autosome has a sub-terminal centromere, the short arm representing perhaps one tenth of the whole chromosome length. Cohen and Pinsky found *two* variant forms, one in which the short arm was effectively absent and one in which the short arm was twice its normal length and had a median secondary constriction. There is a strong implication here of simultaneous origin of the two variants. The authors postulate a reciprocal translocation between the two homologues, but a chromatid intrachange could generate the same two types and is perhaps more likely. Although fifteen animals were studied, neither variant was found in the homozygous state.

The brown rat is of particular interest in that heteromorphic systems have been reported for three different chromosomes. Chromosome 13 has already been mentioned. Hungerford and Nowell [65] found X chromosomes of two types in two different non-inbred laboratory strains. The normal type had a sub-terminal centromere; in the other the presence of a short arm is not obvious. Homozygotes of both types and heterozygotes were found in both populations. Yosida et al. [66] found a similar dimorphism in the third longest autosome in both wild and laboratory populations in Japan. Homozygotes of both kinds and heterozygotes were observed in the wild populations and there was a small but non-significant excess of heterozygotes when compared with expectation from the Hardy-Weinberg rule. This type of dimorphism also occurs in the longest autosome of *R. rattus* and the data of Yosida et al. [67] again demonstrate a small excess of heterozygotes over Hardy-Weinberg expectation in two distinct wild populations.

Good examples of deletion (or duplication) dimorphism of the X chromosome are reported by Matthey [15] in wild populations of African pigmy mice. The one type of X chromosome has a sub-median centromere; the other lacks the short arm of the first and matches its long arm in length. The length difference is therefore decisive. In *M. triton* all of eighteen males had the metacentric X. The twenty-one females examined consisted of eleven metacentric homozygotes and ten heterozygotes; acrocentric homozygotes were not found. The frequency of the normal metacentric X in this sample is therefore $(18 + 22 + 10) \div (18 + 22 + 20) = 0.833$. If the specimens were all truly from the same population the probability of failing to find a single male with the acrocentric X chromosome in eighteen specimens would be $(0.833)^{18} = 0.038$. For males and females combined the probability of the observed sample and all less likely combinations is 0.048. The data therefore give only marginal support to Matthey's suggestion that the deleted X may be lethal to males and homozygous females.

Adaptive significance of chromosome polymorphism. It is now generally accepted that the adaptive significance of the well-known paracentric inversion systems in *Drosophila*, *Chironomus* and other species of Diptera resides in the preservation of co-adapted gene complexes by the exclusion of recombination between the standard and inverted segments. Investigation of the other types of chromosome polymorphism, even in invertebrate species, is less advanced and their significance is still uncertain or unknown. Pericentric inversion systems could serve in the same way as the dipteran paracentric inversions to maintain co-adapted gene complexes through the effective suppression of recombination, though any crossing over that did occur between inverted and standard segments in heterozygotes would result in the production of unbalanced gametic genomes and presumably in a contrary selective disadvantage. It is therefore of interest that the presumptive pericentric inversion systems so far described in mammals involve relatively short chromosome segments within which chiasma formation might be completely suppressed. It is to be hoped that meiotic studies will be carried out in these forms.

No such obvious solution presents itself in the case of the Robertsonian translocation systems. It is possible that the structural differences are attended at meiosis by a difference in the distribution of chiasmata so that, say, crossing-over close to the centromeres is reduced in heterozygotes. This suggestion is entirely speculative, but if correct it would result in a relative preservation of gene combinations nearer the centromeres. Again studies of meiosis in heterozygotes are desirable.

The sytems with supernumerary and heteromorphic chromosomes differ from the two systems just discussed in that both involve a quantitative rather than a positional change in the genome. They are alike in depending on the presence or absence of extra chromosome material but different in that the amount is variable in the first-named, fixed in the second. There must be a strong supposition that the whole chromosomes or segments concerned are heterochromatic and lacking in major structural genes, if only by analogy with cases in plants and invertebrates. Autoradiographic study of DNA replication time is indicated. Presumably the variable part of the genome in these systems has some influence on development or function but at the present time there is no hint of what this influence might be.

In man it is now well-known that there is variation in the length of some chromosome arms in subjects with normal phenotypes. This variation is particularly notable in the long arms of the Y chromosomes and chromosome 16 and in the short arms of at least one member of both the D and G groups. It can therefore be said that there is a variable component of the human diploid genome as there is in the supernumerary and heteromorphic chromosome systems. In man, however, there is continuous variation in the length of the segments concerned. Although this feature makes the human system formally distinct from the heteromorphic chromosome systems as here defined, it may only reflect an insufficient knowledge of the normal range of variation in the chromosomes of other species.

SUMMARY

Abnormal unbalanced karyotypes are much less frequent in juvenile and adult mammals than in man. This might be due to more stringent pre- and post-natal selection against unbalanced types rather than to a difference of primary frequency in zygotic populations. Spontaneous reciprocal translocation appears to occur with approximately the same frequency in mice as in the human populations studied at Edinburgh.

There is very little information on the karyotypes of samples of mammalian embryos (Table 4). Aneuploid, polyploid and mosaic types occur but no useful statement regarding frequencies of occurrence, selective processes or possible inter-species differences can be made at present.

Four types of chromosome polymorphism have been reported in mammals from wild and laboratory populations and from a herd of domestic cattle (Table 5). Evidence is given that the Robertsonian translocations of the common shrew constitute a system of balanced rather than transient polymorphism (Table 6). The adaptive significance of these systems is discussed.

ACKNOWLEDGEMENT

I am grateful to Mr D. G. Papworth for making the exact tests of probability.

REFERENCES

[1] Dobzhansky, T., *Genetics and the Origin of Species*. New York : Columbia University Press, 1951.
[2] Ford, E. B., *Ecological Genetics*. London : Methuen, 1964.
[3] White, M. J. D., *Animal Cytology and Evolution*. Cambridge : University Press, 1954.
[4] Gerald, P. S. and Walzer, S. This volume, p. 143.
[5] Lubs, H. A. This volume, p. 119.
[6] Carr, D. H. This volume, p. 103.
[7] Lewis, K. R. and John, B., *Heredity*, **11**, 11, 1957.
[8] Ford, C. E. In *Cytogenetics of Cells in Culture*, p. 27. Ed. R. J. C. Harris. New York/London : Academic Press, 1964.
[9] Ford, C. E. and Clarke, C. M., Canad. Cancer Conf., **5**, 129. New York : Academic Press, 1963.
[10] Hsu, T. C. and Benirschke, K., *An Atlas of Mammalian Chromosomes*, vol. 1. New York : Springer-Verlag, 1967.
[11] Hsu, T. C. and Benirschke, K., *An Atlas of Mammalian Chromosomes*, vol. 2. New York : Springer-Verlag, 1968.
[12] Benirschke, K., Ed., *Comparative Mammalian Cytogenetics*. New York : Springer-Verlag, 1969.
[13] Gustavsson, I., *Nature, Land.*, **211**, 865, 1966.
[14] Gustavsson, I. and Sundt, C. O., *Hereditas*, **54**, 249, 1966.
[15] Matthey, R., *Rev. suisse Zool.*, **73**, 586, 1966.
[16] Hsu, T. C. and Arrighi, F. E., *Cytogenetics*, **7**, 417, 1968.
[17] Egozcu, J. In *Comparative Mammalian Cytogenetics*, p. 357. Ed. K. Benirschke. New York : Springer-Verlag, 1969.
[18] Fredga, K., *Chromosoma*, **25**, 75, 1968.
[19] Lyon, M. F. and Meredith, R., *Cytogenetics*, **5**, 335, 1966.
[20] Gropp, A. and Citoler, P. In *Comparative Mammalian Cytogenetics*, p. 267. Ed. K. Benirschke. New York : Springer-Verlag, 1969.
[21] Evans, E. P., Ford, C. E. and Searle, A. G., *Cytogenetics*, **8**, 87, 1969.
[22] Russell, W. L., Russell, L. B. and Gower, J. S., *Proc. natn. Acad. Sci.*, **45**, 554, 1959.
[23] Morris, T., *Genet. Res.*, **12**, 125, 1968.
[24] Russell, L. B. and Saylers, C. L., *Genetics*, **46** (7) : 894 (Abstract), 1961.
[25] McLaren, A., *Genet. Res.*, **1**, 253, 1960.
[26] Russell, L. B. and Chu, E. H. Y., *Proc. natn. Acad. Sci.*, **47**, 571, 1961.
[27] Russell, L. B., *Science*, **133**, 1795, 1961.
[28] Court Brown, W. M., *Human Population Cytogenetics*. Amsterdam : North-Holland Publ. Co., 1967.
[29] Ohno, S., Kaplan, W. D. and Kinosita, R., *Expl. Cell Res.*, **18**, 382, 1959.
[30] Carr, D. H., *Am. J. Obstet. Gynec.*, **97**, 283, 1967.
[31] Race, R. R. and Sanger, R., *Br. med. Bull.*, **25**, 99, 1969.
[32] Cattanach, B. M., *Cytogenetics*, **3**, 159, 1964.
[33] Griffen, A. B. and Bunker, M. C., *Proc. natn. Acad. Sci.*, **52**, 1194, 1964.
[34] Griffen, A. B. and Bunker, M. C., *Proc. natn. Acad. Sci.*, **58**, 1446, 1967.
[35] Hamerton, J. L., *Cytogenetics*, **7**, 260, 1968.
[36] Gropp, A. In *Comparative Mammalian Cytogenetics*, p. 247. Ed. K. Benirschke. New York : Springer-Verlag, 1969.
[37] Gropp, A., personal communication, 1969.
[38] Ford, C. E. and Evans, E. P., unpublished observations, 1969.

[39] Ashwood-Smith, M.J., Evans, E.P. and Searle, A.G., *Mutation Res.,* **2,** 544, 1965.
[40] Ford, C.E., Searle, A.G., Evans, E.P. and West, B.J., *Cytogenetics,* **8,** 447, 1969.
[41] Ford, C.E. and Hamerton, J.L., unpublished observations, 1955.
[42] Evans, E.P. and Ford, C.E., unpublished observations, 1964.
[43] Court Brown, W.M. and Smith, P.G., *Br. med. Bull.,* **25,** 74, 1969.
[44] Evans, E.P., Lyon, M.F. and Daglish, M., *Cytogenetics,* **6,** 105, 1967.
[45] Leonard, A. and Deknudt, G.H., *Nature, Lond.,* **214,** 504, 1967.
[46] White, B.J. and Tjio, J.H., *Hereditas,* **58,** 285, 1967.
[48] Fischberg, M. and Beatty, R.A., *J. exp. Zool.,* **118,** 321, 1951.
[49] Carter, T.C., Lyon, M.F. and Phillips, R.J.S., *J. Genet.,* **53,** 154, 1955.
[50] Sparkes, R.S. and Arakaki, D.T., *Cytogenetics,* **5,** 411, 1966.
[51] Zimmerman, E.G. and Lee, M.R., *Chromosoma,* **24,** 243, 1968.
[52] Wahrman, J., Goitein, R. and Nevo, E. In *Comparative Mammalian Cytogenetics,* p. 30. Ed. K. Benirschke. New York : Springer-Verlag, 1969.
[53] McFee, A.F., Banner, M.W. and Rary, J.M., *Cytogenetics,* **5,** 75, 1966.
[54] Sharman, G.B., *Nature, Lond.,* **177,** 941, 1956.
[55] Ford, C.E., Hamerton, J.L. and Sharman, G.B., *Nature, Lond.,* **180,** 392, 1957.
[56] Meylan, A., *Rev. suisse Zool.,* **71,** 903, 1964.
[57] Ford, P.J., unpublished observations.
[58] Ford, P.J. and Graham, C.F., *Bull. Mammal Soc. Br. Isl.,* **22,** 10, 1964.
[59] Fredga, K., personal communication, 1966.
[60] Evans, E.P., personal communication, 1964.
[61] Hamerton, J.L., personal communication, 1958.
[62] Lande, O., *Nature, Lond.,* **131,** 1353, 1958.
[63] Bianchi, N.O. and Molina, O., *J. Hered.,* **57,** 231, 1966.
[64] Cohen, M.M. and Pinsky, L., *Cytogenetics,* **5,** 120, 1966.
[65] Hungerford, D.A. and Nowell, P.C., *J. Morph.,* **113,** 275, 1963.
[66] Yosida, T.H. and Amano, K., *Chromosoma,* **16,** 658, 1965.
[67] Yosida, T.H., Nakamura, A. and Fukaya, T., *Chromosoma,* **16,** 70, 1965.
[68] Cattanach, B.M., *Cytogenetics,* **6,** 67, 1967.
[69] Evans, E.P., Ford, C.E. and Searle, A.G., unpublished observations.
[70] Evans, E.P. and Lyon, M., unpublished observations.
[71] Cattanach, B.M. and Pollard, C.E., *Cytogenetics,* **8,** 80, 1969.
[72] Green, M.M., personal communication.
[73] Lyon, M.F., *Cytogenetics,* **8,** 326, 1969.
[74] Auerbach, C.A. and Slizynski, B.M., *Nature, Lond.,* **177,** 376, 1956.
[75] Charles, D.R., Tihen, J.A., Otis, E.M. and Grobmum, A.B., *Genetic effects of chronic X-irradiation exposure in mice.* University of Rochester Atomic Energy Project. Rochester, New York. Report UR 565, 1960.
[76] Lyon, M.F., Phillips, R.J.S. and Searle, A.G., *Genet. Res.,* **5,** 448, 1964.
[77] Phillips, R.J.S. and Searle, A.G., *Genet. Res.,* **5,** 468, 1964.
[78] Chaganti and Madan, unpublished observations.
[79] Butcher, R.L. and Fugo, N.W., *Fert. Steril.,* **18,** 297, 1967.
[80] McFeely, R.A., *J. Reprod. Fert.,* **13,** 579, 1967.
[81] Hayman, D.L. and Martin, P.G., *Aust. J. biol. Sci.,* **18,** 1081, 1965.

The Operation of Selection

J. H. EDWARDS

Department of Human Genetics
University of Birmingham

SELECTION

¶ IT IS now ten years since the first abnormalities of the human chromosomes were displayed and, in these ten years, the economical but biased studies of abnormal persons, supported by the extensive and unbiased studies of populations, which Court Brown pioneered, have provided complementary sets of data implying that the morbidity due to chromosomal anomalies is of the same order as that due to abnormalities at the level of the gene. When the results of population sampling are extended to those rejected from the uterus, the contribution of chromosomal abnormalities to genetic deaths, or disorders comparable to death in severity, is several-fold increased.

Any study of the role of selection must take account of the origin of the condition selected against, and any practical application to medicine requires study of the effects of compounding natural selection with artificial selection through voluntary sterility or abortion. In order to see this great load of morbidity in perspective it is necessary to consider it against selection as a whole, as there is only a limited scope for selection if adequate proportions are to survive, so that selection at various levels of the hierarchical genetic structure must be considered as competitive. In particular, it is important to consider chromosomal disorders against genic variability, and its morbid consequences.

Unless we keep in sight the problems of genic variation, as they appeared in the fifties, we are likely to repeat the same errors of thought on the chromosomal variants, for it is now clear that, although some genic variants are unconditionally harmful, in most (and man may show common variants at a third of his loci, and has room for several million loci), no meaningful criterion of normality can be advanced either for descriptive purposes, or as an aim for eugenic policies. Before this new vista was opened up by unravelling the nucleus, human genetics, as a subject, was almost wholly dependent on analogies from other species; human variation was seen, almost entirely, as a matter of genic variation, and the differences between species were considered as mere exaggerations of the differences in gene frequencies within species. In the anxieties of radioactive fall-out, the gene, and mutant forms of the gene, dominated most research considered relevant.

Man was assumed to be at the mercy of various allelic variants which were sufficiently small in number and simple in expression for the assumption to be widespread that the genotype could be surveyed by a simple calculus and, given adequate diagnostic facilities, laundered by exhortation through genetic counselling. Other variants, termed polymorphic, were assumed sufficiently few for their presumptive heterozygous advantage to lie within the resolution of anthropological or clinical surveys.

The chromosomes obviously existed, but, notwithstanding observations on insects and flowering plants, were usually regarded as little more than passive vehicles. The major theorists of population genetics had been able to found a powerful calculus without their need. Indeed the chromosomes were a serious bar to a simplicity adequate to conform to algebraic treatment and the orderly florescence of mathematical genetics has been largely undisturbed by the crucial observations of Darlington, Dobzhansky, and White. The two-locus problem, the simplest possible, remains as refractory as the three-body problem of the astronomer.

Matters are now known to be more complicated for the genic variants, and the nature of this variation is most readily considered in terms of a variability for whose existence, and genesis, we are dependent on our origin, and survival, as a species; one of the more recent species. The genetic variability, if considered as a whole, complicates any interpretation still further, and imposes an even more provisional and approximate nature on an accountancy which regards genetic death as an item in the cost of living.

One consequence of this technical advance, and its exploitation for diagnosis and survey, has been the increase in the status of man as a biological organism. His normal and morbid chromosomal variants are now known better than those of any other mammal, and, at the level of the gene, some products, such as haemoglobins, blood group factors, and some enzymes, are also better known.

There is now sufficient data on man to attempt to integrate the findings on those parts of the genetic mechanism open to scrutiny. No doubt in another decade, when the structure of the mammalian chromosome is known, so that we know what we have been looking at, many contemporary problems will be resolved. Meanwhile, there seems little advantage in awaiting the increased numerical precision of further data, which is unlikely to change any accountancy of genetic load by a factor of more than two or so, since neither the theoretical nor the practical implications are likely to be much affected by this increase in accuracy.

Evolutionary arguments relating to genic variability can be summarised, among other ways, in terms of genetic *fitness*, which may be equated with genotypic variance [1], of *load* [2–4] which may be considered as a form of insurance by which established variation is conserved and the potential for future variation maintained, and of *cost*, the cumulative or compound load [5]. Although these lead to paradoxes, and can be criticised as oversimplifications, they allow an approximate audit through which we may compare very roughly, the various components of genetic variation leading to death.

The genetic load has been defined as the proportion by which the mean fitness falls short of the fitness of the fittest possible individual[3]. A dominant lethal contributes a unit deficit; a genotype with a 50 per cent

SELECTION

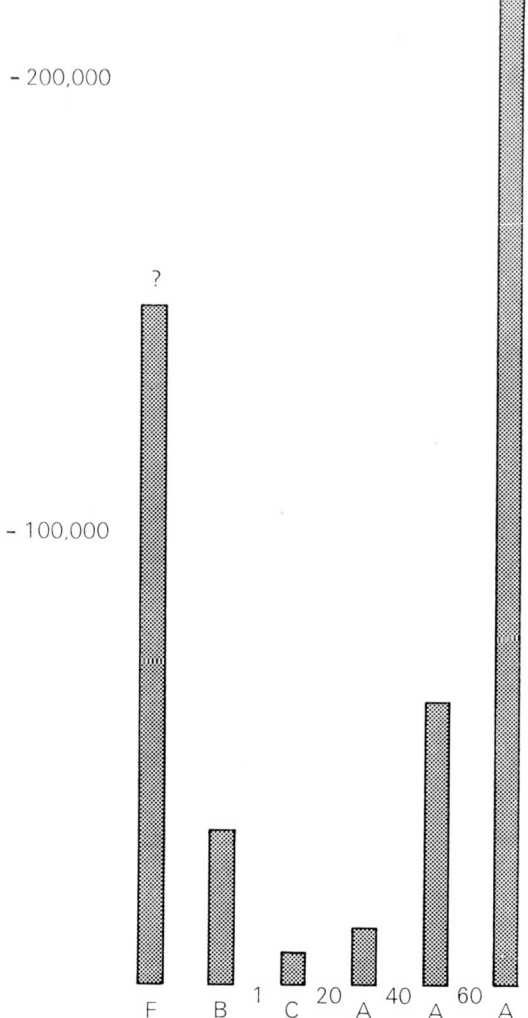

FIGURE 1
Natural Deaths — England and Wales, 1967, with assumed embryonic mortality added

impairment in viability a half unit deficit. If we exclude the proportional argument of impaired viability and restrict ourselves to disorders of simple genetic mechanism with complete genetic death we can make an approximate balance sheet by time of life or by genetic mechanism.

The total deaths by time of life, based on contemporary births and deaths in England and Wales and a very conservative rate of natural or spontaneous abortions of 15 per cent, is given in Figure 1. It is seen to be U-shaped, showing a dip at the period of life when death involves its maximum deprivation in both emotional and economic terms and is the approximate inverse of the distribution of reproductive worth [1].

If the genetic contribution, in terms of simple genetic mechanisms, is imposed on these histograms, it is seen to make a rapidly declining contribution, at least 20 per cent of abortuses, but only about five per thousand later births, dying or becoming sterile or intellectually crippled from chromosomal defects. This contribution to both death and disease is probably about equal to that from genic disease. Genetic causes of infertility are largely unknown, although these may well be present at an incidence of several per cent of the population, perhaps even tenfold the incidence of severe genetic disturbances to the soma.

Before costing this mortality and morbidity in terms of the genetic units involved it is necessary to consider these units and the consequences of their aberrations (Fig. 2). The upper part is largely explored since it relates to units larger than the wavelength of light. The penultimate units, the genes, are also well studied, since some are transcribed and translated to give products which are small enough to dissolve in water and which can be purified and analysed, and from these analyses inferences can be extended to the nucleotide level.

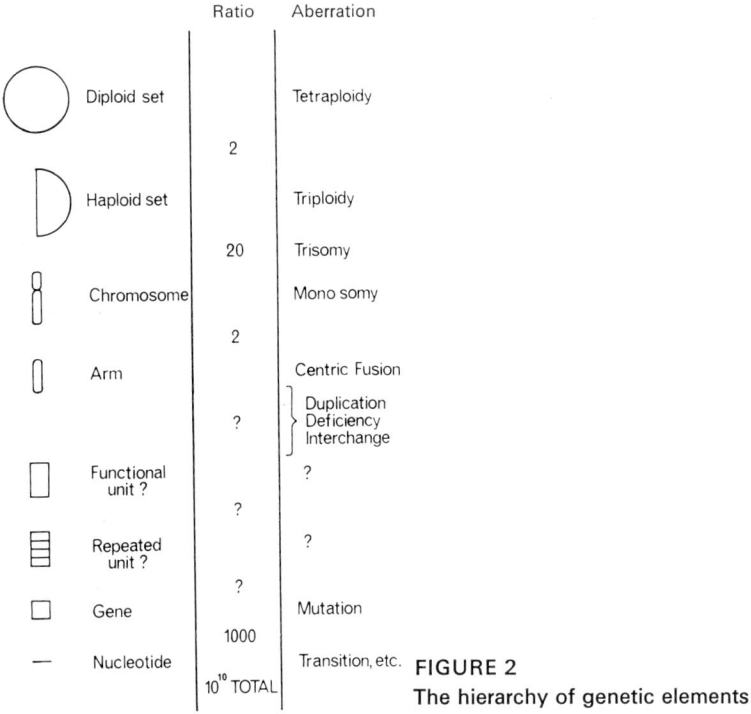

FIGURE 2
The hierarchy of genetic elements

SELECTION

Between those parts which are visible and those whose products are soluble little is known. The human chromosomes are very variable and, if parents are examined routinely alongside their children, chromosome arms of unusual length, or with peculiar satellites, can almost always be seen to have been transmitted unchanged. Chromosomes capable of certain recognition, such as 1, 2, 3 and 16, seem particularly variable, but this is largely because there is no chance of improving on nature by systematic mispairing. The satellites, whatever they are, and they are absent in some mammals, as most rodents, are extremely variable, and as they are directly inherited, and may be conspicuous or invisible on any of the ten acrocentric autosomes in the diploid set, a several fold range of total satellite volume per cell, or per person, is commonly found. A few conceptions must have no satellites, but I have not seen this in the living. These observations are difficult to document by photography and are beyond the resolving power of computer recognition based on a single focal plane.

Presumably this variability is due to variation at various levels in the unknown region, and is constrained and maintained by selective forces with its variability constantly being regenerated by heritable changes or mutations.

In addition to changes in amounts of the various units, structural changes, such as inversions, may be found. However, the meiotic evidence suggests that substantial inversions are rare. Small inversions will remain unobservable until there are technical advances in studying leptotene chromosomes which are readily demonstrable at autopsy [6, 7] in the male.

Sex, the primal genetic variant, has long been regarded as one of man's most costly acquisitions, a substantial excess of male conceptions being weeded out by selective abortion to give the fairly equal proportions at birth. This widespread assertion, based on the extraordinary apparent maleness of the female embryo, and later supported by sex-chromatin studies which failed to apppreciate that loss of the Barr body can be one of the more subtle signs of cellular death, has now been shown to be false. When abortuses have been sexed by caryotyping the proportions are fairly equal, and the proportion of XO cases sufficient to cast doubt on any very elaborate treatment. Any inequality in the sex ratio at conception must now be regarded as unsubstantiated by present techniques and incapable of substantiation by past techniques.

The units, if any exist, at the supragenic and subchromosomal level are unknown. Functional units have been inferred from bacteria, but there is no direct evidence for these in mammals, and the separation of loci coding for sequential enzymes, and parts of tetramers and dimers, suggest such units may not exist. Bacteria, being single cells, have to specialise in time and change their enzyme production rapidly; mammalian cells are usually

TABLE 1. Summary of Carr's (1965) Data showing inferences based on a spontaneous abortion rate of 20%. The data on aneuploidy at birth are estimates based on personal experience in Birmingham

		Data			Inference	
		Abortions		Live births	Conceptions	Embryonic survival
		%	per 10000	per 10000	per 10000 (LB + AB/5) × 5/6	
Total cultured	200					
Normal	156	72				
Abnormal	44	22	2200			
45,X	11	5·5	550	2	93	2%
47,XXY	0	0	0	25	(25)	(80%)
47,XXX	0	0	0			
Trisomy	22	11	1100			
13–15	6	3	300	2	52	4%
16	6	3	300	< 1	50	< 1%
17,18	1	0·5	(50)	3	(11)	30%
21,22	5	2·5	250	15	54	30%
Other	4	2	200	< 1	33	< 1%
Triploidy	9	4·5	450	< 1	75	< 1%
Tetraploidy	2	1	100	< 1	20	< 1%

irreversibly specialised in space by a mechanism which survives or obstructs mitosis and, since they cannot respond in this way, they can exhibit little evidence of such a mechanism. Repeated units may be assumed to exist, although the evidence in man is limited, and the uniformity of DNA content in the mammals is in marked contrast to that in those groups from which duplication was inferred. The extent of this duplication must be maintained at a cost in terms of genetic load. The extreme variation in the proportion of minority DNA fractions in different rodents [8] suggests some very active evolution constrained by intense stabilising mechanisms.

Table 1 attempts to relate the data on population incidence in a newborn population in England with that on a population of spontaneous abortuses in Canada; further data from Carr suggests these estimates of foetal death are conservative. Studies of abortuses in the Birmingham population are entirely consistent with Carr's more extensive findings. Figure 3 shows this flow diagrammatically, the aberrations who survive being derived from a minority of those conceived, plus some with a later origin, the mosaics. Table 2 attempts to express their results, as well as approximate estimates based on experience of the same population, by type of genetic unit. The cost is very great, and imposes serious problems in accountancy on the assumption that all debts have to be costed in terms of renewing

SELECTION

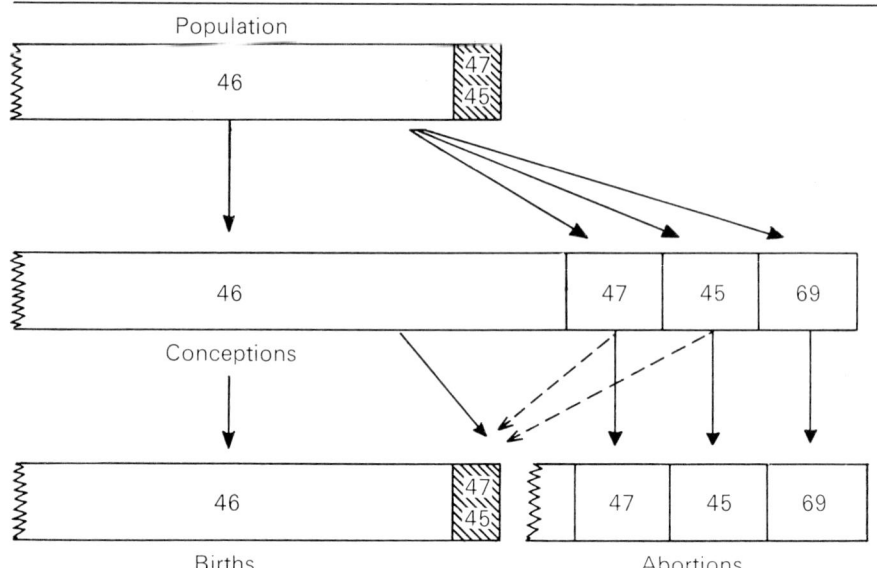

FIGURE 3
Dynamics of intrauterine selection

variability, the mutational load, or of maintaining a variability, the segregational load.

We have considered the *what* of selection, and, since it is very large and evenly spread over the known hereditary units, must continue in terms of *How, When* and *Why*.

How the chromosomal defects lead to death, and why the XO phenotype, which may be so mildly abnormal in childhood, should be so lethal *in utero*, with a mere two per cent survival, while the more disturbed mongol child has a safer passage, may appear a serious problem.

The three trisomies of man which are manifest at birth with an incidence of about 15, 4 and 3 per 10,000 in order of increasing chromosome size, lead to easily distinguished syndromes, each of which is predisposed to malformations which are often observed in isolation, and may be said to have multiple malformations. I think this is a misleading terminology. Multiple malformations, in the true sense of specific malformations in an otherwise normal child, are exceedingly rare in man. These are generalised malformations in which everything is wrong. The whole epigenetic landscape [9] is distorted, and from such distortions numerous evils can be expected. By considering the distortions, rather than their more easily scored, but more variable, consequences an experienced observer can usually make a precise clinical diagnosis. If we invoke Waddington's tent-peg model, [9] a model consistent with later discoveries of gene action,

TABLE 2. Genetic load in man

Mutational unit	Rate per unit	Abortion	Crippling	Infertile only	Carriers
Haploid set					
Triploidy	5×10^{-3}	5×10^{-3}	10^{-5}	—	None
Chromosome					
Trisomy of autosome	5×10^{-3}	10^{-2}	2×10^{-3}	—	10^{-4} (mosaic)
XXY, XXX, etc.	10^{-3}	—	10^{-4}	10^{-3}	None
XO	5×10^{-3}	5×10^{-3}	10^{-4}	10^{-4}	None
Chromosome arm					
Centric fusion	10^{-4}	?	10^{-4}	?	10^{-3}
Reciprocal translocation	10^{-5}	?	10^{-3}	?	10^{-3}
Visible deficiency	?	?	$10^{-3} - 10^{-4}$	10^{-4}	Rare (mosaic)
Invisible deficiency	?	?	?	?	?
Functional unit	?	?	?	?	?
Repeated unit	?	?	?	?	?
Gene					
Mutation	$10^{-5} - 10^{-6}$	?	10^{-3}	$?10^{-3}$	?
Segregation		$?10^{-2}$ (ABO) + ?	10^{-3} (Rhesus) + ?	?	Everyone
Nucleotide	$10^{-8} - 10^{-9}$				
Total due to simple genetic mechanisms		·03	·01	·01	
Total loss		·2	·05	·05	
Mutational load Visible Chromosomal / Genic		·5 – 10	1·0 – 10		

then a duplication of some two to four per cent of a haploid set would lead to distortions which would differ with differing selections of sets of tent-pegs. Further, such distortions in embryonic development would have different consequences at different times, and would be particularly prone to lead to disaster at times of very rapid development when different tissues, including those not partaking in the embryo proper, were in active competition.

An acceptance of this model excludes the possibility of usefully relating pieces of chromosome to pieces of the phenotype. This variability in the time of death and the shape of survivors does not seem to be a real problem, in that the consequences we observe could reasonably be predicted from the nature of the genetic control of embryogenesis; so could the wide variation in the incidence of consequences of this general distortion.

When, we may now ask, do all these errors arise? Here we are badly informed, in that our human data are biased and our analogies based on very distant species.

There are some observations which most experienced observers will probably accept without documentation, and which I shall assume.

1. Tetraploidy is common in tissue culture, and lymphocyte culture, often affecting over one per cent of cells.
2. Triploidy is unknown in human tissue cultures derived from diploid cells.
3. Mosaicism is common in trisomy; in attempting to evaluate my own highly biased data I think the incidence may be as high as ten per cent.
4. Extra chromosomes and abnormal chromosomes are usual in neoplasms and established cell lines.
5. Mosaicism is unusual in structural anomalies of the autosomes and certainly less common than in trisomy. It is usual in structural anomalies of the X chromosome.
6. Structural changes may arise in culture, and are usual in established lines and spontaneous neoplasms.

Further, mongolism [10, 11] and other trisomies are commoner with increasing maternal age and independent of paternal age: the effect is too great to be explained by a changed risk of abortion. No such marked effect is shown in the structural anomalies, or in sex-chromosome anomalies.

If we accept these observations we can attempt to form Table 3, although there is considerable uncertainty; there is suggestive evidence that this costly variability arises from errors at many different stages, and the exact mechanism must await the identification of genetic markers near the centromere in the viable trisomies, and of genetic markers in culture in the

TABLE 3. Common sites of mutation inferred from available data

	Rate	Sites of mutation			
		Interphase	Mitosis	Meiosis	Zygote
Diploid set × 2	10^{-2}	−	+	?	?
Haploid set × 20	5×10^{-3}	−	−	+ or	+
Chromosome × 2	5×10^{-3}	−	+	+	?
Chromosome arm × ?	10^{-4}	(+)	−	+	−
Functional unit ? × ?	?	(+)	−	(+)	−
Repeated unit ? × ?	?	(+)	−	(+)	−
Genic unit × 1000	10^{-6} per unit	+	−	?	−
Nucleotide	10^{-9} per unit	+	−	−	−

TABLE 4. Types of acrocentric imbalance

Possible gametes and incidence, or viability, of product

Nondisjunction	DG	DG	D	G	DGG	D	DDG	G
	N	N	I	I	10^{-3}	I	2×10^{-4}	I
Centric fusion	DG	(DG)	D	G	(DG)G	D	(DG)D Unknown	G
	N	10^{-4}	I	I	10^{-4}	I	$< 10^{-5}$	I
Ratio	−	104	?	?	10	?	> 20	?

N = Normal; I = Inviable

triploids. We cannot even say that mongolism is usually genetic, if by genetic we mean present in the zygotic nuclei.

There is evidence that the sex-chromosome anomalies, and the autosomal anomalies, are very different in action. In the former, structural changes are usually associated with mosaicism, which implies either that abnormal chromosomes are unstable in mitosis, which conflicts with the direct evidence of tissue culture, or that the abnormality and the mosaicism arise at the same time.

Structural abnormalities of the autosomes are rarely mosaic, and therefore more likely to arise at meiosis, when rearrangements are, in any case, occurring. A rough comparison of the incidence of the various numerical and structural anomalies involving the acrocentric chromosomes, on the assumption that only one of the large (13–15) and small (21–22) groups is involved, is given in Table 4. The constancy of the proportions makes it tempting to postulate some single mechanism for these structural and numerical aberrations. However, the frequency of mosaicism in trisomy, especially mongolism, suggests a mitotic non-disjunction may be common. The maternal age effect can exclude male gametogenesis as a usual cause, but cannot distinguish between a predisposition of old ova before or after fertilisation.

The significance of these stages may be clarified by Figure 4, which shows the diploid phase which intervenes between the gametes, accepting two different types, and, on average, conveying an effective pair of one type.

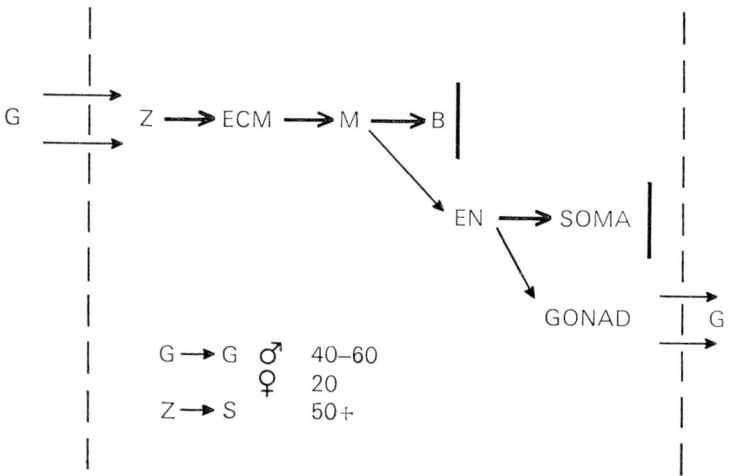

FIGURE 4
Diagram of somatic selection between gametes (G) in the sequence zygote (Z), early cell mass (ECM), morula (M), blastocyst (B), embryonic nodule (EN) and gonad. The number of mitotic divisions between gametes is probably about 20 in the female and 40–60 in the male and in both cases less than the average somatic end cell

The development of the embryonic nodule, at a time when there are of the order of 1,000 cells, from about ten successive divisions, provides strong selection against mosaicism if the extra-embryonic parts remain extra-embryonic, so that, if mosaicism arises in the first few divisions, the embryo would usually be consistently normal or, less commonly, consistently abnormal. Further selection will occur through foetal death and abortion, which seems almost restricted to non-mosaic conditions.

The cell lineage in the diploid phase reveals a mechanism through which the number of mitoses is minimised for the germ cell, especially in the female. This mechanism may lead to aberrations in unusual circumstances, as twinning in cows; in man, as in other primates, chimaerism from twinning appears harmless. It is very rare in man [12]. No case has been observed in 1,200 pairs of twins examined by placental injection and blood and enzyme typing in Birmingham [13].

If we extend our enquiry from *What, How* and *When* to *Why*, we are even further from available data. With such a vast mortality occurring within our species, and the knowledge that it is a recent species and differs in chromosome number, like most species, from its immediate neighbours, it is necessary to consider possible mechanisms, by which I mean mechanisms which might plausibly have assisted evolution in the past or which might assist viability of the species, as a whole, in the present.

If we consider first the present, we may consider each stage in turn. The mechanisms seem uniformly unconvincing.

If we start with female gametogenesis, which, in mammals, usually starts before the male partner's, the recombination fraction may be very variable by woman, and by age, and non-disjunction may arise from absence of recombination [14]. Variability in the female seems to exceed that in the male [15, 16] (Fig. 5), but data on ova from women over 40 are not available. The genetic length is presumably under genetic control and constrained by selection, although little is known of this. There is little evidence of an age effect in recombination frequency in the male, and, although scoring is more difficult, the variance of the male seems less than in the female, and hardly disturbed by age.

Next we may consider the ovum after shedding. If it is not fertilised it will die, and if fertilised just before dying, abnormal behaviour seems likely, and has been demonstrated in the frog [17]. The data in man are suggestive [18]; however, in mongolism, the only common condition capable of accurate clinical recognition, illegitimate births do not seem peculiarly prone, and the incidence in countries reputed to utilise the safe period is not particularly high.

At and after fertilisation mechanisms, such as those of the sterility alleles in plants, may be expected which would reduce inbreeding by making gametes which were similar mutually unattractive. Penrose's observation [19] on mongols resembling their mothers, as opposed to their fathers, over several unlinked markers, remains unexplained, and is well beyond the likely scope of chance (Table 5). It could act against homozygosity. It is unlikely to be explained by selective survival, as the abortion rate of mongol embryos appears, from both translocation families and the pregnancies of mongols, to be fairly low compared to other trisomies. After implantation the reliable aborters, such as triploidy and trisomy 16, might hardly impose any serious load, in that a shortage of conceptions is unlikely to afflict man, and a mechanism encouraging spacing without risk of protracted infertility might well have become established in the past. In particular abortion gives some protection against

SELECTION

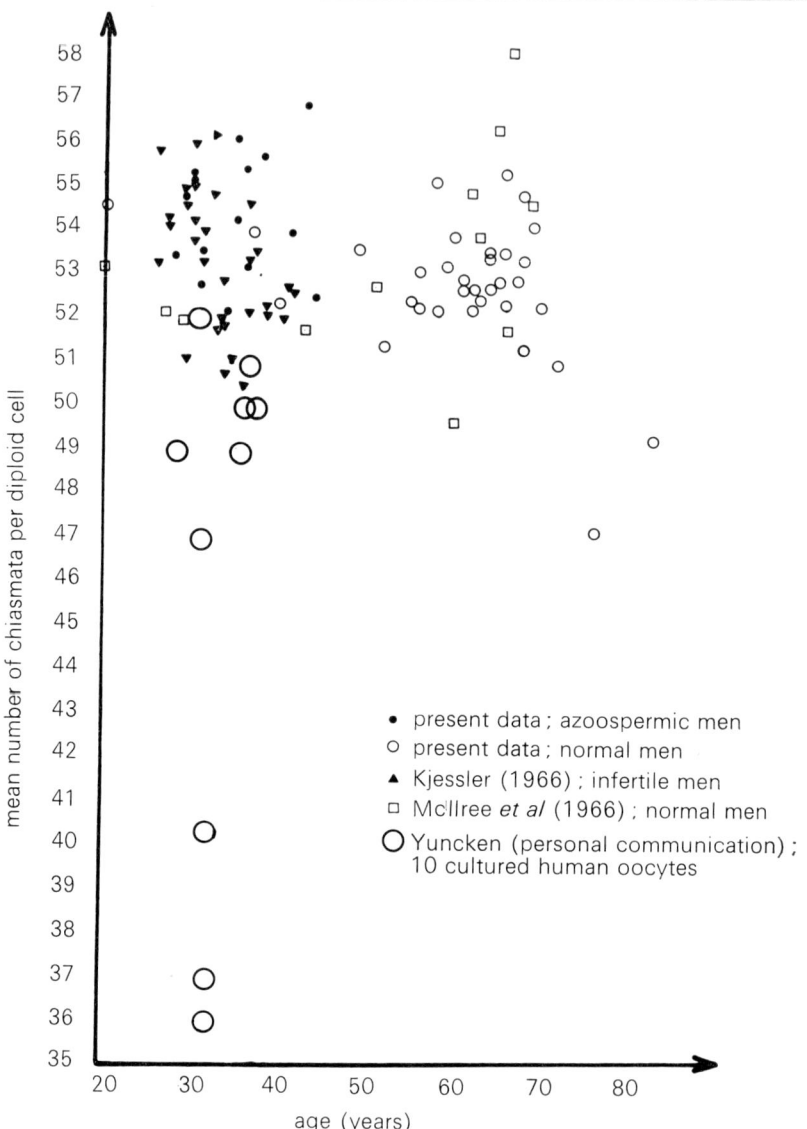

FIGURE 5
Chiasma frequency against age in man and woman

multiple birth : when one in *n* pregnancies is a twin pregnancy, a small abortion rate will prevent two twin pregnancies at a cost of aborting *n* singletons.

After fertilisation the blastocyst may fail to embed, leading to an early and undetectable loss. The incidence of this is difficult to estimate in man. The figure of 15 per cent loss in the cow, which Professor Alan Robertson

TABLE 5. Data on expected and observed maternal similarities of phenotypes in mongolism

System	Expected	Observed	Ratio
ABO	54·8	36	·657
MNS	83·2	82	·986
Rh	83·0	74	·892
P	39·2	26	·663
Lewis	30·7	26	·847
Kell	11·1	5	·450
		Product	0·15

has quoted, would seem consistent with the limited human data. The suggestion that all forms of trisomy are equally likely, but some lead to very early losses, would imply a pre-implantation loss severalfold the rate for post-implantation losses or abortions, so that a majority of fertilised ova would be lost without trace. As the human chromosomes are demonstrably not randomly distributed on the metaphase plate there seems no problem in postulating that some chromosomes might be particularly prone to non-disjunction.

In foetal life some mechanisms for unusual mitosis seem necessary, since some tissues normally have polyploid nuclei, on the evidence of DNA measurements, including some cerebellar cells in man and liver cells in mice. After birth the main hazards relevant to any compensation from mitotic peculiarities are infection and neoplasia. A balanced form of chromosomal aberration, double mitotic non-disjunction involving homologues, would be most advantageous in the lymphatic system in the generation of variation, and would explain the haploid expression of genes coding for myeloma proteins. However, if this were at all common it might be expected to show up as occasional double representation, or non-representation, of distinctive chromosomes and I have never noticed this in observing metaphases in lymphocytes. There is also negative evidence in other tissues. Parents of albinos are without any albino hairs, and, if Negro, are free from any white speckling of the skin. The obvious advantages of chromosomal variability as an aid to cloning variability in lymphocytes, and the extreme rarity (in my experience) of numerical aberrations in lymphocytes, suggests an extreme conformism of mitotic mechanisms in different tissues.

No compensatory advantages are evident and plausible to balance this extreme lethality in early life, and we must consider either that these disorders might be mere accidents of a complicated mechanism which it is impractical, or extravagant, to guard against, or that they are relics of some advantageous evolutionary event : if so we must accept some ascertainment bias from being the first species to discuss its caryotype.

Abnormalities of haploid set number appear to have been devoid of evolutionary significance in most animals, as opposed to plants, and their frequency must be attributed either to chance circumstances, such as the fortuitous spatial relationship of the three pronuclei just after fertilisation, to sharing a mechanism irrevocably tangled with the genesis of other forms of chromosomal variability, or in some way imposing some selective disadvantage at a stage in life when it is relatively free from cost, or may even have a negative cost. Such a method need not be very efficient to be worthwhile : a relatively harmless selective destruction of one per cent of zygotes, even if poorly selected, would compete in costing with a highly efficient destruction after birth, as in the more severe recessive disorders. Such a mechanism would need to be dependent on the male gamete. However, the extraordinary excess of XXY triploids suggests that this may be so [20].

It might also seem plausible to relate congenital and acquired chromosomal disorders, and regard the predisposition to aneuploidy, zygotic or foetal, and neoplasia as distinct and balanced. With 10^6 mitoses a second for red cell production alone, and a number of mitoses per lifetime exceeding the number in direct line from our first nucleate ancestor, and the fatal neoplastic consequences of ill-disciplined clones with distinct caryotypes, the influence of selection for or against determinants influencing mitosis must be strong; a number of disorders which show regular Mendelian segregation and predispose to neoplasia are now known, as well as a few families clearly predisposed to both chromosomal aberrations and neoplasia. Apart from these the familial predispositions to both trisomy and neoplasia appear weak, with a mere doubling of incidence in first degree relatives.

Finally we consider the possibility of chromosomal instability as a major factor in mammalian evolution, the mutations of the large genetic units being similar, in compensatory advantage, to genic mutations. This is hardly a novel concept, and has been widely discussed for other species, especially arthropods [21] and higher plants [22]. White states, 'It is now generally agreed that, as far as the higher animals are concerned, all evolutionary transformations have had their origin in the chromosomes, and that these bodies which constitute the physical basis of heredity also furnish the material source of evolutionary changes' [23]. However, it must be acceded that this agreement has hardly diffused to most theoretical evolutionists. The distinction of the chromosomes of most primates imposes the need either to regard chromosomal changes as a cause or a consequence of evolution [24, 25]. The former is simpler and, if satisfactory, excludes the need for the latter.

Consider a spontaneous reciprocal translocation. The progeny of a mating may have several heterozygous offspring, and these, or their descendants, may mate to produce several homozygotes, who would be capable of breeding as a clone and their genes would be restrained from diffusion by semi-sterility with other matings, and a few complexes temporarily restrained from disruption by recombination. That is, an isolating event would have occurred, as a result of which a clone, starting with the genetic equipment of a mere few dozen individuals, would be capable of competing with its progenitors without the need to postulate any environmental disturbance.

The genetic advantages of this are twofold. Firstly, and obviously, the first translocation would have a 50 per cent chance of occurring in someone of above average fitness, and, independently, the chromosome would have a 50 per cent chance of 'splinting' an above-average set of genes against rearrangement until homozygosity was assured by descendants homozygous for the same chromosome. Secondly, and more important, the clone would be able to evolve with the variability it had at a far greater speed, and, given adequate variability, it could anticipate far more rapid response to any environmental changes, as well as a more rapid riddance of deleterious genes. This follows simply from Haldane's argument of the cost of selection [5].

The cost of changing from a gene frequency of p_1 to p_2, given moderate or mild selection, is $(\log_e p_2 - \log_e p_1)n$ individuals. To allow complete spread of a mutant in a population of n individuals will need $n[-\log_e(1/2n)]$ which is $n \log_e(2n)$ or $\log_e(2n)^n$. This leads to the dilemma that a small population may not have a mutant, and a large population may be unable to afford to exploit a mutation. Recurrent episodes of cloning would appear to allow extreme opportunity for any mutants included. The numbers are not unreasonable. A translocation might have a chance of 10^{-3} of arising, 10^{-2} of leading to several homozygotes, and 10^{-1} of leading to homozygotes whose fitness was in the upper tenth centile. 10^{-6} is by no means impossibly small for a species with 10^{-6} individuals breeding annually for 10^7 years. Indeed it would happen ten million times.

This allows, spontaneously, the development of a restricted population within which complex gene rearrangements can be built up without excessive loss through diffusion, and permits surface-living species the advantage of isolation otherwise restricted to species living on islands or mountain tops, or in lakes or estuaries, the classical sites of rapid evolution involving small numbers.

If this argument is correct, then our high genetical load from chromosomal disorders is to be seen as a curious ascertainment problem which will afflict any species which can evolve with rapidity to the point of discussing its caryotype. Assuming that propensity to translocation is partly heritable, some reduction in liability from the stock which was likely to be particularly prone might be expected.

In evolutionary terms, we have to distinguish between genes spreading by gene diffusion, which is theoretically tractable on the unlikely assumption of zero correlation of parent-child fertility, between chromosomes spreading by a similar diffusion, as in the apparently harmless variations of centric fusion, and between the larger set of the diploid forms, or the descendants of a diploid form, spreading by invasion and ousting the parent stock which lacks the speedy response of the numerically small unit. The effect of the translocation is not to affect the phenotype, as in Goldsmith's argument, but to insulate, and hence isolate, a genotype.

Evolution through this mechanism would be asymmetric, the parent stock maintaining more variety, but the new stock competing in rapidity by such responses as its more limited allelic repertoire allowed.

If we consider the problem of whether artificial selection should be added to natural selection the matter seems fairly simple, whether considered in terms of screening all births, or of screening all mongols, the commonest group of severely deprived survivors.

The population genetics of mongolism is extremely complicated. Figure 6 shows the minimal number of parameters relating to the genesis of balanced, unbalanced and trisomic anomalies (μ_b, μ_u, μ_t) and mosaicism (μ_m) and the selection parameters acting between conception and viable birth (S_b, S_u, S_t), where S_b is probably the normal survival and S_u is probably equal to S_t. The segregation of the balanced form (the relative strengths of the three arrows radiating from B) is also critical. With so many unknowns, precise estimates require extensive unbiased data, and this is only possible from population studies supplemented by studies of abortuses. Even this is inadequate due to the difficulty, or impossibility, of distinguishing chromosomes 21 and 22.

However, the value of most of these parameters is hardly relevant to the practical problems facing both the paediatrician and the administrator, and data are now sufficiently extensive to allay the costly anxieties generated by the biased nature of the earlier data and sustained by irresponsible journalism.

Chromosomal study of all British mongols would identify the 150 annual translocation mongols, and study of the parents would reveal, assuming the data for Denmark [26, 27], which has the largest unbiased series, to be consistent with other data (Figs. 7 and 8), that about forty of them would have a translocation parent. The average parent of a randomly ascertained child would be expected to have a reproductive expectation of about 1·8 for first births and 1·0 for later births, and if the segregation leading to

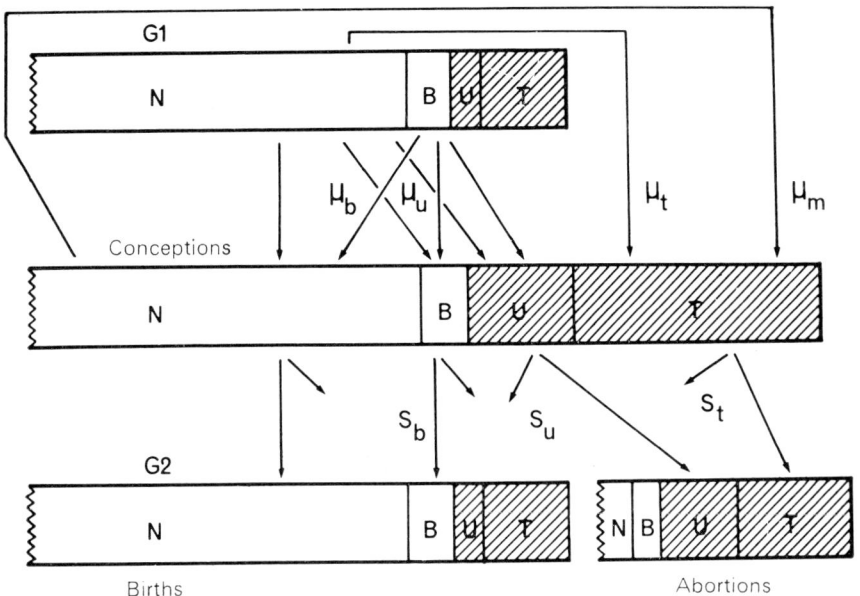

FIGURE 6
Population dynamics of mongolism : b= balanced ; u= unbalanced ; t= trisomic

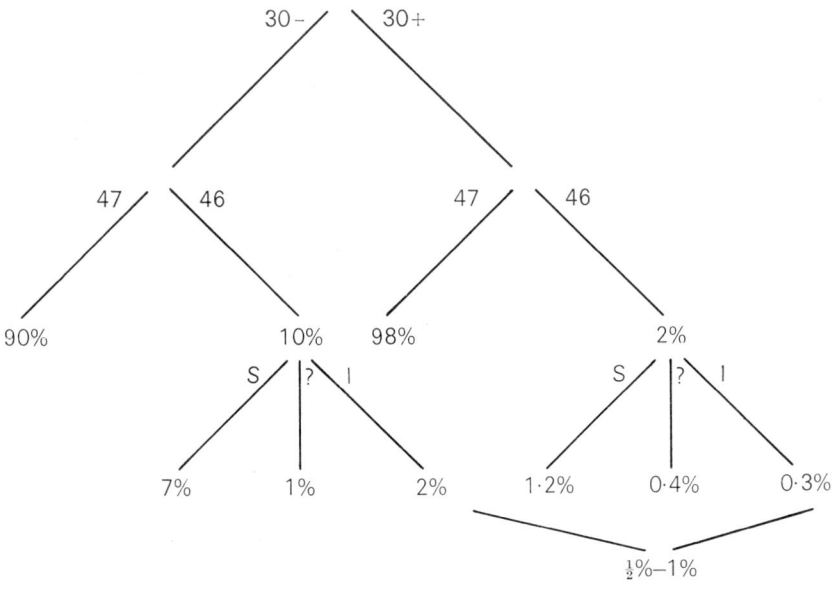

FIGURE 7
Types of mongolism by maternal age (above and below thirty), trisomy and translocation (47 and 46) and spontaneous or inherited (S or I) showing the percentage of inherited forms expected from screening all mongols [26]

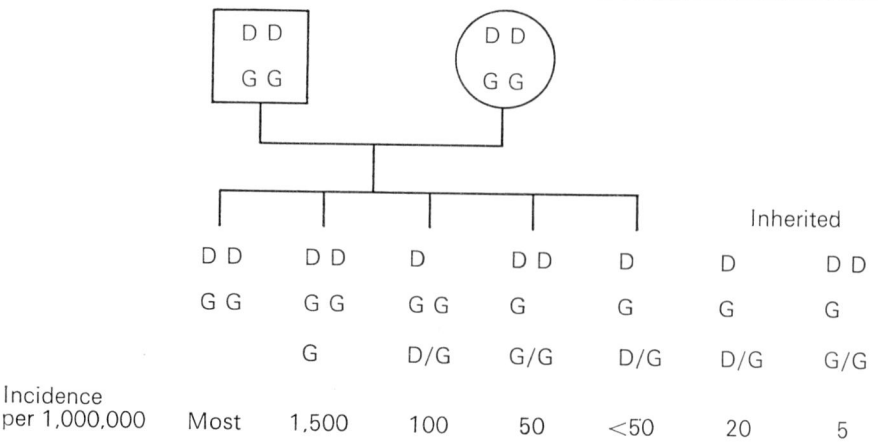

FIGURE 8
Summary of previous figure applied to a population of 1 million births a year

mongolism compared to a normal phenotype were 0·2, this group would have a mean expectation of producing mongols, without benefit of advice, of about 0·3 and, with advice, perhaps 0·15. The contribution of routine chromosomal analysis at a national level would be the non-birth of 0·15 × 40, or about six mongols, at a cost of the non-birth of around 25 non-mongols, some of whom would be carriers. Under stable conditions between ten and twenty fewer mongols might be born per year, a reduction in incidence of about one per cent. Scrutiny of all births would perhaps double this number of non-births of mongols by the detection of potential high-risk parents.

This appears to be the upper limit of immediate benefit from the chromosomal analysis of all mongols. Against this small and costly benefit must be set the need to break the news of the diagnosis immediately after birth, and to tell the parents not only that they have one mongol, but that they should also worry about having another.

The supplementary benefit of more certain diagnosis does not seem very great; in my experience in Birmingham most of the cases confusing to experienced paediatricians were, from the practical point of view, almost as confusing after chromosome study, since they were mosaics or otherwise atypical, and the main practical benefits of chromosomal studies were in the elucidation of intersexes.

A survey of all births would reveal a number of translocations, Turner's, and Klinefelter's; this would be of limited benefit. A few undiagnosed severe trisomies would be saved the indignity of operative intervention and permitted a natural death uncluttered by tube feeding, and one girl in 4,000 could have the offer of breast development for purely cosmetic purposes a few years earlier and increase her chance of a marriage doomed to infertility. Whether it is useful to tell parents of children who are XXY, XXX or XYY that they carry this peculiarity is uncertain, and unless observations are made before this is done, nothing will be known. If, as seems possible, the XYY has a one in a thousand risk of being a murderer, and one in a hundred of being a car-thief, it would seem a little unfair to the majority to make this known. It is at present unclear if psychiatric supervision or guidance of normal children is beneficial, and, until this is studied with controls, it would be unwise and even unethical to submit children with normal behaviour to supervision on grounds of chromosomal abnormalities.

At any rate there are no genetic hazards in aneuploidy, using the term in its strict sense of numerical anomaly without structural change, since the victims are either effectively sterile, or fail to pass on the anomaly, and the benefits of the last decade, which has made man the best studied species from the chromosomal point of view, have no immediate, harmless or economic application on a population scale, except for research purposes in a few selected areas.

REFERENCES

[1] Fisher, R.A., *Genetical Theory of Natural Selection*. Oxford : University Press, 1930.
[2] Muller, H.J., *Am. J. hum. Genet*, **2**, 111, 1950.
[3] Crow, J.F., *Hum. Biol.*, **30**, 1, 1958.
[4] Fraser, G.R., *Ann. hum. Genet.*, **25**, 387, 1962.
[5] Haldane, J.B.S., *J. Genet.*, **55**, 511, 1957.
[6] Edwards, J.H. and Guli, E., *Nature, Lond.*, **199**, 1114, 1963.
[7] Guli, E., *Atti dell'Accademia dei Fisiocritici*, **13**, xvi, 1250, 1967.
[8] Walker, P.M.B., *Nature, Lond.*, **219**, 228, 1968.
[9] Waddington, C.H., *The Strategy of the Genes*. London : Allen & Unwin, 1957.
[10] Jenkins, R.L., *Am. J. Dis. Child.*, **45**, 506, 1933.
[11] Penrose, L.S., *J. Genet.*, **27**, 219, 1933.
[12] Race, R.R. and Sanger, R., *Blood Groups in Man*. Oxford : Blackwell, 1968.
[13] Cameron, H., Edwards, J.H. and Wingham, J., Third International Conference on Congenital Malformations, The Hague, 1969.
[14] Mather, K., *Biol. Rev.*, **13**, 252, 1938.
[15] McDermott, A., Ph.D. Thesis, Birmingham, 1969.
[16] Yuncken, C., unpublished data.
[17] Witschi, E., *Cancer Res.*, **12**, 763, 1952.
[18] German, J.L., *Nature, Lond.*, **217**, 5129, 1968.
[19] Penrose, L.S., *J. ment. Defic. Res.*, **1-2**, 107, 1959.
[20] Edwards, J.H., Yuncken, C., Rushton, D.I., Richards, S. and Mittwoch, U., *Cytogenetics*, **6**, 81, 1967.
[21] White, M.J.D., *Animal Cytology and Evolution* (first ed.). Cambridge, 1945.
[22] Darlington, C.D., *Evolution of Genetic Systems*. Edinburgh : Oliver and Boyd, 1939.
[23] White, M.J.D., *Animal Cytology and Evolution* (sec. ed.). Cambridge, 1954.
[24] Chu, E.H.U. and Bender, M.A., *Science*, **133**, 1399, 1962.
[25] Hamerton, J.L., Klinger, H.P., Mutton, D.E. and Lang, E.M., *Cytogenetics*, **2**, 240, 1963.
[26] Mikkelsen, M., *Ann. hum. Genet.*, **31**, 51, 1967.
[27] Mikkelsen, M., Communication to Conference on Clinical Delineation of Birth Defects, Baltimore, 1968.

Progress on an Automatic System for Cytogenetic Analysis

NIEL WALD, RUSSELL W. RANSHAW
JOHN M. HERRON *and* J. G. CASTLE

Department of Radiation Health and
the School of Engineering
University of Pittsburgh

AUTOMATIC ANALYSIS

¶ THE GOAL of rapid cytogenetic analysis for the study of large populations of cells and individuals is the basis for the design of the automatic cytogenetic analysis system under development in our laboratory [1]. Possible applications include: (1) determining the range of variation and the prognostic value of karyotypes of the newborn, (2) screening of populations whose exposure to deleterious environmental factors can be modified if necessary, and (3) *in vivo* and *in vitro* testing of potentially harmful chemical agents such as food additives, pesticides, and drugs.

The design chosen will supplant some of the more tedious and time-consuming manual steps in our current cytogenetic analysis. The four principal functions within the system that are being automated are:

1. High-speed location of the metaphase cells on batches of six slides which have been prepared for cytogenetic analysis in the usual manner and loaded onto the rotating stage of a special optical microscope;
2. Optical scan, at high resolution and under computer control, for the automatic conversion of the 'image' of each selected cell into binary form useful for the computer karyotype analysis;
3. On-line computer analysis of each selected cell image and the generation of its karyotype; and
4. On-line computer processing of the resultant data concerning the cytogenetic findings and their relationship to the pertinent individual and population data.

The purpose of this paper is to report the progress we have made toward automating these functions. Each function will be discussed separately, after a few words about the developmental history of the complete system.

PRESENT DESIGN OF THE COMPLETE SYSTEM

The automatic system [1] is intended to facilitate several kinds of cytogenetic studies ranging from research in depth on a particular cell or small samples of cells, to the clinical study of problem patients, and the cytogenetic monitoring of large human populations. The successful automatic operation of all four functions mentioned above should lead to the capability of rapid processing of adequate numbers of karyotypes at reasonable expense for these purposes. Successful operation will require obtaining sensitivity to infrequent chromosomal aberrations while maintaining rapid processing.

The equipment assembled to do this analysis is indicated in the block diagram of Figure 1. The block labelled 'Automatic Microscope' represents the cabinet, supplied by the Perkin-Elmer Corporation in December 1967, which contains parts for two optical systems with a common rotatable stage and the associated drive and focus controls. One is the laser optical system for cell location. The other is a high resolution optical system for photographic recording and for image conversion by means of the flying

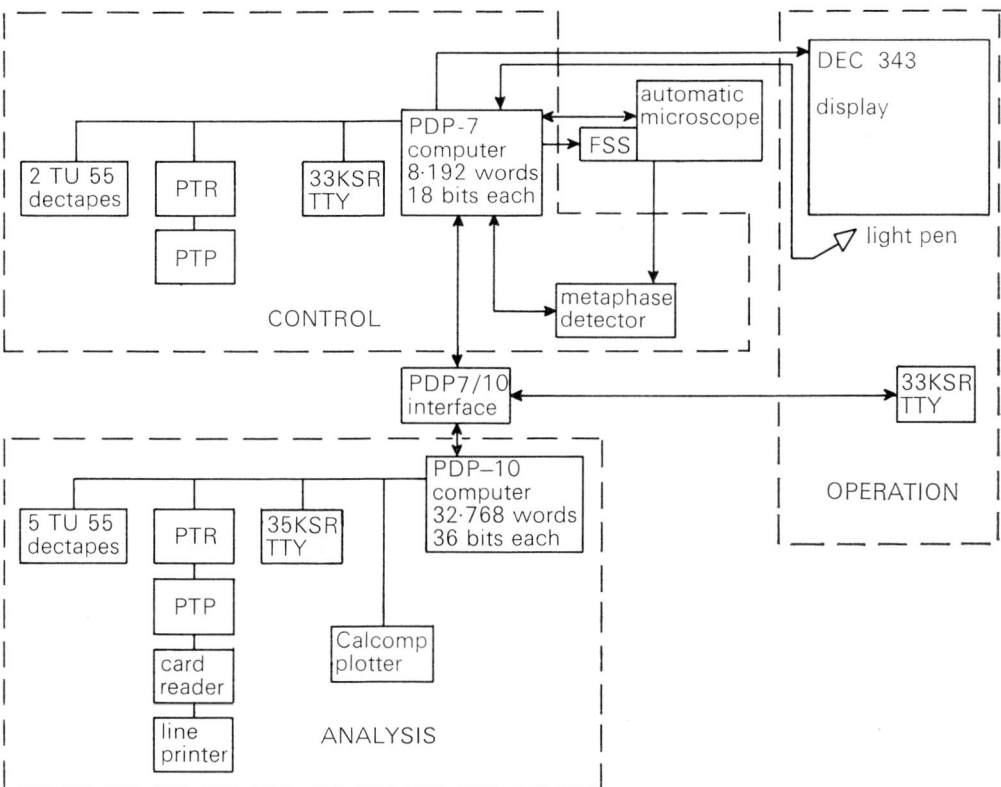

FIGURE 1
Block diagram of the automatic cytogenetic analysis system under development at the Department of Radiation Health of the University of Pittsburgh, showing the installed components as of May, 1969. The automatic microscope block includes the flying spot scanner and the rotatable stage with its two optical systems—a laser system for cell location and a high resolution optical system for scan conversion of the image of each selected cell to binary form

spot scanner. Both have been operated under the command of the PDP-7 control computer.

The control computer, labelled 'Control', has been successfully interfaced to the PDP-10 analysis computer. The interface unit was designed and built in our shop because no standard unit was available; it exchanges 100,000 words per second between the two machines. The primary functions of the interface are to transfer image data in binary form from the automatic microscope to the analysis software in the PDP-10 computer and to transfer control information generated by the PDP-10 analysis programs back to PDP-7 control programs. The interface unit also transfers image data selected by the PDP-10 program back to the PDP-7 for display on the large cathode-ray unit in the operator's station. The display is operational and we expect, as has been suggested by others [2], that display of questionable cell or chromosome images will be a rather frequent occurrence during karyotyping. This may be especially useful when the automatic cell locating system 'locates' a bubble or a piece of dirt, instead of a metaphase cell.

The planned sequence of operations in a karyotype analysis on our automatic microscope system starts with the hand loading of a batch of six slides prepared from cell cultures onto the rotatable stage, shown in the perspective view in Figure 2. Then during steady rotation at about 20 rpm, the slides will pass through the laser beam. The beam will be analysed for scattering by metaphase cells using the two masked photomultiplier tubes shown by dashed lines in the figure. When the decision circuitry indicates that the laser beam is striking a metaphase cell, the position of the platen, as read from the platen (angle) encoder and the radial arm encoder, will be recorded in the control computer. Each recorded position will then be relocated, one at a time, under the high resolution optical system, by driving the platen with its stepping motor. After automatic focusing on each stationary object, the flying spot scanner will convert the image to binary form. The resultant data will be transferred to the PDP-10 computer for inspection and possible karyotyping. Questionable images are to be displayed on the large oscilloscope display for operator decision. Photographs can be taken through the high resolution optics under control of the PDP-7 computer when desired. In addition, pertinent data, such as patient name, date, slide number, and computer-generated karyotypes, can be recorded on the same film by the PDP-7 computer.

A few comments about the real time schedule expected during automatic operation are in order before we discuss the automatic functions separately. The complete sweep of a batch of six slides at 20 rpm for the purpose of cell location is expected to take one hour. With a conservative estimate of 100 usable metaphase cells in a batch, the system should locate about one cell every 30 seconds. Automatic focus may take as much as 60 seconds per selected cell and the subsequent scan of the cell with the flying spot scanner will be delivered to the PDP-10 in about 30 seconds. On the basis of our limited experience with such data processing, we estimate about 90 seconds for the computer karyotype with recording of pertinent data taking another 30 seconds per cell. Adding these steps together and including reloading and some maintenance gives a conservative estimate of the overall rate of processing to be about 200 cells per day; that is about 10 cases per day at 20 karyotyped cells per person, or 100 cases per day at 2 cells per person.

The rate of processing will, of course, be very sensitive to the level of discrimination attainable in the cell locating and in the karyotyping analysis. During the process of locating mitotic cells, the machine will be discrimi-

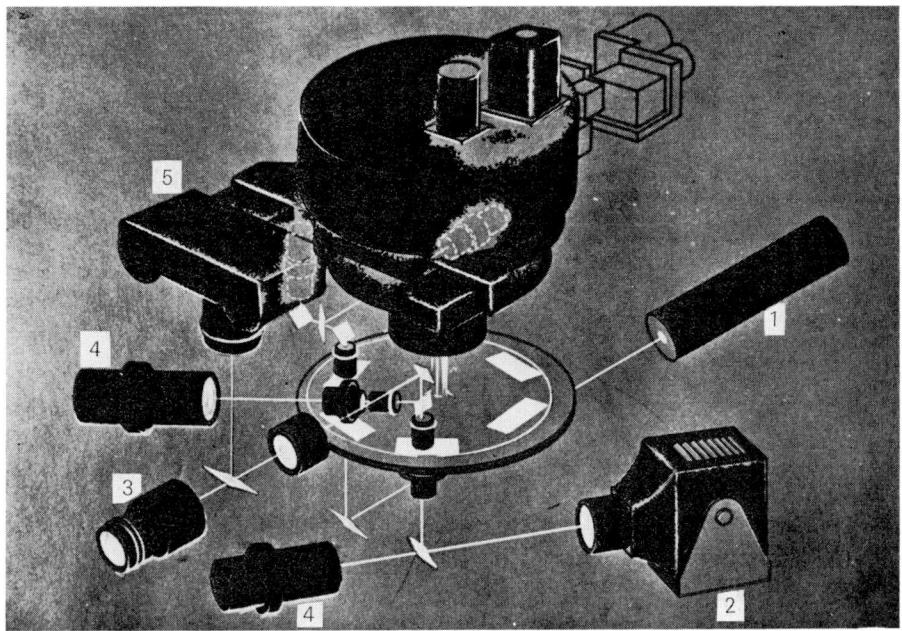

FIGURE 2
Opto-mechanical assembly of the automatic microscope. The rotary platen is shown in the center of the photograph. The other components are : (1) He-Ne laser ; (2) White light source ; (3) Flying spot scanner lens system ; (4) Photomultiplier data collecting system ; (5) 35 mm camera

nating between the light scattered from mitotic cells and from a variety of 'noise' objects, and generating a recognition signal when a cell is 'located'. Early data [1] indicate that recognition of a small fraction of the available mitotic cells can be done with good discrimination (few false recognitions) but that recognition of most of the available mitotic cells will probably involve many recognitions of 'noise' objects, which then must be scanned and rejected in the later steps. During the analysis process, similar discrimination between single chromosomes and other objects will be required. Experience with fully automatic operation is really essential to determine what levels of discrimination are attainable and thereby what rates of processing this system will have for studies of large populations.

In addition to serving as an engineering example of automatic cytogenetic processing of human populations, the automatic system is expected to be capable of performing other cytological analyses for research or routine clinical studies such as peripheral blood leukocyte differential counts. The flexibility required should be available from the control programs of the PDP-7 control computer, the analysis software on the PDP-10 computer, and from the cell location system.

HIGH SPEED LOCATION OF METAPHASE CELLS

Metaphase cells are situated at random on the microscope slides as presently prepared by our cytogenetics laboratory. There are typically a few dozen usable metaphase spreads on each well-prepared slide. Ideally the automatic system should locate all of them, but in so doing, will undoubtedly also locate many 'noise' objects such as hair or dust particles. For example, one of our present problems is the presence of air bubbles in the microscope oil; fortunately, the bubbles do not usually repeat from one revolution to the next. With the 'noise' problem in mind, we shall discuss first some of the design requirements on the cell-locating system, then the means we have chosen to implement the location function, and finally some of the successful steps we have taken toward automatic performance of this task.

Design requirements for location of metaphase cells. The cell-locating system is expected to locate a cell that has been caught in metaphase by producing a recognition pulse and indicating the platen position by reading both the angle and the radial arm encoders. If the recognition pulse is generated falsely by a dust particle or other noise object, then the steps that follow in the automatic chain must do the discriminating. Consequently, the generator of the recognition pulse should discriminate between metaphase cells and other objects as completely as possible within the constraints imposed on machine development by time and money limitations. For reliable recognition, the system must resolve the mitotic cell, usually about 50 microns in diameter, accurately enough to aid in the centering portion of the subsequent automatic focusing operation. At 20 rpm rotation of the platen in our prototype, the resolution time for the generation of a reliable recognition pulse must therefore be a small fraction of 300 micro seconds. With the size of the step in platen position (radial arm) between successive revolutions much smaller than the typical metaphase spread, redundancy on additional passes is available to improve detection and discrimination if necessary.

The discrimination criteria will be adjustable under software control in order to obtain optimal discrimination. For the steady operation envisioned with epidemiological studies of large populations, adjustment of the discrimination threshold may not be used frequently but such adjustments are crucial in the present stage of establishing feasibility and calibrating the instrument's operating characteristics.

Means of detection of metaphase spread. Several years ago the decision was made to use the characteristics of the optical scattering of a metaphase cell to detect its presence within a well-defined optical beam [1, 3]. It is known [4, 5] that every 'edge' or gradient of the optical index of refraction within an illuminated object will scatter light out of the incident beam.

If two or more 'edges' are coherently illuminated by the incoming beam, the scattered light will form a diffraction pattern containing bright spots at certain scattering angles. Each scattering angle, θ, at which a bright spot can be seen corresponds to a particular spacing, d, between 'edges' within the original object. In earlier discussions [1, 3], the term spatial frequency, f, was used to represent the reciprocal of the corresponding spacing, d. The relations between the angle for constructive interference, θ, and the quantities d and f may be written for small values of θ as

$\sin \theta = \lambda/d = \lambda f$,

where λ is the wavelength of the light. In our instrument, the wavelength has been chosen to be 6,328 Å so that a bright spot observed at an angle of 5° corresponds to a spacing in the specimen plane of about 7 microns.

An example of the diffraction pattern generated by the image of a stained mitotic cell is shown in Figure 3. The negative print shows black in each of the many regions where bright red light was observed. The irregularities of the spacings in the metaphase spread are clearly shown by the corresponding irregularities in the bright regions in the diffraction pattern, produced under illumination with 6,328 Å light. The specific correlation of each bright spot with an observable spacing in the image of the stained cell has not been made. For the purposes of rapid data collection, it suffices to integrate over the azimuthal angle; we collect all the light scattered into a given scattering angle by accepting light through the appropriate annular opening. For example, an annulus around the outer edge of the pattern in Figure 3 would pass light scattered into an angle θ of 14° by the stained metaphase spread.

A laser was chosen as the light source because of its high brightness. With small beam diameters, we are not making use of the laser's ability to coherently illuminate large objects, but the brightness seems to be necessary in order to detect a cell image in a few microseconds.

The specific form of locating system being implemented in our laboratory was suggested by the staff of the Perkin-Elmer Corporation early in this project [1, 5]. They proposed the use of a He-Ne laser as the source of the incoming stationary beam, the use of sampling the diffraction patterns continuously through annular openings by photomultiplier tubes (RCA Type 8645), and recognition of a metaphase cell upon the extraction of just two features within scattered angles out to 14°.

Preston based his proposal on static scattering data taken on a few objects placed one at a time in the beam of a laser. We have recently confirmed [6] the plateau (vs. angle of scattering) in the diffraction patterns of about 80 metaphase cells by static surveys on the same laser breadboard out to the same angular limit of about 14°.

Preston's proposal included the collimation of the scattered light with an

AUTOMATIC ANALYSIS

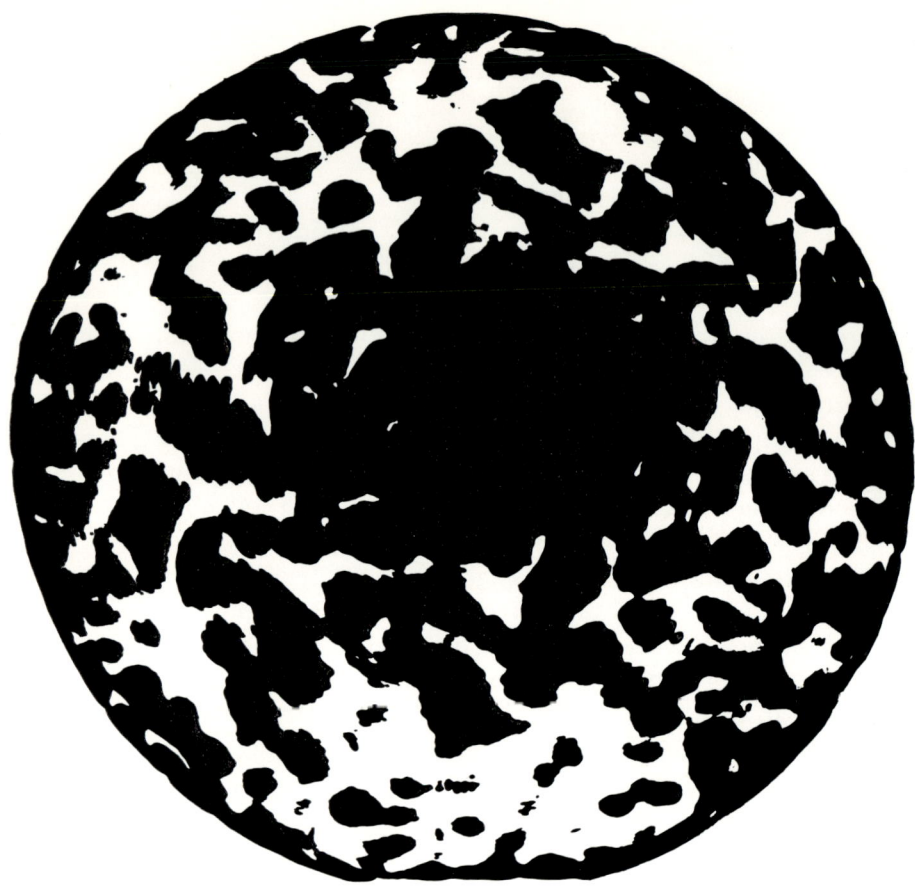

FIGURE 3
The diffraction pattern of a typical metaphase cell usable for karyotyping, as viewed on the laser breadboard. The outer edges correspond to scattering angles in air of about 14°. The photograph is a negative showing black where the observed pattern was bright red

objective lens to get a one-to-one correspondence in the Fourier transform between radius (after the lens) and angle of scattering from the specimen plane, the location of masks with annular openings in front of each photomultiplier tube. By proper imaging in the plane of a single mask near each phototube, we are able to select the two features as recommended [1, 5].
Calibration performed. The laser optical system and the two photomultiplier tubes have been operated in two modes under control of the PDP-7 computer, and one manual mode. The preliminary calibration tests performed to date have the purpose of establishing the magnitude of the signals caused by the images of metaphase cells and typical noise objects on well prepared slides under the conditions which will be present in actual automatic operation.

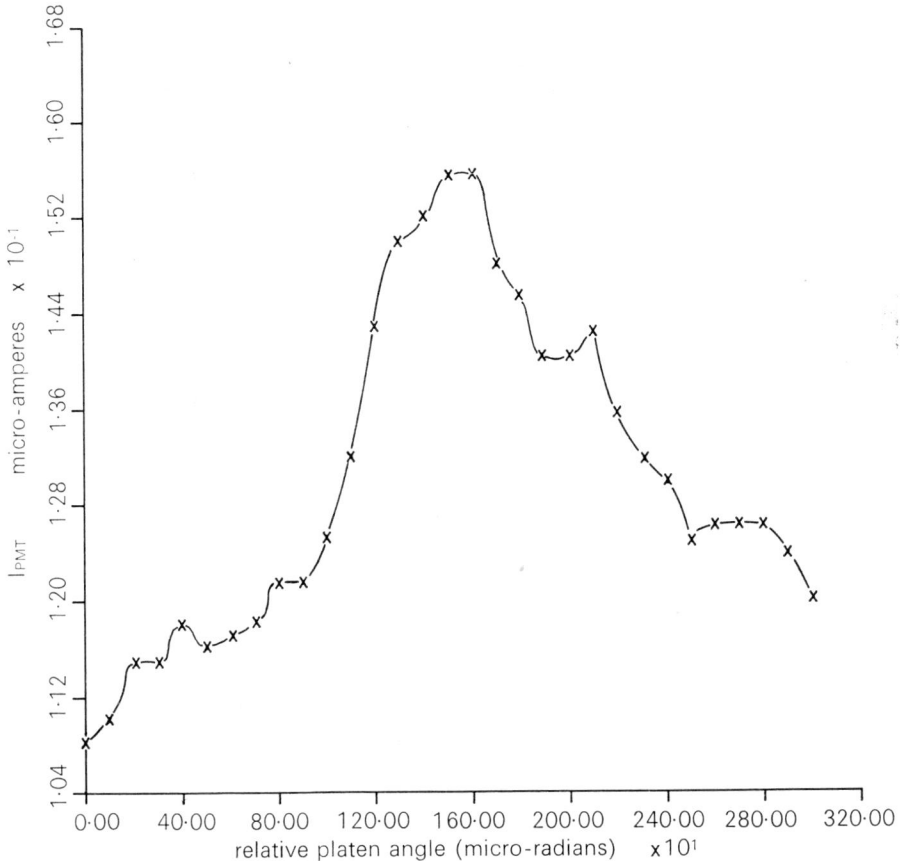

FIGURE 4

Graph of photomultiplier tube current generated by light scattered in angles between 14 and 15° at each of a series of fixed platen positions. The graph was generated automatically from input values of current and platen encoder readings. The peak current represents scattering of the laser light from the metaphase cell whose photo is shown in Figure 5.

1. *Manual operation* consists of reading the photomultiplier current at a fixed angular position indicated by the platen encoder and repeating at a series of other fixed angular positions. A graph of such a series taken with a time constant of about one second for each reading is shown in Figure 4 where the actual plotting has been done by means of the analysis computer. The image of mitotic cell which gave rise to the peak in the graph of Figure 4 can be seen in the photograph in Figure 5, taken under computer control. The peak signal of 0·05 microamperes corresponds reasonably well to the signals observed on the laser breadboard [1, 6] for similar cells.

AUTOMATIC ANALYSIS

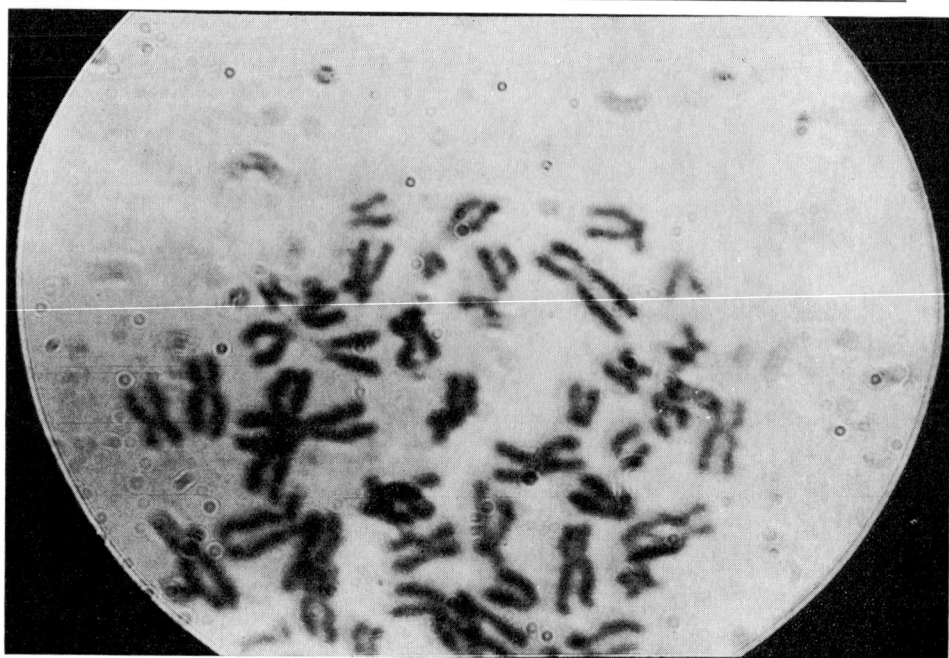

FIGURE 5
Photograph of a stained metaphase cell located on a slide prepared from human blood. The 35 mm photo was taken through the high resolution optics of the automatic microscope by command of the control computer. This cell is the source of the scattered light giving rise to the peak current in the previous figure, as well as to the small peaks indicated by the arrows in the following photographs

2. *Controlled stepping* of the platen causes the two phototube outputs to vary with time. The variation of the light scattered into angles of 5-6° and and 14-15° respectively within the modified cell locating system as the platen was incrementally rotated is shown in Figure 6. Each notch in the upper part of the figure represents 32 of the smallest angular increments possible on the present microscope and were obtained by monitoring the fifth bit of the platen encoder. It is suspected that the peaks in scattered light are mostly from dirt. The small peak over the arrow arises from the same mitotic cell, shown in Figure 5. The patterns are quite reproducible in retracings of the same region. This automatic mode is useful for electrical design purposes and may turn out to be an appropriate one for judging slide quality.

3. *Continuous rotation* of the platen with triggering of the test oscilloscope under control of the PDP-7 gives a record of the two PMT outputs as a function of time. Examples of the two channel outputs are shown in Figure 7 and Figure 8. The test program generates a trigger at a specified angular (platen encoder) reading. We observe that the peaks in the light detected

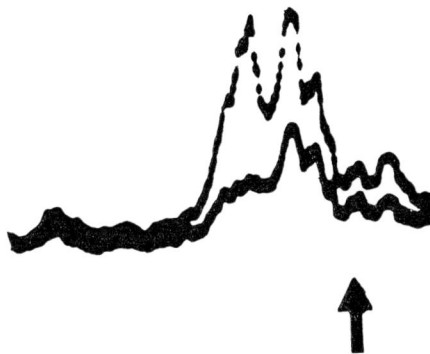

FIGURE 6

Intensity of laser light scattered into the two channels of the modified cell locating system under controlled stepping of the platen at the rate of 32 steps per second. The two photomultipliers indicate the light scattered into the angles of 5–6° and 14–15° respectively. The small peak above the arrow is due to the metaphase cell shown in Figure 5 above, and indicates the same peak current as plotted in Figure 4

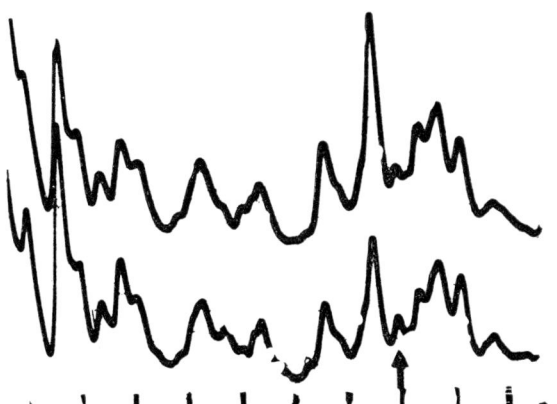

FIGURE 7

Intensity of laser light scattered into the two channels of the modified cell locating system under continuous rotation of the platen at 20 rpm as recorded on a test oscilloscope. The two photomultipliers indicate the light scattered into the angles of 5–6° and 14–15° respectively. The small peak above the arrow is due to the metaphase cell shown in Figure 5 above and indicates the same peak current as plotted on Figure 4. The time scale is 20 milliseconds per division

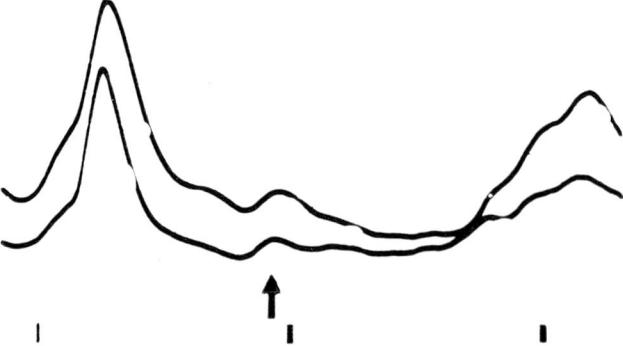

FIGURE 8
Scattered light recorded under continuous rotation as in Figure 9 except the full time span covers the time between the second and fourth marks on the axis in Figure 7

are quite reproducible in position and in magnitude from trace to trace on the test oscilloscope. The small peak near the arrows on Figure 7 and Figure 8 is caused by the same mitotic cell as shown in Figure 5.

These data from our first automatic runs indicate certain discrimination problems in recognising the signals due to metaphase cells in the presence of so many larger signals. They confirm the earlier expectation [1] that recognising mitotic cells at full speed is likely to involve many false recognitions of 'noise' objects. Therefore, considerable inefficiency is expected in the fully automatic operation of our present system.

As a first step toward better discrimination, we have begun to observe more features of the diffraction patterns characteristic of metaphase cells. The engineering analysis under way will also evaluate the investment and the sacrifice in speed of operation necessary in order to obtain improvements in the levels of successful discrimination. To do this well, we need a prediction of the variability of the characteristic features among usable metaphase spreads available with existing slide preparation techniques. Therefore we are considering models of the physical mechanisms by which stained chromosomes scatter the laser light.

Physical models of light scattering. The electronic polarisability of the stained chromosomes by which light is scattered can arise from several different mechanisms. They can be divided into two groups, those due to fixed atomic or ionic sites in the stained cell material and those due to electronic conduction across the surfaces or through the volume of the stained cell material. It is premature to discuss either group in detail. It suffices to say that any contributions from the second group, the conduction processes that are possible in the well-ordered DNA and associated histone protein, would be expected to contribute to the diffraction pattern rather different features from those expected from the absorption due to the staining ions directly. For example, if the conduction were anisotropic as might be expected from the coiling of the DNA [7], the lengths of the chromosome arms would contribute more strongly to the diffraction pattern than with other mechanisms.

Experimental data that facilitate distinguishing between models for the scattering are scarce. We have some preliminary measurements of the amount of light scattered out to wide angles. The graph in Figure 9 shows an excess of the intensity scattered from a metaphase cell over the background intensity present in the incident laser beam out to scattering angles of about 25°. This observed excess scattering indicates prominence of spacings within the cell of around 2 microns, and is a characteristic not previously reported [1, 6]. Clearly much more data are needed.

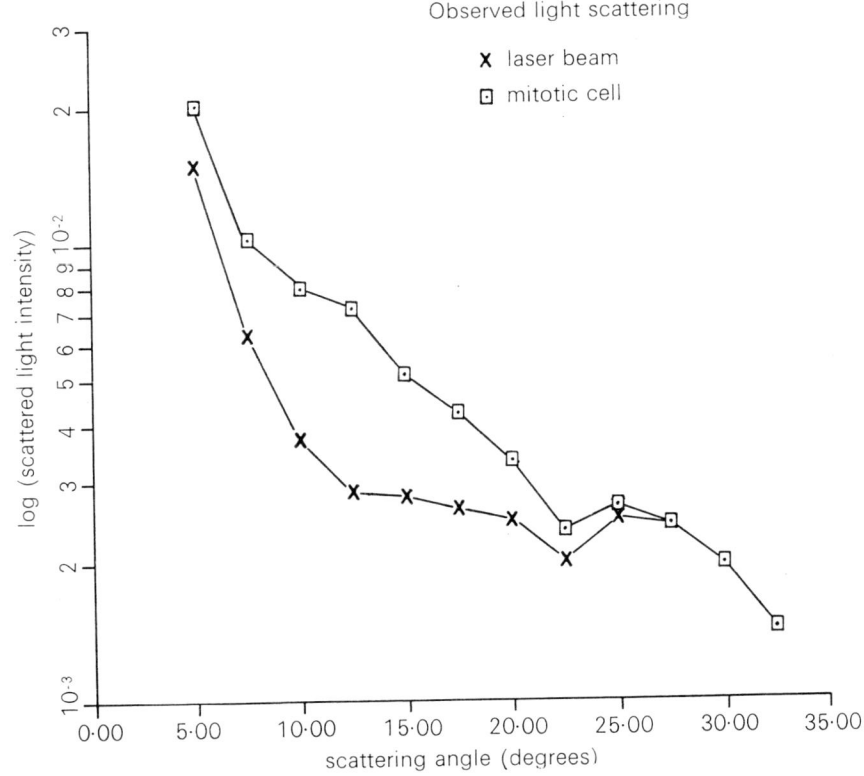

FIGURE 9
Intensity of light diffracted by a typical metaphase cell as a function of the angle of scattering in air. The fall-off of excess intensity with increasing angle may be partly due to the use of a small solid angle for light collection

CONVERSION OF CELL 'IMAGE' TO BINARY FORM

As reported previously [1], the generation of binary information to describe each selected cell is to be accomplished optically in our system by the use of a flying spot scanner under computer control. Recent reports have indicated considerable progress is being made in other ways of converting images to binary form which contain detailed information on the scale of hundreds of Ångstroms. For example, a scanning electron microscope [8] may some day perform this function better than an optical system.

The flying spot scanner that has been installed in the high-resolution optical system within our automatic microscope is of rather conventional design. A typical view of a metaphase cell is shown in Figure 10 where the scan pattern is generated by the control computer. The two images have been superimposed in the camera of the automatic microscope to show the relation of the scan field to the typical metaphase spread of about 50 microns in diameter.

By generating the scan pattern in software in the control computer, flexibility is available for a wide variety of functions ranging from re-scanning portions of the cell under request from the analysis computer to special scanning patterns for other research studies. The scan-converter

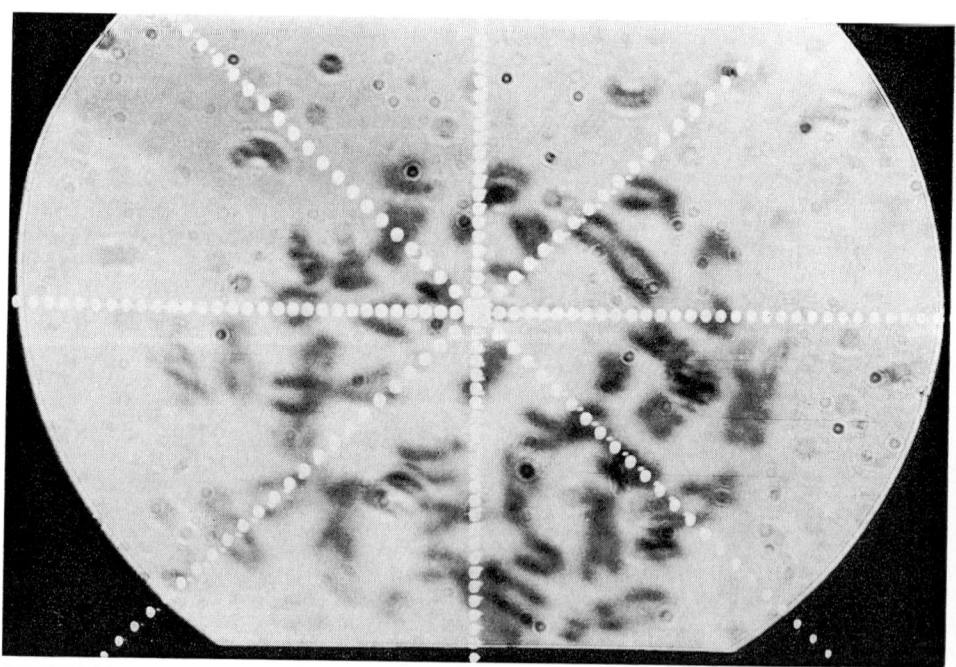

FIGURE 10

Photograph (35 mm double exposure) of a metaphase cell through the high resolution optical system and a scan pattern generated by the control computer. The metaphase spread is approximately 50 microns in diameter.

system is in its final construction stage. Calibration of the shades of the grey scale should begin soon.

The automatic focusing feature of the high resolution optical system is now operable. The depth of focus is, unfortunately, sufficiently small to give rise to some observable variation in focus across the width of a slide. Solution to this problem of variation of focus with stage position within one slide will require some compromise initially with the speed of automatic operation (probably in the control software). Some modification in the microscope hardware may eventually be performed to attain faster operation. The present compromise in the automatic focus of the high resolution optics represents the kind of initial approach being taken to reach fully automatic operation at some speed soon.

IMAGE ANALYSIS AND KARYOTYPE GENERATION

The computer analysis to be performed on-line includes the analysis of the binary 'image' of each selected cell image containing optical density information over the field of view selected by the automatic scanning operations in order to obtain either a definitive karyotype of the cell, or a call for operator help. Initially, the karyotype analysis will be done while the cell is still in focus in the expectation that rescanning or photographing frequently will be desirable.

The computer karyotype analysis, expected to consume about half of the real time needed for automatic karyotyping, is often described in terms of two stages: measurement and pattern recognition. We expect that our initial boundary determination procedures will be very similar to those described in the papers by Mendelsohn and Prewitt [9–11].

The measurement procedures [12] developed by D. Rutovitz and J. Hilditch will be employed to provide some of the necessary measurements, such as arm lengths. In addition, their skeletonising logic for overlap separation will be used as a preparatory phase. The software will resolve the chromosome as far as possible, and the operator will be the final judge of its adequacy. Should he so elect, the operator will be able to over-ride the computer's choice.

The measurement analyses we expect to perform include: integrated optical densities of each connected figure (chromosome?), boundary of each chromosome, locations of the centromere and any secondary constrictions, and various other parameters of each chromosome such as arm length, perimeter, area, etc. Detection criteria for such common puzzles as the single-chromosome-overlapping-arm case and the multiple-chromosome-touching/overlapping cases will be sought with the aid of the operator display unit.

Pattern recognition procedures are being investigated for their applicability to the problems of karyotyping. One promising possibility was suggested to us by Dr C. C. Li of our Engineering School and is being pursued; we are currently translating for the PDP-10 computer an existing program for the Ho-Kashap algorithm [13] as modified [14] for a character recognition problem. In general, we expect to follow the lead of those who have already made a start on these problems.

POPULATION DATA ANALYSIS

The automatic feature of our cytogenetic analysis system which should facilitate the computational aspects of population studies is the time-sharing capability of the PDP-10 computer and its control programs. When, in a few years, the other features of the automatic microscope system work well enough to yield karyotypes rapidly and at a low enough threshold of discrimination to permit classification or at least identification of infrequent aberrations in an individual's chromosomes, the system should be usable for population screening surveys [15–17]. The population analysis will require both the accumulation of pertinent patient data and the statistical comparison of each patients' results with all the recorded histories of related populations.

One population for which accurate karyotype data would be beneficial is the newborn [16]. As soon as fully automatic operation becomes feasible, hopefully next year, slides from our newborn population study will be examined. The population analysis function will include generating population data from the chromosomal aberrations observed in the slides, and from the correlated pathology of the individual patients.

The epidemiological calculations and recording of karyotyping results are not expected to be major time consumers of computer time during automatic operation because the PDP-10 computer is to be run in the time-shared mode. With the control programs now being tested on the PDP-10 computer, we expect to be able to perform the population analysis programs for statistical comparisons between cells and between people as background computer jobs while the analysis computer is primarily engaged with the karyotype analysis on-line.

SUMMARY

The four automatic features of our cytogenetic analysis system will be location of cells in metaphase at full speed, conversion of the cell image of each selected cell to digital form, karyotype analysis of the digital image, and processing of cytogenetic and epidemiological data about the human populations under study. Progress on each has been steady though frustratingly slow. Calibration of many of the steps in each automatic function is in progress.

The metaphase cell-locating function currently promises to be the largest consumer of time during automatic operation under the present plan of extracting just two features of the laser diffraction pattern. Preliminary data on wide angle diffraction by metaphase spreads is under analysis. Eventually modification of the present cell-locating system should yield even faster and more accurate operation.

ACKNOWLEDGEMENTS

This investigation was supported by the National Aeronautics and Space Administration and Contract NAsr-169 and by the U.S. Public Health Service on Research Grant GM 15247 from the National Institute of Health. We are grateful for the help of many people. We wish to thank especially Dr Sylvia Pan and her staff for expert cytogenetic slide preparation and consultation; A. Goldman and W. J. Greger for taking the confirming diffraction data at low angles; S. Shipkovitz for discussion of the analysis section; and Drs George Jacobs, Mortimer Mendelsohn and Denis Rutovitz for their interest and encouragement since the early days of this program. Figure 2 is reproduced by kind permission of *University of California Press*.

REFERENCES

[1] Wald, N. and Preston, K. In *Image Processing in Biological Sciences,* p. 9. Ed. D.M. Ramsey. Los Angeles : UCLA Forum Med. Sci., Univ. of California Press, 1968.

[2] Neurath, P.W., Brand, D.H. and Schreiner, E.D., *Ann. N.Y. Acad. Sci.,* **157**(1), 324, 1969.

[3] Norgren, P.E., *Ann. N.Y. Acad. Sci.,* **157**(1), 514, 1969.

[4] Cf. Becker, H., Meyers, P. and Nice, C.M., Jr., *Ann. N.Y. Acad. Sci.,* **157**(1), 467, 1969.

[5] Izzo, N.F. In *Proceedings of the 18th Annual Conference on Engineering in Medicine and Biology,* p. 180. Eds. P.L. Frommer and G.G. Vurek. Philadelphia : University of Pennsylvania, Conference Committee for the 18th Annual Conference on Engineering in Medicine and Biology, 1965.

[6] Goldman, A., M.S. Thesis, Graduate School of Public Health, University of Pittsburgh, August 1968. Most of the data were taken by W.J. Greger.

[7] Hart, R. In *Advances in Biological and Medical Physics,* p. 139. Ed. J. Lawrence. New York/London : Academic Press, 1968.

[8] Cf. Kimoto, S. and Russ, J.C., *Am. Scient.,* **57**, 11, 1969 and Hayes, T.L. and Pease, R.F.W., *Ann. N.Y. Acad. Sci.,* **157**(1), 497, 1969.

[9] Prewitt, J., *IEEE Trans. on Biomedical Eng.,* BME-12, 14, 1965.

[10] Mendelsohn, M.L., Hungerford, D.A., Mayall, B.H., Perry, B., Conway, T. and Prewitt, J.M.S., *Ann. N.Y. Acad. Sci.,* **157**(1), 376, 1969.

[11] Mendelsohn, M.L., Conway, T., Hungerford, D.A., Kolman, W.A., Perry, B. and Prewitt, J.M.S., *Cytogenetics,* **5**, 233, 1966.

[12] Cf. Rutovitz, D., *J. Roy. statist. Soc.,* A129 # 4, 504-530, 1966 and Hilditch, J. and Rutovitz, D., *Ann. N.Y. Acad. Sci.,* **157**(1), 339, 1969.

[13] Ho, Y.C. and Kashap, R.L., *IEEE Trans. Electron. Computers,* **14**(5), 683, 1965.

[14] Geary, Leo, Ph.D. Thesis, Electrical Engineering Dept., University of Pittsburgh, April 1969 (unpublished).

[15] Turner, J.H., Wald, N., Li, C.C., and Borges, W. In *Proceedings of the Conference on Research Methodology and Needs in Perinatal Studies,* p. 176. Eds. S.S. Chipman, A.M. Lillienfield, B.G. Greenberg and J.D. Donnelly. Springfield : Charles C. Thomas, 1966.

[16] Turner, J.H. and Wald, N. This volume, p. 153.

[17] Cf. Wald, N., Koizumi, A. and Pan. S. In *Human Radiation Cytogenetics,* p. 183. Eds. W.M. Court Brown, H.J. Evans and A.S. McLean. Amsterdam : North-Holland Publ. Co., 1967.

Instrumentation and Organisation for Chromosome Measurement and Karyotype Analysis

D. RUTOVITZ
J. CAMERON
A. S. J. FARROW
RUTH GOLDBERG
D. K. GREEN *and*
C. JUDITH HILDITCH

MRC, Clinical and Population Cytogenetics Research Unit, London

AUTOMATIC INSTRUMENTATION

¶TECHNIQUES for automatic recognition and measurement of chromosomes are gradually edging out of the domain of speculative pattern recognition studies and into the world of practical laboratory cytogenetics. This paper describes what is intended to become a working installation for automatic or semi-automatic karyotyping, able to deal with both substantial populations of human individuals in which the principal task might be that of screening normal from abnormal subjects or with large cell populations in which the accumulation of statistics of normal and aberrant cells and chromosomes is the objective.

It is not our purpose here to discuss the logic of chromosome recognition and classification or the problems of accurate measurement (see p. 297). However, as an indication of the present state of the art and of the kind of contribution that machine measurement is beginning to make, we have included a table (Table 1) comparing the separability of ten groups of chromosomes on the basis of size alone as determined by: (i) manual measurement, with length as the size criterion; (ii) machine measurement, with (a) length, (b) area, (c) integrated optical density as the size criterion.

It can be seen at once that in almost every case neighbouring groups are better separated by machine measured area than by any other parameter. Manual measurements of length are better than machine measurement; integrated densities are somewhat better than machine measured length but more variable than manually measured length and machine measured area. Now, integrated optical density should be the best measurement because if all the variables involved were properly controlled it would be an estimate of the mass of DNA in the chromosome, which is independent of the configuration and of the state of contraction. The fact that with our present equipment (Fidac : see [1]) this measurement is less consistent than area, shows that we still have many problems to overcome in measurement of optical densities of objects as small as chromosomes. The greater variability in our estimates of length as compared with the operator's shows that we are still struggling with the problems of identifying the ends of chromosomes (although our group has not concentrated on this measurement to any extent – others have done better).

Area on the other hand presents no recognition problems additional to the difficult one of correctly segmenting objects. Though manual extraction of areas would be exceedingly tedious, their calculation by machine is fast and almost trivial once the segmentation problem has been overcome. The fact that it now gives consistently better separation between neighbouring groups than does hand measured length is an example of the contribution that machine analysis and measurement is beginning to make.

INSTRUMENTATION

The equipment required for chromosome analysis falls into the following categories : (1) machinery for mechanical positioning and transport of microscope slides, and associated electronic controls; (2) an optical train to present the information contained in microscopic objects on a macroscopic scale; (3) scanning devices to convert enlarged images to time-dependent analogue or digital electrical signals; (4) analogue electronic logic for the rapid evaluation of certain overall characteristics of entire optical fields; (5) a means of presenting visual field information to an operator and of receiving operators' graphic 'comments' [3]; (6) a digital computer or computers for controlling machinery, carrying out the measurement, recognition, analysis, logging and statistical tasks set, and communicating with operators on the one hand and cytologists on the other.

Our answers to these requirements are as follows cf [4]:

(1) A microscope with a mechanically propelled stage capable of a 320 mm × 32 mm movement in a horizontal plane (which allows for scanning of a 10-slide magazine without reloading), and 0·41 mm travel in the vertical direction for focus control. The specification calls for marked points on the same slide to be relocatable to within 4 microns on the same, and 20 microns at different, loadings, the maximum speed of movement being 1 cm per second. The focus motion is in steps of 0·1 μ, but repeatability is not asked for. To meet these specifications, Imanco (Image Analysing Computers Ltd., Melbourn, Herts) have designed an unconventional microscope with a stepping-motor-propelled, air-bearing glass stage in which the objective turret and condenser are fixed to a four-pillared inner frame connected to a larger four-pillared outer frame by leaf springs allowing small vertical displacements of the inner assembly in relation to the stage (Fig. 1). The interface to the controlling computer is to be kept as simple as possible, but will accept a command to move a specified number of steps forward or backward in a selected direction, the effective position being continuously logged by the computer itself.

(2) The optics of the instrument will consist of a selected Leitz planapochromat X100, 1·32 NA oil immersion lens with anti-reflection coatings, together with matching eye pieces and a number of mirrors and prisms, some half-silvered or beam-splitting and some under computer control. Dr W. T. Welford, of the Applied Optics Department, Imperial College, has kindly agreed to act as a consultant in this regard. The instrument is designed to be used for either source or image-plane scanning, with provision of the following optical channels:

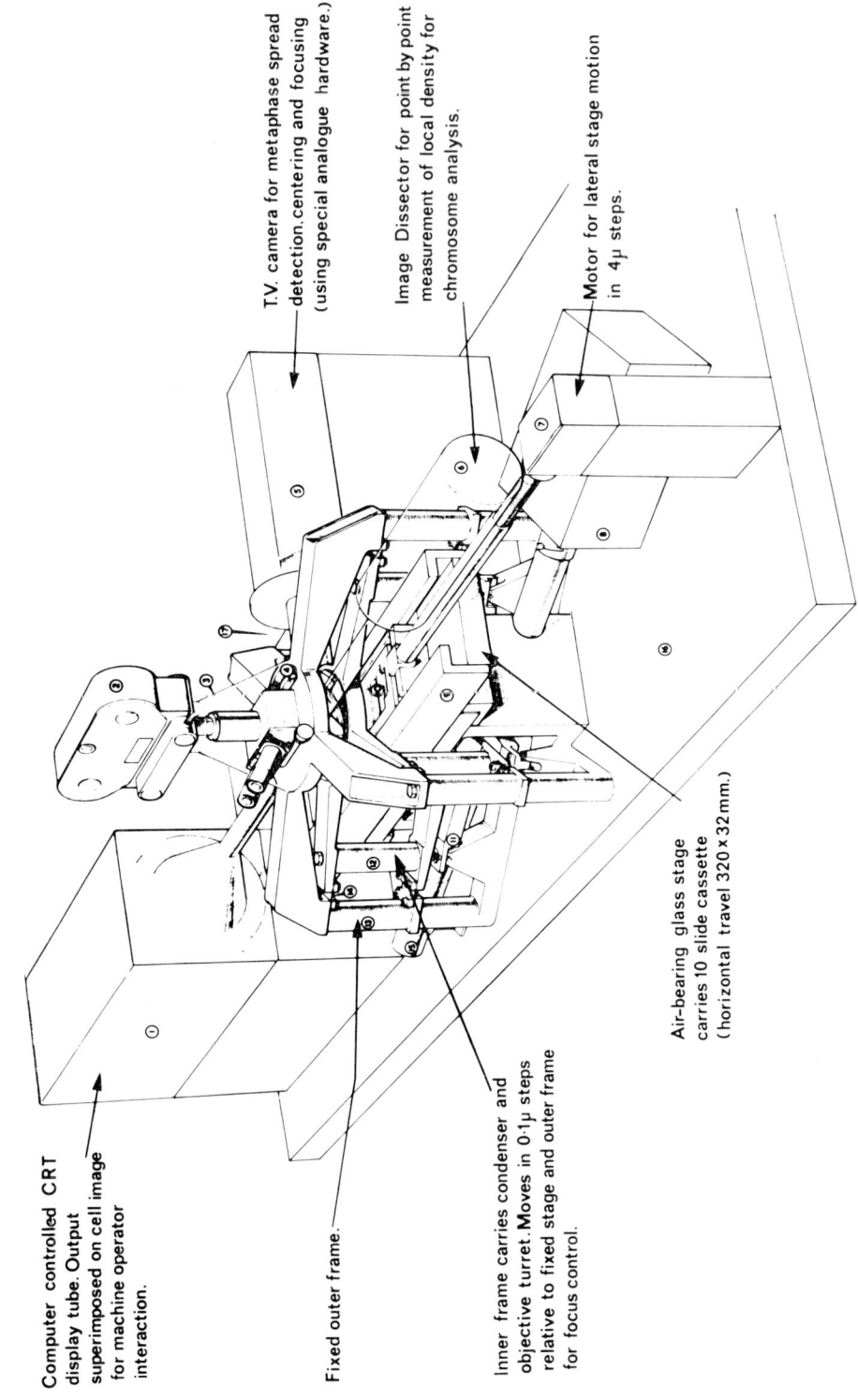

Computer controlled CRT display tube. Output superimposed on cell image for machine operator interaction.

Fixed outer frame.

Inner frame carries condenser and objective turret. Moves in 0·1μ steps relative to fixed stage and outer frame for focus control.

Air-bearing glass stage carries 10 slide cassette (horizontal travel 320 x 32mm.)

T.V. camera for metaphase spread detection, centering and focusing (using special analogue hardware.)

Image Dissector for point by point measurement of local density for chromosome analysis.

Motor for lateral stage motion in 4μ steps.

284]

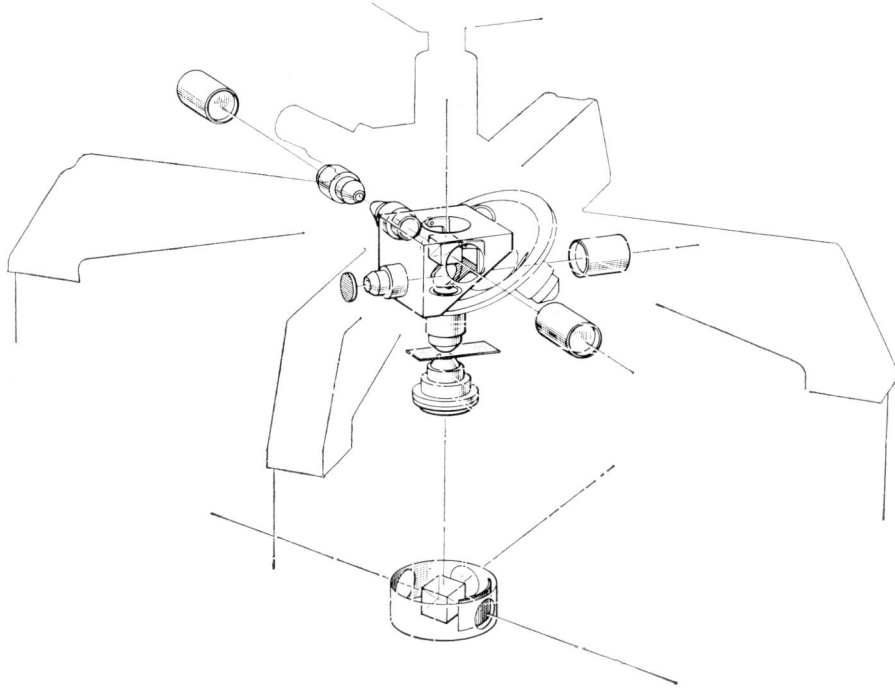

FIGURE 1b
The optical system of the microscope

FIGURE 1a (facing)
Diagram of the computer controlled microscope under construction by Imanco
(1) Flying Spot Scanner, write on film and manual input ; (2) Computer controlled camera ; (3) Camera support (not shown) ; (4) Objective turret ; (5) TV Camera for fast metaphase spread detection, centering and focusing ; (6) Image dissector ; (7) 'X' Actuator (Y and Z not shown) ; (8) Constant light source for photo/visual use ; (9) Air bearing X-Y stage ; (10) Ten slides in cassette ; (11) Focusing levers move the inner frame ; (12) Microscope inner frame carries condenser and objectives in fixed relationship ; (13) Microscope outer frame carries all other components ; (14) Flexible leaf springs to allow focusing motion ; (15) Light pipe from image dissector ; (16) Shock free mounted table ; (17) Flashing light source for TV Camera (Hidden behind stage)

(a) light input for image-plane scanning (if a flying spot device is substituted for an image-plane scanner, this becomes a photomultiplier channel);
(b) a reference channel for monitoring the strength of the light source;
(c) an auxiliary light channel to be used in conjunction with a vidicon scanner;
(d) two channels for projecting the image on to camera tubes. One of these would possibly be associated with a flashing light source, the idea being to employ a storage tube such as a vidicon or orthicon capable of retaining an image for at least one frame scan to register a 'snapshot' of the field over an interval of about 50 μsec. During the next frame scan the temporarily recorded image would be analysed by special purpose analogue hardware and because of the 'snapshot' effect, the image analysed will not be affected by stage motion. The second channel for image-plane scanning will accommodate a scanner producing accurate point-by-point digital encodings of density measurements, for subsequent computer analysis of visual field content;
(e) a channel for superimposing the output of a CRT on the image seen by an operator, a camera or one of the image-plane scanners. Selected points on the CRT face will be brightened up by computer instruction, either because of a program's calling for such display during pattern analysis, or as a result of a concurrent operator command to the machine. The reason for this provision is that it enables an operator to monitor the computer's actions in relation to the image and to communicate with the machine when necessary, without necessitating the reconstruction of the original optical scene by computer action as would be the case with the normal type of display screen. This frees the central processor of the computer in question from a rather heavy burden and also allows direct viewing of the original specimen;
(f) a channel for photographic attachments;
(g) a channel for operator viewing.

SCANNERS

As remarked above, two types of scanning devices are envisaged. One would be used for obtaining point-wise digitisations of the local image intensity, the other would be associated with analogue scan processing hardware for assessing certain overall field characteristics at a rate consistent with the frame time of the device, which should be about 25 frames per second. Characteristics which could be measured in this way are such things as overall variation along scan lines, total number of line segments over a threshold, total number of convex objects in a field, centroids of area or mass and so on. Measurements of this type can be applied to metaphase spread location, focus control and cell-centering. Our suppliers, Imanco, have built a mock-up of a system of this kind which responds correctly when connected to a special television camera trained on a large-scale sketch of a microscope field containing metaphase cells and other objects; theoretical assessments and some simulation studies done on Fidac also tend to confirm the feasibility of this approach. Vidicons, orthicons and similar tubes are well suited to applications not calling for photometric uniformity of any high degree, and not requiring random access to picture points, as this enables the integrating and storage features of the tubes to be used to advantage in overcoming signal-to-noise problems. For programmable and densitometrically accurate high resolution image-sampling however, the foremost devices for consideration are the CRT and the image dissector tube. The image dissector resembles a rather large photomultiplier with a long drift tube and a conducting plate with a central aperture inserted between the photocathode and the dynodes. When external magnetic focus coils are employed a secondary electron image of the optical field can be produced on the plate, and the output of the tube then depends only on the light impinging on the small area at the centre of the photocathode which is imaged onto the aperture. If deflection coils are added, it is possible to scan the entire electron image across the aperture plate; the anode current will then depend on the output from whatever portion of the photocathode is currently focused onto the aperture. In such tubes modulation at a given spatial frequency is very close to the theoretical estimate for the size of circular or other shaped aperture used (Table 2). Although such a comparison is not easy to make precise, it does seem that the image dissector gives better resolution than a CRT for an equivalent spot size, perhaps because the image tube's aperture is much more sharply cut off than are the edges of the spot. The dissector also has some disadvantages in comparison with a CRT. The first is that the signal-to-noise ratio depends on the energy received during the actual dwell time at each point sampled (as with a CRT) but there is no usable light source comparable in brightness with a CRT. In order to make the best use of the light, Information International Inc. (from whom we are procuring an image tube system) have devised a signal processing unit which gives constant signal-to-noise ratio measurements, the output signal being in effect the time taken to accumulate a specified number of photons. Thus on the bright parts of the image the dwell time is comparatively short whereas on the darker parts, such as the centre of a chromosome, the dwell time may be twice as long. A cut-off is provided for spots below a brightness threshold selectable by the computer. Nevertheless with available light sources integration times may be as long as 150 μsec for a $\frac{1}{8} \mu$ equivalent aperture diameter sample point (Table 3). Another problem is that with an image tube it is not possible to take concurrent measurements of the sensitivity of the portion of the photocathode from which the reading is being taken, whereas with the flying spot system one can certainly take parallel measurements of spot brightness. This does not matter providing there are no, or at least very few, rapid changes of sensitivity in the tube. Whether this can in reality be achieved still seems to be an open question, although we have been offered an as yet unfulfilled guaranteed specification by the instrument suppliers, based on results of the examination of some tubes of this type.

COMPUTER ORGANISATION

In automatic karyotyping there are several levels of control function which can be distinguished. At the lowest level there are the details of device control; e.g. in scanning an optical field for light or dark regions, the search limits, the step length and the threshold level, each has to be defined and an appropriate action taken when a point above or below threshold is encountered or a line-end or the scan end is reached. At this level there is often a difficult choice between building special hardware to deal efficiently with each type of situation as against the alternative of keeping the specialised control equipment to a minimum and placing a greater workload on the computer. Again one must keep track of the positions of all moving parts and of the states of electronic equipment; data must be channelled to and from scanners, display units and parametric input devices such as analogue logic units for evaluating the likelihood of a field being a metaphase, or the direction of the cell centre; or operator communication units such as push buttons, instrument dials, the light pen and so on. At the next level of complexity there are tasks of storing, packing, retrieving and unpacking optical density measurements, and to this might possibly be added smoothing or other local operations which can be defined irrespective of context. Next, one is concerned with field segmentation, and analysis logic for an individual component of a field; and then with control and classification procedures for a single cell and finally with the evaluation of a composite karyotype for an individual based on analyses of a number of separate cells, and also with decisions as to whether to examine more cells from the same individual or not.

In common with Dr Wald's group in Pittsburgh (see p. 263), our proposed solution to the problems posed by these multiple and overlapping requirements is to employ two computers, one a machine with a small memory and limited analytic repertoire for carrying out the less complex control and image processing functions, and the other a larger machine in direct communication with the smaller for simultaneous performance of the higher recognition, decision and statistical functions.

The smaller computer is already acquired and is a PDP-9 which has been interfaced by International Information Inc. to an operator's console, display screen and image dissector scanner. The display and scanner have symmetrical control arrangements permitting point, vector and data channel mode operations. The console is equipped with four parameter dials, 16 push buttons, six switches, two foot pedals and a light pen which seem enough to provide for operator-machine communication! We plan that during metaphase spread searching the scan limits will be computer set and controlled, and that the analogue metaphase cell detecting equipment will interrupt the computer when a possible metaphase has been located (but focus control will not go through the CPU). Once a cell has been located and centred, digital scanning will commence, in a coarse raster mode to begin with: i.e. at about $\frac{1}{2}\mu$ spacing of sample points. Present calculations show that the image dissector dwell time for a signal-to-noise ratio of 32 : 1 (deemed adequate for the coarse scan) will be about 30 microseconds, and therefore the total time for a cell of about 70 microns diameter will be of the order of $140^2 \times 30$ μsec $<$ 1 sec for the coarse scan. The PDP-9 will then transmit to the larger computer, not the entire 20,000-odd density values found, but only a list of end points of line segments above the background noise level. This computer will rearrange the segment lists into separate ones describing different connected regions of the image. These will then be sent back to the small computer, one at a time. On receipt of one of these lists, the PDP-9 will commence a detailed higher signal-to-noise ratio scan of the area referenced by the list, with raster spacing now equivalent to $\frac{1}{8}$ micron on the slide. While the PDP-9 is obtaining the density information for one object, the larger computer will be analysing the preceding one, so that a complete overlap should be achieved between scan time and analysis time. Since at $\frac{1}{8}$ micron sampling density metaphase cells average about 25,000 points of chromosomal material, this will mean that with 120 μsec minimum dwell time (darkest points 150 μsec) giving 64 : 1 signal-to-noise ratio, the total scan time will be about 3 seconds which is compatible with a projected analysis time of over 20 seconds.

All recognition procedures carried out in the main computer will have confidence criteria associated with them. Wherever the confidence levels attained are inadequate one of the following alternatives will be taken depending on the circumstances:

(1) A more elaborate (more time and storage consuming) program will be called in to try to resolve the ambiguity automatically;

(2) The recognition failure will be noted but no further action taken (in some statistic-gathering tasks it is not necessary to resolve every difficulty);

(3) Operator assistance will be requested. The main computer will transmit a short message specifying the whereabouts of the object and the nature of the difficulty to the smaller one which will interrogate the operator. Meanwhile to avoid time-wasting delay the larger machine will proceed with the next object and only come back to the questionable one when the operator's response has been obtained and transmitted by the PDP-9.

The action taken by the small machine will generally be to alert the operator (e.g. by ringing the teletype bell!), to identify the problem object by displaying a light circle superimposed on the operator-viewed microscope image and to communicate the nature of the query by means of a suitable ideograph : for example, $\boxed{\text{C?}}$ near but outside the light circle could indicate a request for centromere identification : at the same time a short line segment could be shown at the centre of the pointing circle. Using the parameter input knobs provided, the operator would steer the line until satisfactorily positioned, then press a button or foot pedal to tell the machine to send the current position of the line segment to the main computer as being that of the centromere line. Of course, it may happen that the assistance requested is incompatible with the nature of the object identified, and the operator will be provided with a means of signalling such errors.

Another mode of operation that will be allowed for is one in which object-segmentation and centromere-placing are displayed as they occur, so that although the weight of calculation and classification is still borne by the machine, human judgements replace statistical confidence tests; it may turn out that greater efficiency can be achieved by a combination of man/machine intelligence of this kind than in any other way.

Incidentally, it is well within the capacity of a small machine simultaneously to maintain a display of a diagram of limited size and complexity, to service interrupts from the larger machine or operator, and to attend to the scanning and information transmission tasks described above. Thus the total time per cell should not be greatly increased by occasional calls on the operator as his response time will be overlapped with scanning and machine analysis of other objects.

Later it is hoped to add a second scanner to the system so that metaphase cell location on one scanner could proceed in parallel with cell analysis on another; if this is achieved, and if the estimated analysis time is right, then the total time per cell, including metaphase cell searching, will be limited by the analysis time. In this connection it should be remarked that total time per cell, using Fidac retrieved images stored on magnetic tape, is at present around $2\frac{1}{2}$ minutes. A fair amount of this is taken up in magnetic tape transport, there is no overlap of scanning and analysis and the program is written in unoptimised Fortran. This, together with the prospect of considerably faster arithmetic on a 1970s machine, would seem to provide some basis for a timing assumption in the vicinity of 0·5 minutes per cell. However the present system makes no use of operator intervention which, despite the overlapping with machine analysis, could slow proceedings down considerably. Here again, one would hope that with two or even three microscope-scanner combinations one of the machines could usually proceed with aspects of the work while an operator was pondering some difficulty in a field being scanned by another.

TABLES/

AUTOMATIC INSTRUMENTATION

TABLE 1a. Mean and standard-deviation by group

Group	Machine measured AREA			Machine measured MASS *		
	Mean	Std.	$\frac{\text{Std.}}{\text{Mean}} \times 100$	Mean	Std.	$\frac{\text{Std.}}{\text{Mean}} \times 100$
1	84·24	3·83	4·55	91·71	8·51	9·28
2	83·20	3·70	4·44	92·48	7·46	8·07
3	66·52	3·48	5·23	72·00	6·98	9·69
4–5	62·00	3·30	5·33	65·04	6·21	9·56
6–12 X	48·47	5·74	11·84	48·53	5·79	15·65
13–15	34·08	2·71	7·94	32·46	4·23	13·05
16	29·58	2·28	7·72	27·08	2·95	10·90
17–18	26·62	1·91	7·18	23·19	2·98	12·84
19–20	19·59	2·35	12·00	15·51	2·84	18·34
21–22 Y	14·38	2·06	14·32	11·33	2·37	20·94

* Integrated optical density

TABLE 1b. Mahalonobis separation between adjacent groups

Groups	Length * (manual)	Area (machine)	Integrated optical density (machine)	Length (machine)
1/2	1·651	0·076	0·009	0·514
1/3	17·634	23·650	6·535	8·487
2/3	10·591	21·601	8·039	4·463
2/4–5	29·420	37·981	16·993	7·544
3/4–5	3·290	1·807	1·154	0·304
3/6–12 X	15·945	10·664	9·730	7·369
4–5/6–12 X	6·974	6·446	5·075	4·891
4–5/13–15	81·152	88·631	40·269	18·375
6–12X/13–15	11·899	7·992	5·525	3·243
6–12X/16	13·349	12·044	8·885	8·247
13–15/16	0·634	2·998	1·872	1·436
13–15/17–18	3·216	9·646	6·092	1·883
16/17–18	1·284	2·105	1·712	0·047
16/19–20	21·951	18·399	16·118	5·204
17–18/19–20	5·301	10·780	6·957	3·050
17–18/21–22 Y	2·229	37·783	19·672	6·417
19–20/21–22 Y	10·751	5·605	2·580	1·030

* After J. Lejeune [2]

Machine measured LENGTH			Manually measured LENGTH (after J. Lejeune [2])			No. of objects measured by machine
Mean	Std.	Std./Mean ×100	Mean	Std.	Std./Mean ×100	
3·90	5·48	7·42	88·68	5·38	6·93	21
9·96	5·51	7·88	82·36	4·41	5·35	25
9·16	4·68	7·91	67·91	4·47	6·58	25
6·65	4·47	7·90	60·59	3·80	6·27	49
6·22	4·78	10·34	47·68	5·14	10·78	185
7·83	4·33	11·45	31·89	2·70	8·47	71
3·00	3·03	9·19	29·88	1·90	6·35	26
2·25	3·67	11·39	27·04	2·76	10·21	52
6·65	2·65	9·95	21·83	1·62	7·42	51
3·72	3·06	12·92	15·56	2·12	13·62	58

Table 1a and b. The separation between chromosome groups as given by four different measures of size; machine measured area, integrated optical density and length, and manually measured length. The calculations for the manual measurements are based on Lejeune's figures [2].

In each case the distance between two groups is given by the formula

$$\frac{(\bar{x}_i - \bar{x}_j)^2}{\{\sum_k (x_{ki} - \bar{x}_i)^2 + \sum_k (x_{kj} - \bar{x}_j)^2\}/(n_i + n_j - 2)}$$

where $\bar{x}_i$ and $\bar{x}_j$ are the means of the parameter for groups i and j, x_{ki} is the value for object k in group i, and n_i and n_j the number of chromosomes in groups i and j.

TABLE 2. Observed vs theoretical modulation in an image dissector

	Full cycles/diameter *	Theoretical † modulation	Observed modulation
Center Field	341	100	92
	481	92	72
	630	71	26
Edge of Circle	315	100	90
	390	99	79
	409	97	70

* The number of t.v. lines per diameter is twice this figure.
† For a uniform disc scanning a square wave the theoretical modulation is $\frac{2}{\pi}(p\sqrt{1-p^2} + \sin^{-1}p)$, where p = (aperture diameter)/($\frac{1}{2}$-cycle width)

Table 2. Digital readings from a three-inch image dissector with ·002 diameter circular aperture were recorded with the dissector programmed to scan various size resolution targets both at the centre of the tube and at the edges of the 'quality circle' (a circle of $\frac{1}{\sqrt{2}}$ × tube diameter). Modulation was interpreted as the ratio of the maximum difference between readings over an individual target to the maximum difference observed when scanning a very wide reference target. The table shows the figures obtained, and compares them with the theoretical modulation for a uniform circular aperture of the same size scanning over a perfect square wave bar chart.

TABLE 3a. Image dissector dwell time and signal-to-noise ratio

No of signals processed in two seconds	Dwell time per point (mean signal processing time less fixed delay and computer instruction time) = Microseconds =	Data for 1600 consecutive readings at the field center		
		Mean *	Standard * deviation	Signal-to-noise ratio
6,204	339	35·4	0·8	44
11,357	192	33·9	0·9	37
21,986	107	46·8	1·8	26
35,857	72	33·0	1·6	22

(Quartz-halogen tungsten filament lamp, Leitz 1·3 N.A. oil immersion objective lens, 0·9 N.A. condenser, clear portion of orcein stained metaphase cell, S20 photocathode, gain of 2·6 at first dynode, photocathode aperture geometrically equivalent to $\frac{1}{15}\mu$ diameter circular area on the microscope slide)
*Adjusted for truncation error

Table 3a. Scanning rates in microphotometry are limited by the time taken to accumulate sufficient photons at each point examined. Whilst the luminous energy available with various microscope and lamp combinations is in other respects fairly easy to estimate it is difficult to be sure of the transmission losses through the whole system, including the projection lens, the slide, immersion oil and stained specimen.

The overall transmission factor can be inferred from the measurements presented in Table 3a, as follows:

On comparing the observed standard error $\frac{1}{\rho}$ with $\frac{1}{\sqrt{t}}$ (where t is the time in microseconds), we obtain the regression line

$$\frac{1}{\rho} = \frac{\cdot 34}{\sqrt{t}} + 0 \cdot 005;$$

thus signal-independent noise contributes 0·05 to the standard error, and the photo-electron noise component is $\cdot 34/\sqrt{t}$. It follows that if the photo-electron rate is N per μsec, $\frac{1}{\sqrt{Nt}} = \frac{\cdot 34}{\sqrt{t}}$, and N is of the order of $\left(\frac{1}{\cdot 34}\right)^2$, ie, about 10 per μsec for the $\frac{1}{280}$ sq. μ area scanned, which tallies quite well with estimates of the rate derived from the formula.

$$N = \text{photocathode current} \times \frac{rk(\sigma - 1)}{\sigma e}$$

where r is the ratio of the photometric aperture to the photocathode area, k is the mesh transmission factor, σ is the secondary emission factor at the first dynode, and e is the electron charge.

We conclude that with the quartz-halogen tungsten filament lamp employed, an average image dissector photo-electron rate of about 2,800 per microsecond per square micron at the slide can be achieved, compared with an estimate of 10,500 for a microscope with 100 per cent transmission, a 5,000 lumens/cm² /ster lamp, and the condensor, objective, photocathode and photomultiplier combination employed, thus indicating an overall transmission factor of 27 per cent.

For pattern recognition purposes we consider a signal-to-noise ratio of about 16 : 1 for individual point readings adequate, but the 'signal' must be interpreted as the difference between the average centre-chromosome reading and the background reading. Since this seems to be only about one quarter of the mean background value recorded in a metaphase cell, a 64 : 1 signal-to-noise ratio is necessary in the actual light measurement. As it is intended to use a photocathode aperture size equivalent in geometric optics to an $\frac{1}{8}$ μ diameter circular area in the slide, the figures show that for a 64 : 1 signal-to-noise ratio (4,096 photons) a dwell time of about 150 microseconds at each chromosome point is needed, and 120 microseconds at each background point.* With the planned sample spacing of $\frac{1}{8}$ μ, this should give a signal-to-noise ratio of well over 100 : 1 in determining integrated optical densities of areas of 1 μ × 1 μ and above.

* Assuming that the signal-independent noise can be eliminated.

TABLE 3b

Lamp	Brightness (lumens/cm²/ster)	Integration Time (μsecs)
Normal Tungsten filament	1,000	525
Overrun Tungsten filament	3,500	178
Quartz-Halogen Tungsten filament	5,000	125
High pressure mercury arc	20.000	31

Contrast could be improved to some extent by using filters, but with those tried so far the gain is disproportionate to the increase in signal processing time.

Table 3b. This shows how integration times would vary with different light sources. Although Orcein, Fuelgen and other stains fade over a period of about 10 minutes with the quartz-iodine lamp tested, brighter lamps could possibly be used to advantage with filters, but the spatial instability of a high pressure arc lamp renders it unsuitable for this application.

REFERENCES

[1] Ledley, R.S., Rotolo, L.S., Golab, T.J., Jacobsen, J.D., Ginsberg, M.D. and Wilson, J.B. In *Optical and Electro-Optical Information Processing,* p. 591. Ed. J.T. Tippet. Cambridge, Mass. : MIT Press, 1965.

[2] Turpin, R. and Lejeune, J., *Les Chromosomes humains.* Paris : Gauthier-Villars, 1965.

[3] Neurath, P.W., Brand, D.H. and Schreiner, P.D. *Ann. N.Y. Acad. Sci.,* **157** (1), 324, 1969.

4] Rutovitz, D. In *Human Radiation Cytogenetics,* p. 58. Eds. H. J. Evans, W. M. Court Brown and A.S. McLean. Amsterdam : North-Holland Publ. Co., 1967.

The Principles of a Software System for Karyotype Analysis

C. JUDITH HILDITCH*

MRC, Clinical and Population Cytogenetics Research Unit, London

---

* This paper describes the work jointly developed by D. Rutovitz, P. Balakumaran, A. S. J. Farrow, Ruth Goldberg, C. Judith Hilditch, K. Paton and B. Stein

¶ OVER the past few years the MRC Clinical & Population Cytogenetics Research Unit has been working on a system for automatic chromosome analysis. This is intended initially for use in screening the few abnormals in predominantly normal populations and referring these to the cytogeneticist for further investigation. Although it is hoped and fully expected that additional advantages will accrue from such a system, for example a more precise knowledge of the nature and variation in 'normal' karyotypes, the primary aim is simply to relieve the cytogeneticist of the need to examine enormous numbers of normal cells in order to be able to investigate a very small number of interesting abnormals.

This paper covers the principles of the software side of the system presently being developed by the Pattern Recognition Section of the Unit.

The input to the system to date has been a Fidac film scanner. This has been extensively described elsewhere [1] and details will not be given here. The use of a film scanner has necessarily required that suitable cells be pre-selected and photographed before being presented for analysis, so although an automatic means of locating cells is an essential part of an efficient high throughput system, this aspect of the problem is not discussed here. The reader is referred to Dr Wald's paper in this volume for one interesting approach to this problem. A second aspect which we feel will be an essential component of a successful system but which is also not discussed here, is that of man-machine interaction. Not having the necessary equipment, we have not done any investigations in this field to date, but the power of the system will undoubtedly be enhanced by supplementing the measuring power of the computer with man's tremendous pattern recognition ability in those cases too complex for the machine to handle alone.

The work of Dr Neurath's group [2] for example has been greatly concerned with this aspect of the problem, and our own approach to it is described more fully in an accompanying article in this volume (p. 281).

DESIGN CRITERIA

There are three main conditions which we feel the system must meet in order to be useful:

(a) It should be fast and efficient.

(b) It must have a very low rate of false negatives. As the rate of incidence of abnormals is low, it is important that as few as possible of these rare occurrences are missed.

(c) It must not have a high rate of false positives, otherwise it will not succeed in its task of appreciably reducing the number of normal cells the cytologist must examine to find any abnormals.

The system which we have been implementing and which is described here, is designed to meet these three conditions, and the way in which they have influenced it are indicated.

OVERALL VIEW OF THE SYSTEM

Very briefly the system has as input the digitised images of a number of metaphase spreads, and its output takes the form of a message to the effect that the individual from whom the cells were obtained has a normal karyotype, or is suspected of having an abnormal one, with as much information as possible as to the nature of the suspected abnormality.

To achieve this end the input information passes through a series of distinct operations:

(1) Segmentation. This extracts the interesting parts of the picture from the background, and presents them as a number of distinct objects, most of which will be individual chromosomes, to the next stage.

(2) Preliminary measurements and normalisation. The absolute size of many of the measurements obtained for the objects retrieved from a cell are of limited use in identifying the various chromosomes, or even differentiating between whole chromosomes and pieces and aggregates of several chromosomes, because they are dependent on variable factors, such as the degree of condensation of the cell, staining etc. Normalisation can be achieved either by forming dimensionless combinations of the raw variables, or by normalising with respect to some aggregate value for the cell. In order to do the latter, measurements must first be obtained for all the objects in the cell, so that the aggregate value can be calculated.

(3) Recognition of chromosomes and calculation of centromeric index. The two processes of recognising an object as being a chromosome and locating its centromere (and hence calculating its centromeric index) are so closely interrelated and interdependent that we consider this as one single phase.

(4) Classification of chromosomes. When all the chromosomes in a cell have been recognised the next stage is to classify them into groups according, for example, to the Denver Classification.

(5) Karyotyping. A karyotype can be obtained as a result of the classification of the chromosomes in one cell, but in general the results of a number of cells are combined in order to produce a reliable karyotype.

SEGMENTATION, PRELIMINARY MEASUREMENTS AND NORMALISATION

The process of segmentation, both as used by ourselves and other workers, has been extensively described elsewhere [1, 3, 4] so it will suffice here to say that the result of the process is a number of connected components, each of which represents a non-background area of the picture. With a reasonable metaphase spread most of these will correspond to individual chromosomes, but some may represent pieces of chromosomes when, for example, the centromere is particularly faint, or if there is disjunction at the centromere, and others may be collections of touching or overlapping chromosomes. Some examples of objects retrieved as a result of this initial segmentation are shown in Figure 1.

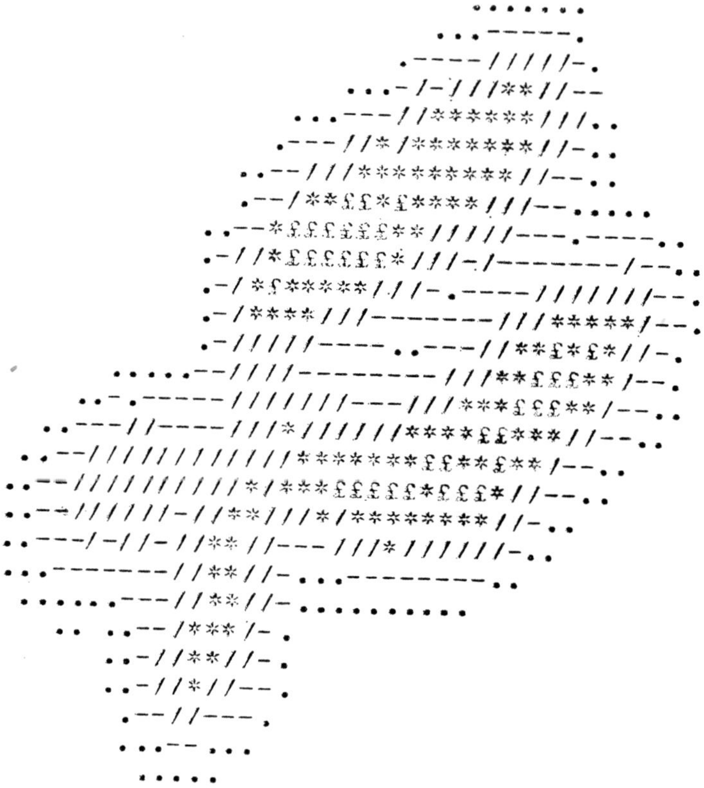

Figure 1a and 1b: ASCII-art representations of chromosome objects retrieved by initial segmentation.

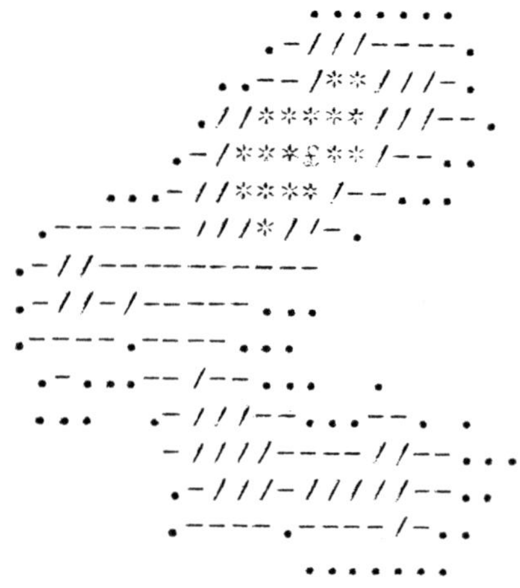

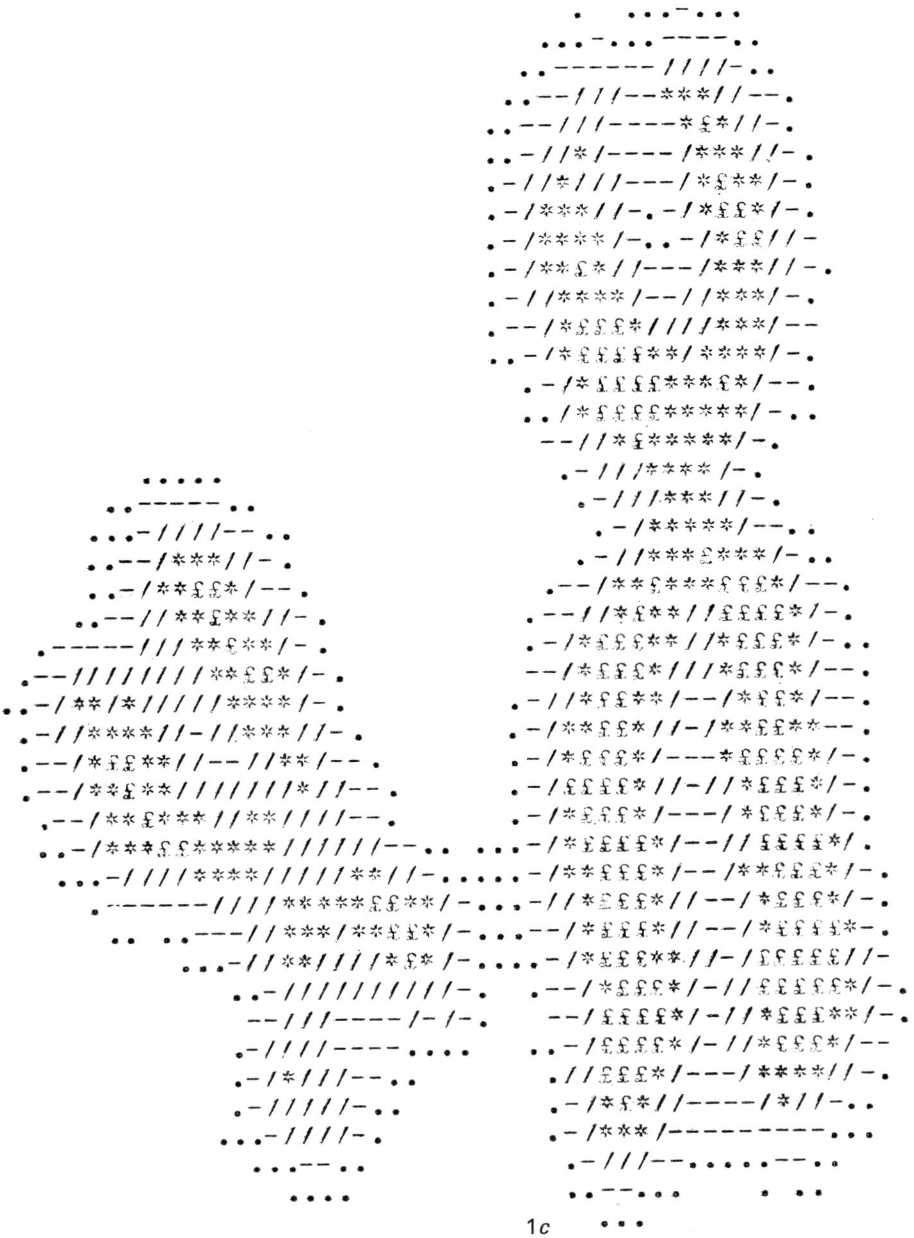

FIGURE 1

The segmentation process results in a number of objects, most of which represent single chromosomes, as in 1a and 1b. Some however represent overlapping or touching chromosomes as in 1c, and a few may represent pieces of chromosomes which appear disconnected as a result of the scanning process

(Each point of the picture is assigned a grey value between 0 and 7 depending on the darkness of the corresponding part of the photograph. In these computer printouts characters have been chosen such that their darkness when printed corresponds approximately to the darkness of the original)

The more complicated configurations sometimes require elaborate and time-consuming pattern processing techniques, but to apply these to every object would be extremely inefficient since the great majority of the objects are single chromosomes and, what is more, most of these are fairly straightforward in that they lie straight with arms more or less parallel. Such objects are comparatively easy to identify as chromosomes and to measure, and can therefore be dealt with rapidly by quite simple techniques. As something like 90 per cent of the objects retrieved are of this type [5] it would be exceedingly inefficient to deal with them in any other than the most straightforward way possible. Nevertheless, there still remain the more difficult 10 per cent; and as over half the cells will contain at least one of these [5] it is necessary to have the capability of dealing with the majority of these more difficult configurations as well.

We have therefore devised a system which initially assumes that each object is a single, whole, straight chromosome, that is, that it is an 'easy' object. At each subsequent stage of the program this assumption is checked and a measure of confidence attached to the results. If it becomes apparent that the assumption was wrong or the degree of confidence becomes too low to be acceptable, an alternative assumption is made or more sophisticated programs brought into use. It is of course necessary that these checks do not become as time consuming as the programs which they are intended to avoid, and this is done by basing them on parameters which are easily calculated, or are a natural result of the recognition technique being used.

In order to identify and find the centromere of an 'easy' chromosome the first thing that can be done is to determine its orientation. Various techniques have been described for doing this [6, 7, 8], but we have chosen to define it as the line of best fit to the density distribution in the least squares sense. The first phase of the recognition (as opposed to segmentation or classification) part of the program therefore consists of determining this line of best fit or principal axis of each object, and at the same time obtaining a number of preliminary measurements as a check on the type of object. As mentioned above, these are either simple to obtain – area, integrated optical density, length and width of the object in the direction of its axis, etc., or proceed naturally from the technique being used – moments about the axis, and the perpendicular line through the centre of gravity, and so on. Most of these measurements are dependent on variable factors such as staining or the degree of condensation of the chromosomes, and thus require normalisation before they can form a useful basis for decisions as to the nature of an object. These measurements are therefore calculated for all the objects and an aggregate value determined for each measure, for comparison purposes, before any decisions are taken (except for discarding very small noise objects and large undivided nuclei). For some of the measurements, for example area and mass, the total summed over all the objects can be quite useful as this aggregate measure, though it will be badly disturbed by the presence of an unrecognised undivided nucleus. On the other hand measures such as moments are not additive and here some alternative to a sum is required. Perhaps the most obvious possibility is the mean, but this is liable to distortion not only by extraneous objects but also by one or two composites of touching chromosomes, or a few small noise objects. We have found that the median, determined after discarding the six largest objects, and any objects smaller than 2 per cent of the remaining total, is a stable aggregate measure, and this has been used generally as the normalisation factor in subsequent parts of the program.

The results obtained by this stage of the program for a particular cell are summarised in Figures 2 & 3.

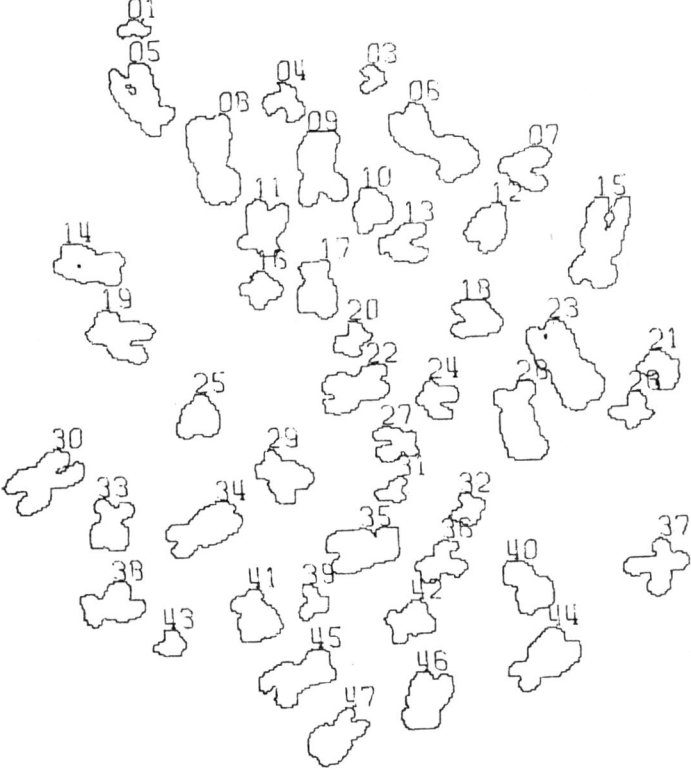

FIGURE 2

The outlines of all the objects retrieved as a result of applying the segmentation procedure to a cell. The objects have been numbered sequentially by the computer (object 2 is missing because it was rejected as a noise object)

SOFTWARE SYSTEM

OBJECT SERIAL,	NUMBER HAS	AREA	MASS	LNGTH	WDTH	MUU
1	1	74	299	13	7	3634
2	3	83	344	11	11	3177
3	4	167	752	18	14	15675
4	5	447	2063	33	18	147396
5	6	570	3650	41	21	438318
6	7	258	1401	19	21	32097
7	8	598	3139	38	19	331642
8	9	478	2788	30	20	186176
9	10	226	1155	19	16	22282
10	11	337	1764	23	18	61338
11	12	253	1156	21	15	28420
12	13	235	1080	20	17	25738
13	14	371	1942	28	16	109697
14	15	595	3328	39	19	368586
15	16	171	757	18	15	11761
16	17	316	1645	23	16	58019
17	18	251	1414	21	16	33885
18	19	361	1784	27	18	83909
19	20	138	564	15	13	8884
20	21	189	972	17	13	20604
21	22	389	2045	29	17	108335
22	23	639	3473	40	19	386754
23	24	186	774	17	15	13140
24	25	244	1230	19	17	24986
25	26	157	689	17	14	11963
26	27	191	816	19	14	17055
27	28	475	2515	33	17	193910
28	29	310	1533	24	18	51102
29	30	463	2527	32	19	177169
30	31	88	310	14	9	3613
31	32	127	486	14	12	6004
32	33	285	1427	21	17	45017
33	34	421	2041	32	15	139094
34	35	449	2414	30	18	143895
35	36	222	957	20	16	24876
36	37	316	1713	26	22	60327
37	38	313	1612	25	17	61043
38	39	115	472	14	12	5805
39	40	303	1766	25	17	73141
40	41	311	1685	22	18	53908
41	42	234	1097	18	14	25450
42	43	102	435	14	11	4186
43	44	432	2419	30	19	136388
44	45	415	2224	32	18	144213
45	46	376	1984	26	18	80129
46	47	340	1791	26	16	78732

TOPNUM IS 47 TOPSRL IS 46 ASUM IS 14021 MSUM IS

MEDIANS	NMA	NMM	NML	NMB
	316	1713	23	17

FIGURE 3

Computer output giving the measurements obtained by the first phase of the recognition procedure for the cell illustrated in Figure 2. The second column of figures gives the object numbers corresponding to those given in Figure 2; subsequent columns give area, mass (integrated optical density), mass, length (in direction of principal axis), width, Σmu^2, Σmv^2, $\Sigma m|u|$, $\Sigma m|v|$ (where m is the optical density of a point of the object and (u, v) its coordinates relative to the principal axis and perpendicular axis through the

V	MAU	MAV	APR	SDRD	ACIR	IX	IY	ANGLE
801	885	400	66	93	36	46	200	1.535
1909	913	664	59	86	54	70	297	-0.285
8153	2976	2029	49	72	53	79	260	1.174
3443	15053	8460	42	39	39	75	202	0.546
4477	35881	12732	49	58	37	97	321	0.919
8962	5826	5254	55	55	48	109	360	-0.605
8835	27848	13032	61	57	43	99	232	0.204
3154	20006	11400	57	44	47	106	274	0.014
6987	4297	3683	79	43	50	124	295	-0.487
8931	9089	7370	59	36	54	130	250	-0.323
7768	4831	3930	73	58	55	133	343	-0.493
9157	4355	4154	52	46	51	139	309	-1.354
9908	12572	6802	60	60	46	143	178	1.335
5992	30364	14216	35	53	39	141	387	-0.296
8621	2412	2103	69	40	54	156	249	1.547
6146	8140	5654	65	57	59	157	272	0.063
1808	5789	4839	62	54	55	171	336	1.401
6590	10656	7255	48	51	44	173	190	1.134
4723	1924	1374	61	62	43	178	285	-0.902
9857	3954	2619	64	53	49	194	410	0.814
5334	12866	7486	52	39	42	198	286	-1.243
7219	32432	14428	57	48	46	190	372	0.530
1502	2698	2698	54	47	48	204	319	1.177
0024	4874	4167	79	53	53	209	221	-0.924
7527	2462	1937	59	74	50	211	398	-1.251
1214	3132	2703	45	49	42	222	301	1.478
3146	19100	8858	58	50	41	214	353	0.203
8456	7549	5954	63	55	50	234	255	0.978
8202	18481	10919	38	43	45	232	157	-1.024
1590	881	593	63	56	57	241	300	-1.089
4153	1504	1189	66	82	63	250	330	-0.753
4458	6904	5146	55	38	53	252	184	-0.270
3409	14421	7327	56	56	46	254	223	-1.129
9949	15959	9825	58	53	49	265	287	-1.449
4371	4208	3221	51	46	49	272	319	-1.124
0832	8365	7172	43	56	51	277	407	-1.285
5912	8233	5478	57	57	44	285	184	-1.508
4056	1426	1152	61	77	57	287	265	-0.319
1674	9995	5350	69	54	49	283	353	0.684
9542	8056	6106	68	49	58	292	242	0.577
5511	4519	3622	65	46	51	296	306	-1.237
2731	1189	883	75	63	50	302	206	-1.271
0847	15274	9745	62	53	47	313	360	-0.951
9440	15507	8284	52	49	45	318	259	-1.139
8319	10755	7691	73	50	61	329	310	-0.484
8909	10053	6255	59	43	48	344	273	-0.892

| 2432 | LSUM IS | | | 1093 | | | | |

JU	NMMVV		NMMAU		NMMAV			
29	29908		10053		6255			

e of gravity of the object), area/(perimeter)2, standard deviation of the radius of the
t from its centre of area, area of object/area of circumscribing circle ; the last three
ns give the position of the centre of gravity of the object and the direction of the
ipal axis. The second to last line gives the total number of objects retrieved by
entation, the number retained after rejecting noise and undivided nuclei and the
area, mass and length of these. The final line gives the median (see text) values for
rst seven parameters

IDENTIFICATION OF CHROMOSOMES
AND CALCULATION OF CENTROMERIC INDEX

Having obtained the medians, the program is in a position to normalise the parameters, both by dividing individual measurements by the corresponding aggregate value, thus giving, for example, values for mass, second moments about the principal axis, etc., relative to the cell as a whole, and also by forming dimensionless combinations of measurements for each object. For example, area divided by perimeter squared gives an indication of how nearly circular the object is, area divided by width times length shows how well it fills the circumscribing rectangle with sides parallel to its principal axis, width divided by length indicates how nearly square this rectangle is, and so on. An indication of how these derived parameters can be used is given later.

These measurements are now used to check the assumption, implicit in the determination of orientation, that the object is an 'easy' chromosome, and a confidence measure is also calculated. The main aim is to distinguish between single whole chromosomes, and, on the one hand, composites made up of several touching or overlapping chromosomes, and on the other, pieces of chromosome. The parameters also provide information as to the reliability of the orientation determined in the preceding phase of the program and this determines how the next stage, the identification of the centromere, will proceed.

(*A*) *Touching chromosomes*. Let us consider first the situation where the values of the parameters indicate that the object is not in fact a single chromosome but made up of a complex of several chromosomes. This may have been shown for example by a high value for the mass of the object, or by its being much wider than the median width for the cell.

If this is so, it is next assumed that the object is made up of touching chromosomes (this is assumed, rather than that it consists of overlapping chromosomes, both because we find that touches occur more frequently than overlaps and because they can be dealt with much more easily). An attempt is then made to separate them by finding any 'thin necks' at which the object can be disconnected by cutting through at a place where it is only a few points wide, with the points in this region having low density values. If any such cuts are possible they are implemented and the two or more resulting objects are re-input to the system and thereafter treated just as if they had appeared as separate objects after the initial segmentation. It should be noted that this will provide a check on the validity of the splitting process, for each of the resultant objects will have to satisfy the same criteria as any other before being accepted as chromosomes. Figure 4 shows an instance of the separation of touching chromosomes.

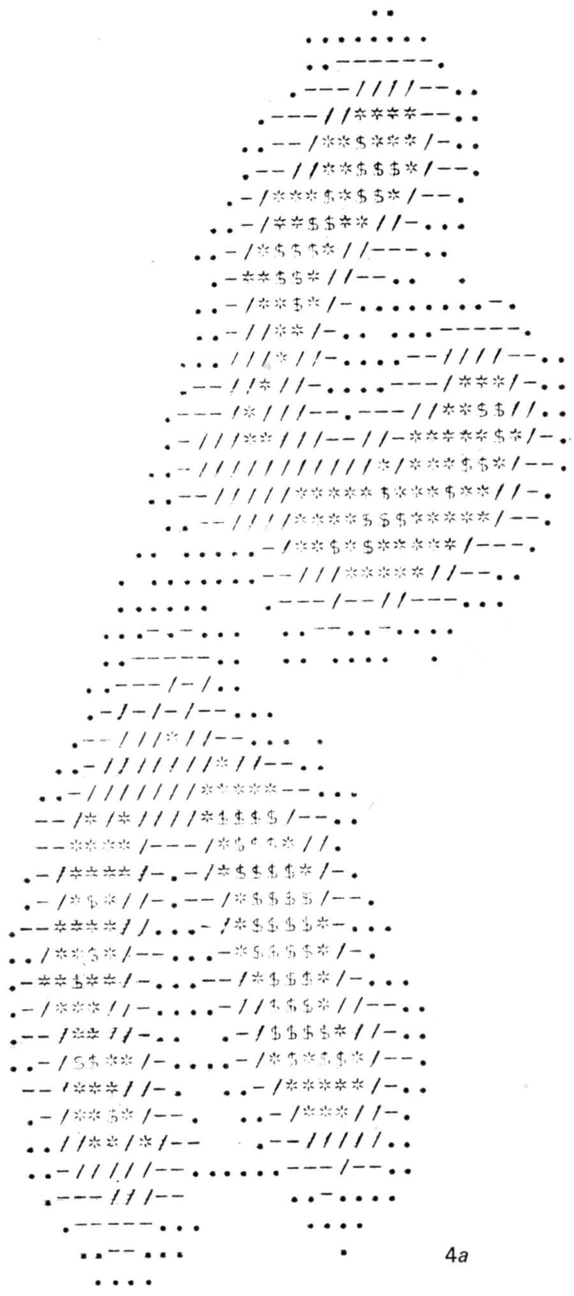

FIGURE 4

Splitting touching chromosomes: Figure 4a is the computer output for an object consisting of two touching chromosomes. 4b and 4c show the two objects obtained after this object has been split by cutting through at any narrow connections where the density level is low (see p. 308)

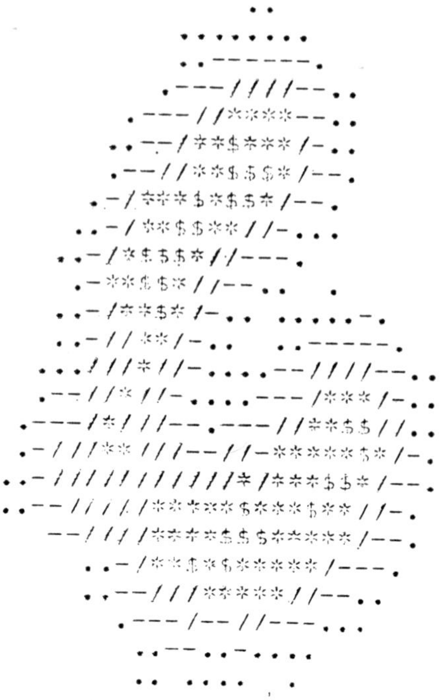

SOFTWARE SYSTEM

4b

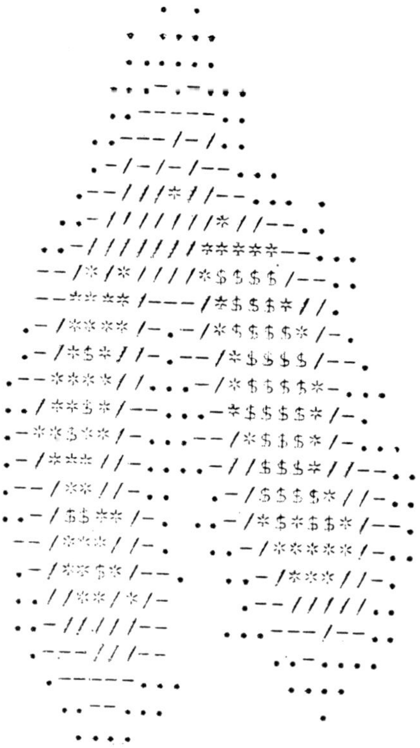

4c

Caption p. 307

In cases where splitting is not successful, the identity of the object is reconsidered in the light of this fact, and may be treated thereafter either as a single chromosome or as an assumed overlap. Methods for dealing with overlapping chromosomes are being studied but are not at present incorporated in the system and so are not described here; the reader is referred to [9].

(B) *Pieces of chromosomes.* At the other end of the scale, the parameter values may indicate that the object is somewhat less than a whole chromosome, for example an arm that the scanning and segmentation process has artificially disconnected from the remainder of the chromosome. In this case it is tested to see whether it lies close to any other pieces which have already been recognised as such. If so the two are considered together as one object which is re-input to the system and thereafter treated like any other object. If there is no sufficiently close piece to combine it with, it is added to the list of pieces for checking against any others which occur later.

(C) *Single chromosomes.* If all the measurements obtained have values compatible with an object being a single chromosome, the next step is to identify the position of its centromere. The way in which this is done further depends on the values of particular parameters.

If the ratio of the width of the object to its length in the direction of the principal axis is small, the axis direction is usually correct. If in addition the object fills the circumscribing rectangle it is probably also true that it is a reasonably straight chromosome with parallel arms, and the program continues as if this were in fact so. The chromosome should be reasonably symmetrical with sister chromatids lying on either side of the axis, and the next step is to check that this symmetry is sufficiently good. If it is, then the position of the centromere is ascertained by a comparatively simple technique pioneered by Mendelsohn [10], namely by considering the 'density profile', that is the profile obtained by plotting the sum of the densities across strips perpendicular to the axis against distance along the axis, as illustrated in Figure 5. In the case of a non-acrocentric chromosome the position of the centromere is given by a well-defined dip in the profile, in the case of an acrocentric there is usually not a dip, but a flattening out, or shoulder, at the end of the profile where the short arms are. (At present satellites do not prove a complicating factor, for the simple reason that our equipment is not sufficiently sensitive to pick them up.) Having located this dip or shoulder we take as centromeric index the area under the profile to one side of it, divided by the total area. (This is adjusted later if it represents the long rather than the short arms over the whole.) This simple calculation gives the centromeric index in terms of the integrated optical densities of the arms and not more conventionally in terms of length.

If instead of considering the density profile of the chromosome as a

SOFTWARE SYSTEM

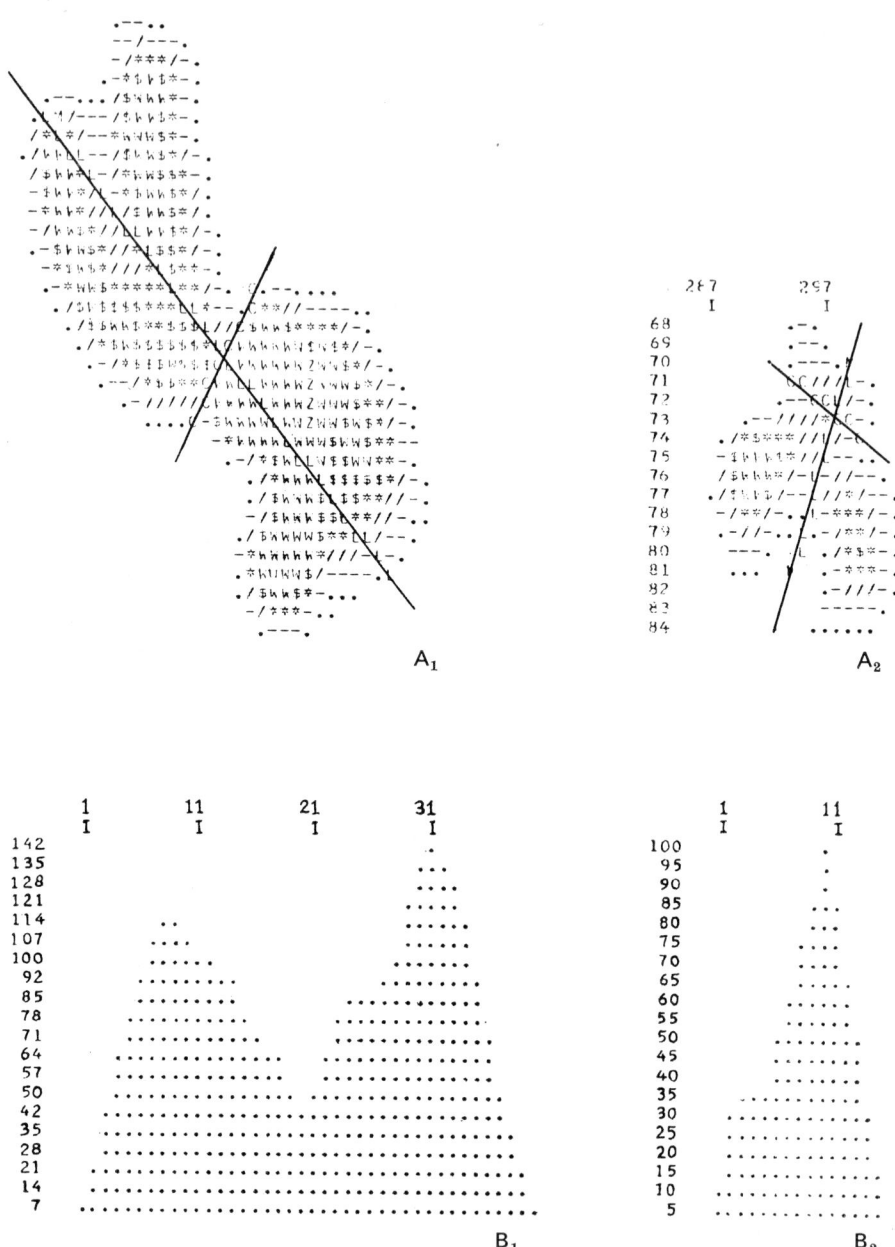

FIGURE 5

Chromosomes and their density profiles. Figures 5A₁ and A₂ are line printer reconstructions of the optical density distributions for two chromosomes. A line has been superimposed on the axis of best least squares fit shown by the printed symbol L, and the line perpendicular to this through the centromere given by the symbol C (the lines do not appear to be at right angles because of distortion produced by line printer spacing). B₁ and B₂ are 'whole profiles' for each of these chromosomes, that is graphs of the sums of the optical density values along lines perpendicular to the axis. For the first object the centromere is located at the line corresponding to the principal dip in the profile, for the second object it is located at the shoulder seen to the left of the profile. C_1, C_2, D_1 and D_2 give the 'half profiles' obtained by considering the half of the chromosome lying on each side of the axis separately. For the axis to be correct the two half profiles must correspond in number of dips and centromere position. The second object is more symmetrical about its axis than the first, and its two half profiles are a correspondingly better match

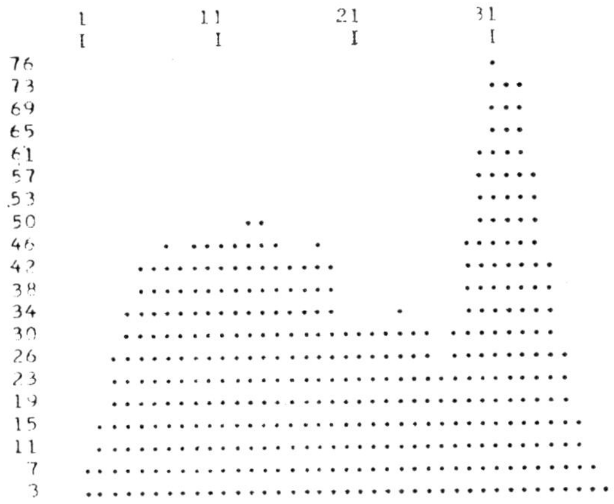

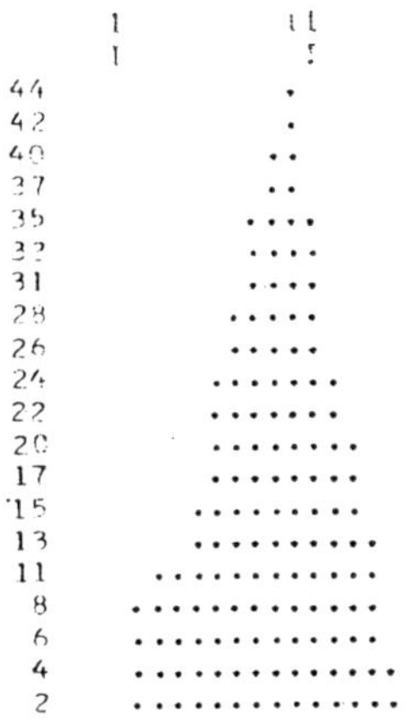

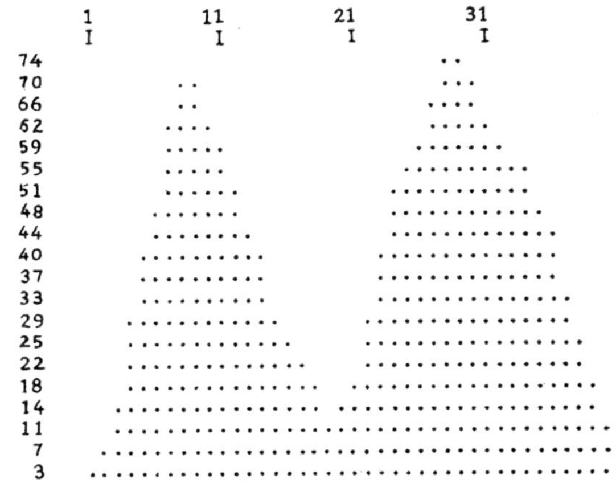

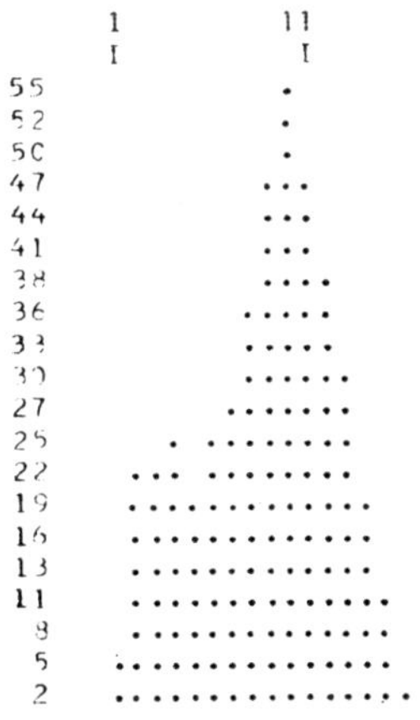

```
OBJECT NO      19

AREA OF OBJECT     397

LINE PRINTER RECONSTRUCTION OF AN OPTICAL DENSITY DISTRIBUTION

LINES     156 TO     182,        COLUMNS     139 TO     166.

            139           149           159
             I             I             I
    156                           .---..
    157                         -/*$*/.
    158                         -/$$W$/-
    159                         ./$WhH$/.
    160                        ./$WWhh*-.
    161                        ./$WWWh$/-
    162                       .-$hhWWh*-.
    163                      -*hhWWW$/-./.---..
    164                     -*hWWWW$/-./.  /*$$*-.
    165                    -/$hhhh$/-./ -/$Wh$*/.
    166                   -*hhhh$*---/-*WWWW$/-.
    167                  -$hhhh*/---/ /*$WWWW*/..
    168                  -$hhW*---/ /*WWWW$$/-.
    169                 .*hh*/.. /* $hWWW$*/-.
    170                .-**/-./ /hhhWW$*/-.
    171              ...-////-/ /$WhhW$*/-.
    172             .////////// /*hWhhW*/-..
    173            .-$h$$**//   /$hhhhW$/-..
    174           -/$hWW$$* /// $Whh*/-..
    175          -$$$$**$** /---/**/-.
    176         -///--/ *♦  ..  ......
    177       ......./*/-.
    178        /.-*$$-.
    179       ./$h$/.
    180      -*$$*/.
    181     -/*//-.
    182    .--...
```

FIGURE 6a (caption facing)

whole, the two halves on either side of the axis are considered separately a density profile can be produced, the centromere located and the centromeric index calculated for each half alone. A final check on the assumption that the object is a chromosome lying along the axis can then be obtained by comparing the results from the two halves, and this is in fact what is done. Instead of producing a whole profile, two half profiles are produced, symmetry is calculated in terms of the sum of the absolute differences between these profiles and, if this is satisfactory, the dip or shoulder representing the centromere is located for each. If in each case just one such centromere is found, the two values for the centromeric index are calculated. These are compared and if the difference between them is sufficiently small the object is finally accepted as being a chromosome, and its centromeric index taken as the average of the two half profiles (subtracted from 1, of course, if it is greater than $\frac{1}{2}$, in order that it should represent the short arms over the whole). If the difference between the two

```
 1. FROM TAPE F 46 , FILE IDENTIFIERS      0, A25 ,    1,    2, REFLST, OBJECT    19 READ IN
 2. COPY ALL PARAMETERS FOR OBJECT
 3. DISTINGUISH BETWEEN CHROMOSOMES  PIECES AND COMPOSITES USING ALL AVAILABLE PARAMETERS
 4. CONFIDENCE 52 PER CENT  THAT OBJECT IS CHROMO
 5. GET PROFILES
 6. OBJECT NEITHER SMALL NOR FAT SO TRY PRINCIPAL AXIS
 7. LPT,KPT,AA,BB=   159    163   69   -71
 8. SYMMETRY=  9 GIVING CONFIDENCE  90
 9. THE TWO INDICES ARE   77   78
10. THE DIP POSITIONS    21   20
11. THE TWO CONFIDENCES  68   74
12. RESULTANT CONFIDENCE  85
13. CONFIDENCE  85 PER CENT THAT INDEX IS  23
```

FIGURE 6*b*

In Figure 6a, the principal axis determined by the computer program has been drawn in on the computer output representation of a chromosome. Figure 6b shows the output from the program during the process of locating the centromere of this chromosome. Line numbers have been added for convenience ; the significance of the lines is as follows : line 1 identifies the object which has been input ; line 2 indicates that the parameters calculated during the first phase of recognition (mass, orientation etc) have been copied ready for use ; line 3 shows that these parameters are being used to decide whether the object is more likely to be a single whole chromosome, a piece of a chromosome or a composite of several chromosomes ; line 4 says that it is a chromosome with 52% confidence in this result ; line 5 therefore shows that the next step is to obtain density profiles ; and line 6 says that since the object is neither small nor has a high width to length ratio the principal axis is probably the axis of symmetry of the chromosome and should be taken as such ; line 7 gives the position and direction of this axis ; line 8 gives the symmetry about this axis, this symmetry is high and gives a good measure of confidence in the axis ; line 9 gives the values of the centromeric index obtained from each half profile (multiplied by 100 for convenience) ; line 10 gives the position in each half profile at which the centromere was located ; line 11 gives the confidence associated with the centromere as found from each half profile, this depends on such factors as how well defined the dip in the profile at the centromere is ; the two half profiles give sufficiently similar results and lines 12 and 13 give the resultant confidence in the final value of the centromeric index, which is the average of the half profile values (subtracted from 100 so that it represents the short and not the long arms divided by the whole) Note that this computer output is only for monitoring and demonstration purposes and would not normally be produced during an analysis halves is acceptable, but not small enough to ensure complete confidence, the whole profile is also considered, and the result accepted only if this too has just one centromere, which lies sufficiently close to the average obtained from the two half profiles. Some computer output for such a Chromosome is illustrated in Figure 6.

Less confidence can be placed in the orientation determination in cases where the object is small, or where its width is almost as great as its length. The method by which the axis is calculated is such that the calculated axis is frequently perpendicular to the true axis of the object. In such cases therefore both the chosen axis and its perpendicular through the centre of mass are considered, and half profiles and symmetry are calculated in each direction. If the symmetry in one direction is very much better than that in the other, this gives an immediate indication of which of the two possible lines should be taken as axis. The small acrocentrics usually produce this sort of result and an example is shown in Figure 7. With the larger acrocentrics and the small metacentrics this is less often the case, and here it is often necessary to consider, as well as symmetry, the apparent centromere position in each of the two directions and then choose the one for which the two half profiles give the better match. Having chosen the orientation, the calculation of the centromeric index can proceed just as before.

If at any stage in either of the processes described above it becomes apparent that the assumption that the object is a chromosome with the chosen axis as its axis of symmetry is incorrect, or values are such that confidence in its correctness becomes very low, e.g. if the symmetry is poor, then an alternative must be tried. In the case of thin objects the axis angle is changed slightly (by $\pi/16$ radians) in each direction to see whether either of these gives an improvement. In the case of the small, wide chromosomes the original axis is more likely to be in error, and all possible orientations may be tried (in steps of $\pi/16$ radians). This is naturally very time consuming, and a more reliable technique for the initial determination of orientation is clearly required for the smaller chromosomes. It appears that a study of the boundary of the object, as pioneered by some other workers [1], may be useful here, and programs are being developed to this end.

If all possible orientations have been tried and have failed, or if the original values of the parameters indicated that the object could not possibly be a straight chromosome, one further, more sophisticated, and much more time consuming technique is employed. This consists of thinning the object down to its 'skeleton', where each arm of the chromosome is reduced to a thin line of points, as illustrated in Figure 8. The density of the chromosome can then be concentrated onto the skeleton, by adding the density at each point of the object into the nearest skeleton point. The arms

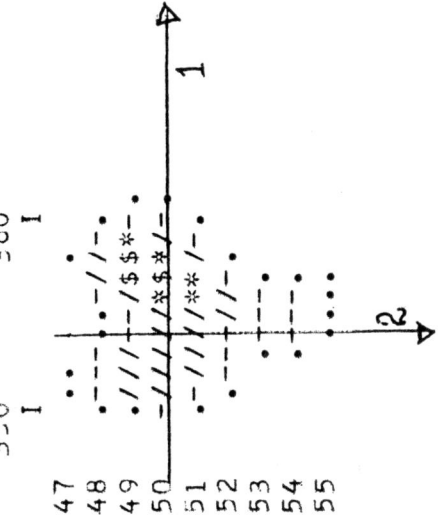

FIGURE 7a (caption p. 316)

SOFTWARE SYSTEM

```
 1.  FROM TAPE F 46 , FILE IDENTIFIERS          0, A25        1,       2, REFLST,   OBJECT     1 READ IN
 2.  COPY ALL PARAMETERS FOR OBJECT
 3.  DISTINGUISH BETWEEN CHROMOSOMES   PIECES AND COMPOSITES USING ALL AVAILABLE PARAMETERS
 4.  CONFIDENCE 73 PER CENT   THAT OBJECT IS CHROMO
 5.  GET PROFILES
 6.  OBJECT SMALL OR FAT AND FILLING SO TRY BOTH AXES
 7.  LPT,KPT,AA,BB=     50    361     2   -99
 8.  SYMMETRY=  4 GIVING CONFIDENCE  30
 9.  LPT,KPT,AA,BB=     47    354    99    2
10.  SYMMETRY=  9 GIVING CONFIDENCE  90
11.  SYMMETRY
12.  INDICATES PARALLEL SECOND AXIS
13.  THE TWO INDICES ARE  92  93
14.  THE DIP POSITIONS         8    8
15.  THE TWO CONFIDENCES      73   70
16.  RESULTANT CONFIDENCE  81
17.  CONFIDENCE  81 PER CENT THAT. INDEX IS   8
```

FIGURE 7b

In Figure 7a both the principal axis determined by the program (axis 1) and the perpendicular to it through the centre of gravity (axis 2) have been superimposed on the representation of a chromosome. 7b shows the output given while locating its centromere. Line numbers have been added for convenience, and the significance of the lines is as follows: lines 1 to 5 are the same as lines 1 to 5 in Figure 6; line 6 states that as the object is small, not only the principal axis but the perpendicular axis should also be considered; line 7 gives the position and direction of the principal axis; line 8 gives the symmetry about this axis; lines 9 and 10 give the same information for the perpendicular axis; lines 11 and 12 say that since the symmetry about the second axis is very much the greater this will be taken as the axis of the chromosome; lines 13 to 17 then correspond to lines 9 to 13 of Figure 6

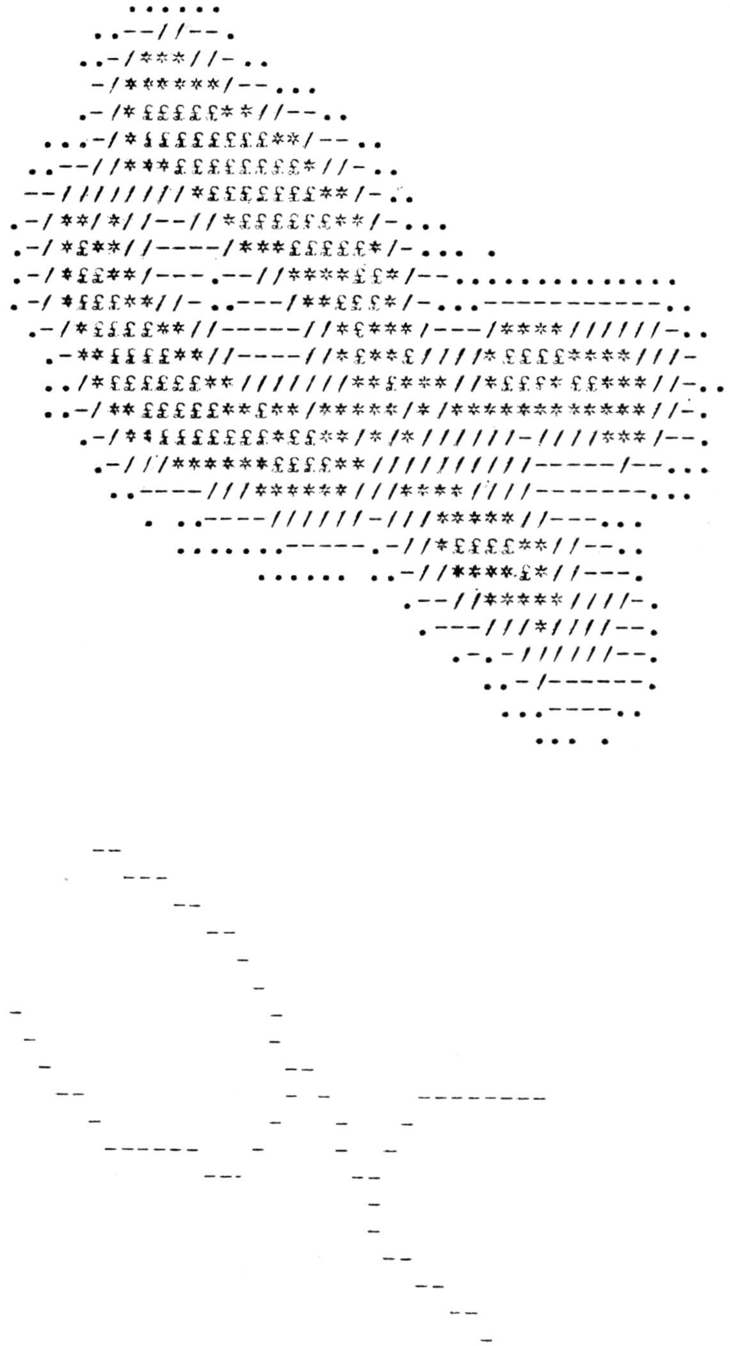

FIGURE 8
A computer output representation of a chromosome, and the 'skeleton' obtained from it by thinning the object until only lines which are one point wide remain

of the chromosome can be much more readily identified in the skeleton than in the original digitisation and by considering the density sum at each point of the skeleton as one moves along it, density profiles can be obtained analogous to those got by the principal axis technique. The program can then proceed exactly as before in determining the centromeric index, with the same criteria to be satisfied if it is to be accepted. In some cases, even this fails to locate the centromere with sufficient confidence to be acceptable. The assumption that the object was a single chromosome is then reconsidered, and if it appears possible that it could consist of more than one chromosome an attempt is made to separate out its components; otherwise it is accepted as being a chromosome whose centromere could not be identified, and the program proceeds to deal with the next object. This continues until every object in the cell has been dealt with.

CLASSIFICATION OF CHROMOSOMES

At this stage of the program, about 46 objects will have been identified as being chromosomes, and for each of these two parameters will have been calculated, namely size and centromeric index. At present the measure of size which we use is area, as this has been found to give the best discriminating power between the chromosome groups. The more conventional measure of length could be used, but it is much more difficult to estimate accurately by automatic means than is area.

With better techniques of staining and improved densitometry, we expect to be able to use integrated optical density as the measure of size, as with a DNA specific stain, this should be an estimate of the DNA content and hence invariant under different degrees of condensation of the chromosomes etc. Centromeric index is already calculated on the basis of integrated optical density since the variations making i.o.d. less useful than area as a size measurement seem to be on rather a large scale, varying from one chromosome to another, rather than within the length of a chromosome itself. As mentioned earlier, there are inevitably some objects which appear to be chromosomes but for which it is impossible to identify the centromere with any confidence. Rather than leave these out altogether they are later classified on the basis of the one parameter, size, alone. Assigning the chromosomes to groups on the basis of the two parameters raises several problems. Firstly: absolute size has very little significance and as mentioned earlier must be normalised before it can be used as a parameter for classification. Conventionally, this is done by dividing the size of each chromosome by an 'adjusted total cell size' – that is the total size of the cell (sum of areas, lengths or integrated optical densities of all chromosomes in the cell) adjusted to allow for any missing or additional chromosomes. For example, if the female cell is taken as standard the size of a male cell must be multiplied by about 1·02 to compensate for the fact that the Y chromosome is smaller than a second X chromosome.

Of course, in order to be able to normalise in this way it is necessary to know the sex and normality, or otherwise, of the cell – i.e. to have already classified all the chromosomes; an impossible situation. However, at this stage of the program all extraneous matter, such as other nuclei, should have been rejected, and the total size of the cell is unlikely to differ too much from the norm (even a 47,XXX cell differs by only 2 per cent from a normal diploid cell).

An initial attempt at normalising the chromosome sizes is therefore made using the unadjusted size of the cell, the size of each chromosome being multiplied by 2,000 and divided by this total size.

The chromosomes are then assigned to groups on the basis of their preliminary normalised sizes and centromeric indices, as follows:

Initially a sample set of normal cells was taken, the chromosomes in these cells classified manually, and their normalised sizes and centromeric indices calculated automatically. When plotted on a size vs. centromeric index chart all the number one chromosomes fall into a cluster in one section of the plane, the C group cluster in another part of the plane and so on, as illustrated in Figure 9.

Given the measurements of an unclassified chromosome it is possible, by examining its position in the size-centromeric index plane to decide to which group it is most likely to belong. In some cases this will be unambiguous, for example if a chromosome has size 85 and index 49 (see Fig. 9) one can be very confident that it is undoubtedly a number one. In other cases there may be some doubt, for example at (60, 28) a chromosome could be a group C chromosome, but is almost as likely to be a member of the B group. In other cases where the chromosome's measurements do not fall within any of the clusters it is more likely to be abnormal.

This intuitive procedure has been formalised by generating a probability distribution of the measurements for each group from the chart of discrete values by treating each point value as the center of a small normal distribution. By applying Bayes' inversion formula to these probability distributions a table has been generated giving, for each pair of measurements, the three most likely groups and their associated probabilities. It is then straightforward to classify each chromosome by assigning it to the most likely group from this table, and thus very rapidly produce a preliminary karyotype, such as is shown in Figure 10.

A second problem now appears. As mentioned above, it is possible for a chromosome to have ambiguous measurements and indeed a B group chromosome in one cell may have the same measurements as a C chromosome in another. Nevertheless cytologists are found to be in general

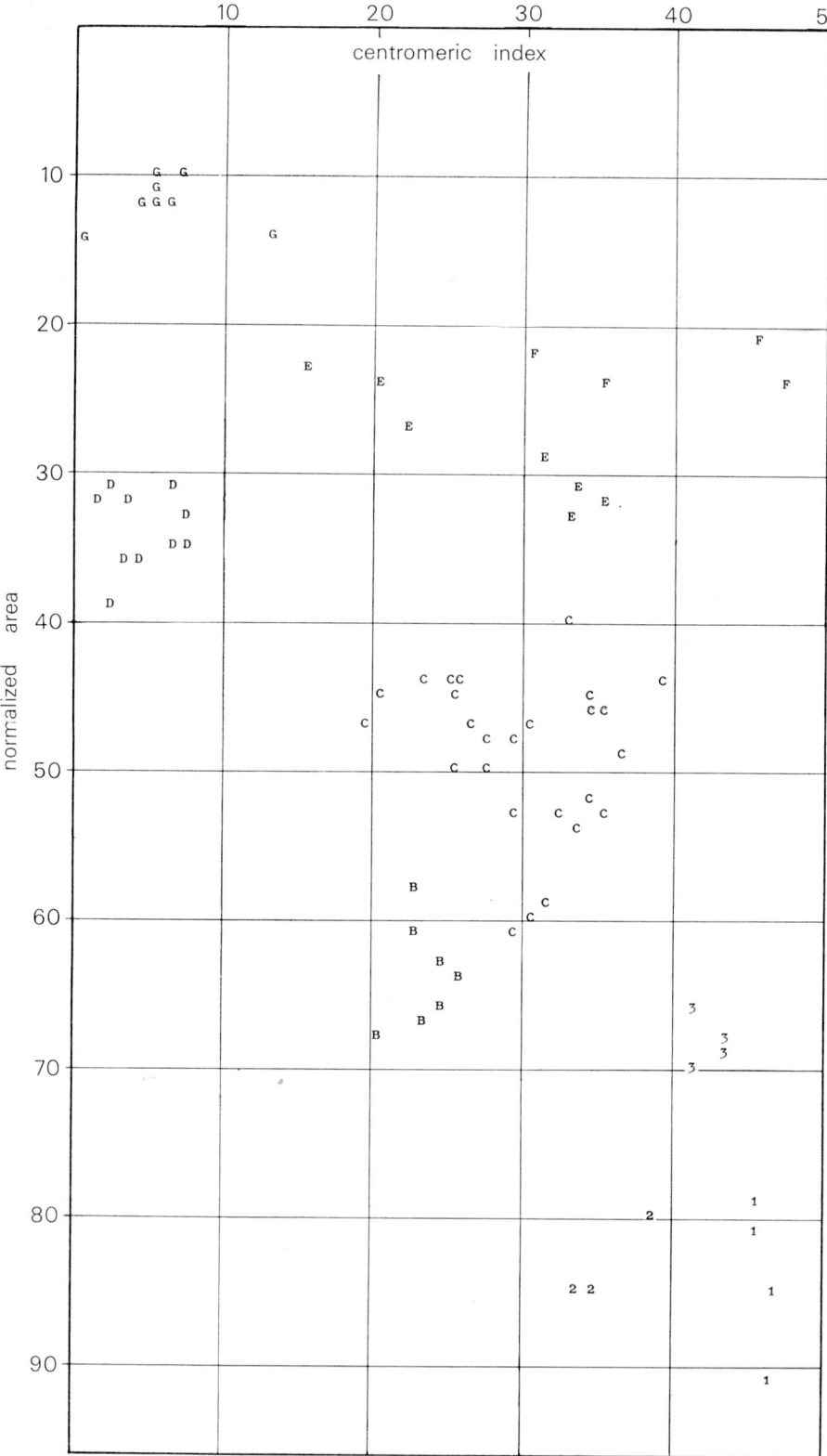

agreement when classifying such chromosomes, and this is no doubt due, at least in part, to using their prior knowledge of the expected karyotype. The program can make use of this same knowledge in the following way: the karyotype produced by the peliminary classification of chromosomes is compared with the expected normal karyotype, and if some group (or groups) have more than the expected number of chromosomes and others have less, a check is made to see if any of the chromosomes in the excessive group(s) has a reasonably good 'fit' in one of the deficient groups. If so, this chromosome is reassigned to the deficient group. At present 'fit' depends entirely on the probability (as given in the look-up table) of the chromosome belonging to the two possible groups, but could also be extended to cover, for example, similarity with other chromosomes in the group.

Another possible source of confusion can occur when two small chromosomes lie close together in such a way that they appear very similar to one larger chromosome (this occurs particularly with the satellite association of acrocentrics). This may well result in an apparent extra large chromosome and a deficit of two acrocentrics. When this occurs, an attempt is made to split the most likely large chromosome to remedy this imbalance.

It should be noted that a chromosome is split or reclassified at this stage if, and only if, this is both necessary in terms of producing a more nearly normal karyotype, and plausible in terms of the prior probability of the alternative classification. Nevertheless, it is quite possible that this technique of always attempting to produce a normal karyotype could hide genuine abnormalities. This is avoided by recording the size and centromeric index of any chromosome which is reclassified or split, for further consideration when the results from the complete set of cells are combined, as described later.

Returning to the individual cell, it should be remembered that the classification so far has been based on measurements which have been normalised only in a rather inaccurate manner. Now that the chromosomes have been classified, a more accurate normalising factor can be calculated. This is done as follows: if a chromosome has been classified for example as a 1, we know from our previous sample what proportion of the total

FIGURE 9 (facing)
The chromosomes in two cells were classified manually, and their normalized areas and centromeric indices calculated automatically. This chart was then produced by plotting the letter representing the group to which each chromosome was manually assigned (or number in the case of chromosome pairs 1, 2 and 3) at the position given by its measurements. As can be seen the chromosomes in different groups cluster in different parts of the plane. By applying this operation to a larger number of cells, and replacing each point value by a small normal distribution a probability density function can be produced for each group

```
OBJECT  1 ALLOCATED TO GROUP 21-22Y .ITS MEASUREMENTS ARE... INDEX =  8 RELATIVE AREA = 10
OBJECT  2 ALLOCATED TO GROUP 21-22Y .ITS MEASUREMENTS ARE... INDEX =  6 RELATIVE AREA = 12
OBJECT  3 ALLOCATED TO GROUP 13-15  .ITS MEASUREMENTS ARE... INDEX =  8 RELATIVE AREA = 33
OBJECT  4 ALLOCATED TO GROUP 6-12X  .ITS MEASUREMENTS ARE... INDEX = 30 RELATIVE AREA = 53
OBJECT  5 ALLOCATED TO GROUP 4-5    .ITS MEASUREMENTS ARE... INDEX = 25 RELATIVE AREA = 63
OBJECT  6 ALLOCATED TO GROUP 6-12X  .ITS MEASUREMENTS ARE... INDEX = 35 RELATIVE AREA = 46
OBJECT  7 ALLOCATED TO GROUP 13-15  .ITS MEASUREMENTS ARE... INDEX =  7 RELATIVE AREA = 31
OBJECT  8 ALLOCATED TO GROUP ABNRML .ITS MEASUREMENTS ARE... INDEX = 16 RELATIVE AREA = 14
OBJECT  9 ALLOCATED TO GROUP 2      .ITS MEASUREMENTS ARE... INDEX = 39 RELATIVE AREA = 80
OBJECT 10 ALLOCATED TO GROUP 17-18  .ITS MEASUREMENTS ARE... INDEX = 16 RELATIVE AREA = 23
OBJECT 11 ALLOCATED TO GROUP 17-18  .ITS MEASUREMENTS ARE... INDEX = 21 RELATIVE AREA = 24
OBJECT 12 ALLOCATED TO GROUP 13-15  .ITS MEASUREMENTS ARE... INDEX =  2 RELATIVE AREA = 32
OBJECT 13 ALLOCATED TO GROUP 13-15  .ITS MEASUREMENTS ARE... INDEX =  2 RELATIVE AREA = 26
OBJECT 14 ALLOCATED TO GROUP 6-12X  .ITS MEASUREMENTS ARE... INDEX = 36 RELATIVE AREA = 53
OBJECT 15 ALLOCATED TO GROUP 1      .ITS MEASUREMENTS ARE... INDEX = 50 RELATIVE AREA = 79
OBJECT 16 ALLOCATED TO GROUP 16     .ITS MEASUREMENTS ARE... INDEX = 32 RELATIVE AREA = 29
OBJECT 17 ALLOCATED TO GROUP 19-20  .ITS MEASUREMENTS ARE... INDEX = -1 RELATIVE AREA = 22
OBJECT 18 ALLOCATED TO GROUP 19-20  .ITS MEASUREMENTS ARE... INDEX = 48 RELATIVE AREA = 24
OBJECT 19 ALLOCATED TO GROUP 4-5    .ITS MEASUREMENTS ARE... INDEX = 23 RELATIVE AREA = 56
OBJECT 20 ALLOCATED TO GROUP ABNRML .ITS MEASUREMENTS ARE... INDEX = 24 RELATIVE AREA = 17
OBJECT 21 ALLOCATED TO GROUP 13-15  .ITS MEASUREMENTS ARE... INDEX =  3 RELATIVE AREA = 31
OBJECT 22 ALLOCATED TO GROUP 16     .ITS MEASUREMENTS ARE... INDEX = 34 RELATIVE AREA = 33
OBJECT 23 ALLOCATED TO GROUP 13-15  .ITS MEASUREMENTS ARE... INDEX =  3 RELATIVE AREA = 39
OBJECT 24 ALLOCATED TO GROUP 13-15  .ITS MEASUREMENTS ARE... INDEX =  7 RELATIVE AREA = 35
OBJECT 25 ALLOCATED TO GROUP 16     .ITS MEASUREMENTS ARE... INDEX = 36 RELATIVE AREA = 32
OBJECT 26 ALLOCATED TO GROUP 6-12X  .ITS MEASUREMENTS ARE... INDEX = 27 RELATIVE AREA = 47
OBJECT 27 ALLOCATED TO GROUP 21-22Y .ITS MEASUREMENTS ARE... INDEX = 14 RELATIVE AREA = 14
OBJECT 28 ALLOCATED TO GROUP 6-12X  .ITS MEASUREMENTS ARE... INDEX = 37 RELATIVE AREA = 49
OBJECT 29 ALLOCATED TO GROUP 6-12X  .ITS MEASUREMENTS ARE... INDEX = 20 RELATIVE AREA = 47
OBJECT 30 ALLOCATED TO GROUP 6-12X  .ITS MEASUREMENTS ARE... INDEX = 40 RELATIVE AREA = 44
OBJECT 31 ALLOCATED TO GROUP 6-12X  .ITS MEASUREMENTS ARE... INDEX = 31 RELATIVE AREA = 47
OBJECT 32 ALLOCATED TO GROUP 3      .ITS MEASUREMENTS ARE... INDEX = 44 RELATIVE AREA = 69
OBJECT 33 ALLOCATED TO GROUP ABNRML .ITS MEASUREMENTS ARE... INDEX = 49 RELATIVE AREA = 47
OBJECT 34 ALLOCATED TO GROUP 6-12X  .ITS MEASUREMENTS ARE... INDEX = 28 RELATIVE AREA = 50
OBJECT 35 ALLOCATED TO GROUP 6-12X  .ITS MEASUREMENTS ARE... INDEX = 30 RELATIVE AREA = 61
OBJECT 36 ALLOCATED TO GROUP 6-12X  .ITS MEASUREMENTS ARE... INDEX = 26 RELATIVE AREA = 50
OBJECT 37 ALLOCATED TO GROUP 3      .ITS MEASUREMENTS ARE... INDEX = 42 RELATIVE AREA = 70
OBJECT 38 ALLOCATED TO GROUP 4-5    .ITS MEASUREMENTS ARE... INDEX = 25 RELATIVE AREA = 66
OBJECT 40 ALLOCATED TO GROUP 19-20  .ITS MEASUREMENTS ARE... INDEX = 46 RELATIVE AREA = 21
OBJECT 41 ALLOCATED TO GROUP 2      .ITS MEASUREMENTS ARE... INDEX = 34 RELATIVE AREA = 85
OBJECT 42 ALLOCATED TO GROUP 1      .ITS MEASUREMENTS ARE... INDEX = 47 RELATIVE AREA = 85
OBJECT 43 ALLOCATED TO GROUP 6-12X  .ITS MEASUREMENTS ARE... INDEX = 36 RELATIVE AREA = 46
OBJECT 44 ALLOCATED TO GROUP 4-5    .ITS MEASUREMENTS ARE... INDEX = 21 RELATIVE AREA = 68
OBJECT 45 ALLOCATED TO GROUP 21-22Y .ITS MEASUREMENTS ARE... INDEX =  7 RELATIVE AREA = 12
OBJECT 47 ALLOCATED TO GROUP 6-12X  .ITS MEASUREMENTS ARE... INDEX = 34 RELATIVE AREA = 54
OBJECT 48 ALLOCATED TO GROUP 6-12X  .ITS MEASUREMENTS ARE... INDEX = 26 RELATIVE AREA = 45
```

FIGURE 10

The preliminary karyotype for a cell. This computer output for the cell shown in Figure 11 gives the values for size and centromeric index calculated for each object and the most likely group for each object based on these values alone. Subsequently, chromosomes may be moved from one group to another if they have measurements which make this possible where doing so will produce a more nearly normal karyotype diploid cell size it should have, so from the actual size of this particular chromosome it is possible to estimate the total expected diploid cell size. This can be done for every chromosome in the cell, and by appropriately weighting the size estimate given by each chromosome depending on the variance of the group to which it belongs, the confidence in its measurements, etc., an average estimate of cell size can be calculated. If this estimated size differs from the previous cell size by less than 1 per cent the classifications are accepted as final. Otherwise the size measurements are renormalised using this new estimated cell size as factor, and the process of classification repeated. This continues until the cell size stabilises, which generally happens very rapidly—usually on the first pass. The last classification, the shifts of chromosomes from one group to another, the splits which took place during it, and the resulting karyotype, are taken as definitive for the particular cell.

FIGURE 11
This computer output shows the outlines of all the objects found in a cell by the segmentation process. The objects are numbered sequentially and on each chromosome the computer has superimposed a line in the position where it has located the centromere and the final group to which the chromosome has been assigned

SOFTWARE SYSTEM

THE COMPOSITE KARYOTYPE

As mentioned earlier, the main aim of the programs described here is for screening purposes; eliminating normals and referring abnormals to the cytogeneticist for further investigation. In survey work it is general to examine at least two cells from each individual (and frequently many more) in order to reduce the number of abnormalities which are missed. With an automated system the need for scanning several cells from each slide becomes even more obvious for, without impossibly complex and time consuming recognition procedures (or improvements in slide preparation techniques such as to produce metaphase spreads which are perfect for machine analysis every time), there will inevitably be a significant number of misclassifications owing to incorrect segmentation, incorrect centromere location, inability to locate the centromere, or merely measurements with a low confidence value. If, however, several cells are examined and a composite karyotype built up, errors in individual cells cease to be so important, and suspicious cases can be either confirmed or rejected.

In our programs, the results of the various individual karyotypes are combined in the following way:

Firstly, the number of chromosomes allocated to each group is examined for all the cells. If most cells have the expected number of chromosomes in each group, and there are not a significant number of cells exhibiting an anomalous number of chromosomes in any group then the group distribution is accepted as normal. If however the modal number of chromosomes in a group is not the expected number, or if there is a bimodality, with a significant proportion of cells exhibiting a different number of chromosomes in the same group, this is classed as an abnormality to be referred to the cytogeneticist.

Secondly, the various suspect classes of chromosomes recorded during the individual karyotypes are examined. These are : chromosomes with apparently abnormal measurements; chromosomes which have been shifted from one group to another, after their preliminary classification, in order to try and produce a normal karyotype; and chromosomes which have been split, or resulted from a split, at this same stage. These are all cases which may appear as artefacts in individual cells, but which, if they represent a genuine abnormality, will appear in a large proportion of the cells examined. Thus the measurements of those due to artefacts will be scattered randomly in the size/centromeric index plane, but genuine abnormalities will appear again and again with very similar measurements.

For this reason all chromosomes in each suspect class are pooled and examined together. If their measurements are scattered widely in the plane they are ignored, but if clusters containing points from several different cells appear, each of these clusters is taken to represent a genuine abnormality and this is again referred to the cytogeneticist for further consideration.

It should thus be possible initially to tolerate quite a significant error rate in the recognition and measurement of individual chromosomes, without either missing genuine abnormalities, or inundating the cytogeneticist with large numbers of suspect cells which then turn out to be from a perfectly normal individual. This should be achieved simply by considering a sufficiently large number of cells per subject. As improvements increase the accuracy with which individual chromosomes are analysed, the number of cells needed for a reliable karyotype can be steadily decreased giving a corresponding increase in the overall efficiency of the system.

REFERENCES

[1] Ledley, R.S., Rotolo, L.S., Golab, T.J., Jacobsen, J.D., Ginsberg, M.D. and Wilson, J.B. In *Optical and Electro-Optical Information Processing,* p. 591. Ed. J.T. Tippet. Cambridge, Mass. : MIT Press, 1965.
[2] Neurath, P.W., Brand, D.H. and Schreiner, E.D., *Ann. N.Y. Acad. Sci.,* **157**, 324, 1969.
[3] Butler, J.W., Butler, M.K. and Stroud, A. In *Proceedings of the Conference on Data Acquisition and Processing in Biology and Medicine,* p. 261. New York : Pergamon Press, 1963.
[4] Hilditch, C.J. and Rutovitz, D., *Ann. N.Y. Acad. Sci.,* **157**, 339, 1969.
[5] Hilditch, C.J. In *Proceedings of the Summer School on Automatic Interpretation and Classification of Images.* Academic Press (in press).
[6] Gallus, G., Montanaro, N. and Maccacaro, G.A., *Computers Biomed. Res. 2* (in press).
[7] Neurath, P.W., Bablonzian, B.L., Warms, T.H., Serbagi, R.C. and Falek, A., *Ann. N.Y. Acad. Sci.,* **128**, 1013, 1968.
[8] Rutovitz, D. In *Human Radiation Cytogenetics,* p. 58. Eds. H.J. Evans, W.M. Court Brown and A.S. McLean. Amsterdam : North-Holland Publ. Co., 1967.
[9] Hilditch, C.J. In *Machine Intelligence 3,* p. 325. Ed. D. Michie. Edinburgh : Edinburgh University Press, 1968.
[10] Mendelsohn, M.L., Mayall, B.H. and Prewitt, J.M.S. In *Conference on Biological Image Processing UCLA Forum in Medical Sciences 9* (in press).